Advances in

# Nutritional Research

Volume 9

## Nutrition and Osteoporosis

Advances in

# Nutritional Research

A Continuation Order Plan is available for this series. A continuation order will bring delivery of each new volume immediately upon publication. Volumes are billed only upon actual shipment. For further information please contact the publisher.

Advances in

# Nutritional Research

## Volume 9

## Nutrition and Osteoporosis

Edited by Harold H. Draper

*University of Guelph*
*Guelph, Ontario, Canada*

Plenum Press • New York and London

The Library of Congress cataloged the first volume of this title as follows:

Advances in nutritional research. v. 1–
New York, Plenum Press, c1977–
1 v. ill. 24 cm.
Key title: Advances in nutritional research, ISSN 0149-9483
1. Nutrition–Yearbooks.

QP141.A1A3 613.2'05 78-640645

ISBN 0-306-44893-9


A Division of Plenum Publishing Corporation
233 Spring Street, New York, N. Y. 10013



Printed in the United States of America

# Contributors

**Lindsay H. Allen,** Department of Nutrition, University of California at Davis, Davis, CA 95616

**John F. Aloia,** Department of Medicine, Winthrop–University Hospital, 259 First Street, Mineola, NY 11501

**John J. B. Anderson,** Department of Nutrition, Schools of Public Health and Medicine, University of North Carolina, Chapel Hill, NC 27599-7400

**Susan I. Barr,** School of Family and Nutritional Sciences, University of British Columbia, Vancouver, British Columbia, Canada V6T 1Z4

**E. C. H. van Beresteijn,** Department of Nutrition, Netherlands Institute for Dairy Research (NIZO), P.O. Box 20, 6710 BA Ede, The Netherlands

**Mona S. Calvo,** Department of Health and Human Services, Public Health Service, Food and Drug Administration, Washington, DC 20204

**H. H. Draper,** Department of Nutritional Sciences, University of Guelph, Guelph, Ontario, Canada N1G 2W1

**Takuo Fujita,** Calcium Research Institute, Kishiwada, Osaka, Japan

**Catherine Gilligan,** Department of Preventive Medicine, University of Wisconsin, Madison, WI 53706

**D. M. Hegsted,** New England Regional Primate Research Center, Harvard Medical School, Southborough, MA 01772-9102

**Isabelle F. Hunt,** School of Public Health, UCLA, Los Angeles, CA 90024

**Jukka A. Inkovaara,** University Central Hospital, PL 2000, 33521 Tampere, Finland

**Jane E. Kerstetter,** School of Allied Health Professions, Department of Nutritional Sciences, University of Connecticut, Storrs, CT 06269

**E. M. C. Lau,** Department of Community and Family Medicine, Lek Yuen Health Centre, Shatin, N. T., Hong Kong

**Richard A. Lazenby,** Anthropology Programme, University of Northern British Columbia, Prince George, British Columbia, Canada V1N 4Z9

**P. Lips,** Afdeling Endocrinologie, Academisch Ziekenhuis, Vrije Universiteit, Postbus 7057, 1007 MB Amsterdam

**Allan G. Need,** Division of Clinical Biochemistry, Institute of Medical and Veterinary Science, Adelaide, South Australia

**B. E. Christopher Nordin,** Division of Clinical Biochemistry, Institute of Medical and Veterinary Science, Adelaide, South Australia 5000, and Department of Pathology, The University of Adelaide, Adelaide, South Australia

**Susan K. Pfeiffer,** School of Human Biology, University of Guelph, Guelph, Ontario, Canada L8S 4L9

**William S. Pollitzer,** Department of Cell Biology and Anatomy, School of Medicine, University of North Carolina, Chapel Hill, NC 27599-7400

**Jean-Michel Pouillès,** Unite Maladies Osseuses et Metabolique, Centre Hospitalier Universitaire, Hopital de Purpan, Place du Docteur Baylac, 31059 Toulouse, France

**Jerilynn C. Prior,** Department of Medicine, University of British Columbia, Vancouver, British Columbia, Canada V5Z 1M9

**Claude Ribot,** Unite Maladies Osseuses et Metabolique, Centre Hospitalier Universitaire, Hopital de Purpan, Place du Docteur Baylac, 31059 Toulouse, France

**Everett L. Smith,** Department of Preventive Medicine, University of Wisconsin, Madison, WI 53706

**Lorri J. Tommerup,** Department of Preventive Medicine, University of Wisconsin, Madison, WI 53706

**Florence Trémollières,** Unite Maladies Osseuses et Metabolique, Centre Hospitalier Universitaire, Hopital de Purpan, Place du Docteur Baylac, 31059 Toulouse, France

**J. Woo,** Department of Medicine, Chinese University, Hong Kong

# Contents of Earlier Volumes

# Preface

## Nutrition and Osteoporosis: Seeing Through a Glass, Darkly (1 Cor. 13:12)

This volume of *Advances in Nutritional Research* deals with the present state of knowledge relative to the role of nutrition in the etiology of osteoporosis, one of the most serious degenerative diseases in the aging population. As a backdrop for subsequent chapters on specific nutrients, **Chapter 1** provides a comprehensive account of the gain and loss of bone throughout the life cycle, with emphasis on the architectural changes in later life that predispose to osteoporotic bone fractures. **Chapter 2** documents the occurrence of aging bone loss throughout human archeological history and **Chapter 3** extends this documentation to all non-human vertebrate species so far examined, including primates living in the wild. It is apparent that a progressive loss of bone tissue is a normal accompaniment of aging among higher vertebrates. Whether it is a cause of bone fractures in animals, as it is in humans, is still unknown. It has also been established that there are significant differences in the frequency of osteoporotic fractures among human families, ethnic groups, national populations and diet cultures. Numerous studies have been carried out in an effort to explain these differences, and many of these deal with the possible effect of nutrition.

Protracted controversies over the role of nutrition in the etiology of osteoporosis are reflected in the contents of several of the ensuing chapters. These controversies have arisen for a number of reasons: the slow progression of the disease over a period of years makes accurate assessment of its dietary correlates extremely difficult; methods of assessing current nutrient intake by free living individuals are notoriously inaccurate and retrospective assessments are even less reliable; a capacity to adapt to a range of intakes of some nutrients implicated as risk factors for osteoporosis creates uncertainty regarding the significance of effects observed in short term laboratory studies; a strong influence of non-nutritional factors, particularly heredity and lifestyle, on the risk of osteoporotic bone fractures generates a high level of background variability against which the influence of nutritional factors must be appraised.

Despite these complications, recent epidemiological studies on several diverse populations have produced a general consensus on the most contentious question pertaining to the role of nutrition in the prevention of postmenopausal osteoporosis, namely the dietary requirement for calcium. A 10-year prospective study on two groups of perimenopausal Dutch women revealed no difference in bone loss or bone mineral density after they were matched for bone mass index, despite a marked difference in calcium intakes (564–796 versus 1354–2580 mg per day) (**Chapter 4**). In a large study on U.S. women, no response to calcium supplementation, assessed in terms of a reduced rate of bone loss, was found in the perimenopausal or early postmenopausal periods, and only in those with an habitual calcium intake of less than 400 mg per day in the late postmenopausal period. These estimates are compatible with the long-standing WHO recommendation of 400–500 mg per day, which recognizes the low calcium content of cereal-based diets. Large differences in the recommended intake of calcium in different countries reflect the influence of Parkinson's Law, i.e., the intake recommended increases to meet the supply available. A committee of the U.K. Department of Health has estimated the average calcium requirement of British women to be 525 mg per day and has proposed a Reference Calcium Intake (equivalent to the American RDA and the Canadian RDI) of 700 mg per day. This was the RDA and RDI a decade ago, before they were raised to 800 mg as a token response to a claim (since disproved) that postmenopausal bone loss could be prevented in the average woman by a calcium intake of 1500 mg per day. Modest responses to calcium supplements have been observed in some healthy and osteoporotic postmenopausal women, but these supplements often have included vitamin D, which recent studies indicate may be a more important consideration.

The low incidence of hip fractures among Japanese women, despite their low intake of calcium, has been a focus of recent research interest. **Chapter 5** contains a proposed explanation for this discrepancy which, if correct, has widespread significance for the aging populations of industrialized societies. It is based on simple differences in lifestyle between Japanese and Western women that affect the incidence of falls and the severity of the trauma caused by falls. These differences include frequent getting up and down associated with the traditional Japanese habit of sitting on the floor, which serves to strengthen the hip musculature, improve motor function, agility and body balance, and thereby to reduce the incidence of falls, the main immediate cause of hip fractures. The Japanese lifestyle also provides more weight-bearing exercise, which has a stabilizing effect on bone. It is noteworthy that the difference in the incidence of vertebral fractures between Japanese and Western women is much smaller than it is for hip fractures. The mean intake of calcium by free living Japanese adults of different ages has been estimated at 500–600 mg per day. The intake of institutionalized elderly was found to be much lower. The Japanese RDA has been set at 600 mg per day, a figure that, if adjusted to the same bone mass index, equates with an RDA of about 700 mg per day for Western adults.

In studies on Hong Kong postmenopausal women, little or no increase in bone mineral density in response to calcium supplementation was seen in subjects with a calcium intake of 545 mg per day or more (**Chapter 6**). The range of intakes beyond which no significant benefit could be expected was estimated to be 500–600 mg per day, a value that exceeds the intake of many Hong Kong women. Urbanization in recent years has been associated with a significant increase in the incidence of hip fractures in elderly women in Hong Kong, Singapore and other Asian countries.

The reported global epidemiology of hip fractures is directly, rather than inversely, related to calcium intake (**Chapter 7**). Evidence that the incidence of vertebral fractures in Japan and other Asian countries is no greater than it is in Western societies further indicates that the higher calcium content of the national food supplies of these societies is not protective against fractures at this site. This does not negate the probability that a very low calcium intake is a risk factor for osteoporotic fractures at both sites and in all countries.

Ethnic and racial differences in bone mineral content, bone density and fracture rates are discussed in **Chapter 8**, with particular reference to American blacks versus whites.

Recent evidence, reviewed in **Chapter 9**, indicates that low vitamin D status may be a more prevalent problem than calcium deficiency among elderly men and women, particularly those who are housebound or confined to institutions where they receive little exposure to solar radiation. Vitamin D supplements have been reported to reduce the rate of bone loss and incidence of fractures among such women in the U.S., France and Finland. Such findings, in addition to evidence for low vitamin D intakes and a decrease in the efficiency of vitamin D synthesis in the skin and in its conversion to active metabolites during aging, have prompted several government health agencies to recommend that elderly persons confined indoors consume a vitamin D supplement.

**Chapters 10** and **11** describe the basis for the contention that the high protein and high phosphorus content of the "Western" diet constitute risk factors for osteoporosis by increasing the requirement for calcium. This contention is consonant with the higher rate of fractures in industrialized societies, an exceptionally high rate of bone loss among Arctic semi-carnivores, and by the adverse effect of a large excess of either nutrient on calcium homeostasis observed in experimental subjects. Oxidation of excess sulfur amino acids generates an endogenous acid load that is excreted in the urine. The efficiency of the parathyroid hormone-dependent reabsorption of calcium from the renal tubules decreases with increases in urine acidity, resulting in a greater loss of calcium in the urine. Excess dietary phosphate causes a depression of serum calcium and a consequent increase in the synthesis of parathyroid hormone, which acts to restore normocalcemia by increasing the reabsorption of calcium from the filtered urine as well as the resorption of bone. When consumed individually in large excess, each nutrient increases bone resorption and bone loss. Because of the natural association between these nutrients in foodstuffs, however, in high protein diets they are consumed in excess simultaneously. Under these conditions, the calciuretic action of excess protein is, at least in part, counteracted by an increase in the reabsorption of calcium from the filtered urine resulting from the increased synthesis of parathyroid hormone induced by excess dietary phosphate. Thus, a high protein intake increases the phosphorus requirement for calcium homeostasis.

It seems unlikely that the hypocalciuric effect of excess dietary phosphorus and the hypercalciuric effect of excess protein are always fully offsetting. **Chapter 10** reviews the evidence that there is an increase in urinary calcium excretion on high protein diets that is not counteracted by phosphorus and is not due to an increase in calcium absorption (i.e., that reflects bone loss). In any event, the ratio of phosphorus to protein, rather than the more frequently cited ratio of phosphorus to calcium, is the most important nutrient relationship affecting calcium homeostasis. This ratio is relatively uniform in diets composed of unprocessed foods. However, it has been disturbed by the widespread unilateral addition of phosphorus to the diet in the form of phosphate food additives. These compounds currently appear to be the source of about one-third of the phosphorus in the U.S. diet (**Chapter 11**). The phosphorus present in high protein diets composed of common commercial foodstuffs has been shown to be sufficient to cause an increase in serum parathyroid hormone and a consequent increase in bone resorption, which was cited in the report of the 1984 NIH consensus conference on nutrition and osteoporosis as a risk factor for bone loss. While there is undoubtedly some level of phosphate intake beyond which bone loss occurs (as demonstrated in several species of animals), it is unclear whether the increase in bone resorption observed in human adults consuming high protein, high phosphorus diets results in an increase in bone loss or only in bone turnover. Why the increase in parathyroid hormone-induced bone resorption seen on a low calcium diet is widely perceived as indicating a need for more calcium, whereas the increase induced by the same hormone on a high phosphorus diet is viewed as clinically benign, has not been explained.

**Chapter 12** presents evidence that a high salt intake can significantly increase bone resorption and loss of calcium in the urine, particularly on low calcium intakes and in the postmenopausal state. How the sodium load signal is transmitted to bone is unknown. Salt restriction has been observed to decrease bone resorption following the menopause, but an effect on bone loss has not been demonstrated. The low incidence of hip fractures among women in Southeast Asia indicates that a high salt intake is not a major risk factor for this disease.

There is no consistent relationship between fluoridation of drinking water and the incidence of osteoporotic fractures (**Chapter 13**). Fluoride treatment of osteoporosis is counterindicated by an increase in hip fractures arising from the increased brittleness of fluorotic bone as well as by gastrointestinal disturbances and arthralgias.

Densitometric evidence for a rapid rate of bone loss among Arctic Inuit consuming a semi-carnivorous diet and a cross-cultural association between a high protein intake and the incidence of hip fractures have prompted comparative studies on the bone mineral content of vegetarians and omnivores. The evaluation of current evidence presented in **Chapter 14** indicates that there is no significant difference in the bone mineral content of postmenopausal vegetarians and omnivores when calcium intakes are similar and protein intakes are within the range usually reported for women in the United States. Also, no significant relationship was found between the current intake of calcium or phosphorus and bone mineral content in either group.

The protection against osteoporotic fractures, particularly of the hip, afforded by obesity is documented in **Chapter 15**. Postmenopausal women who are overweight have a greater bone mass than women of normal weight. This difference is attributable to several factors, including the inhibition of bone loss by estrogen synthesized in adipose tissue, the homeostatic effect on bone of a higher level of weight-bearing stress, and the cushioning effect of a fat deposit on the physical impact associated with a fall.

The potential of exercise to decrease fracture risk by reducing bone loss, increasing muscle strength, improving agility, and thereby reducing both the incidence and severity of falls is described in **Chapter 16**. Cross-cultural studies indicate that the risk of falls and hip fractures is related to the level of exercise. The fact that the incidence of vertebral fractures among Japanese women approaches that in the West, even though exercise has a stabilizing effect on the spine, indicates that their much lower hip fracture rate is related to a difference in the frequency of falls. The Japanese custom of keeping fracture patients in bed in order to reduce the risk of additional falls, as opposed to the American practice of getting them out of bed as soon as possible to prevent the rapid bone loss associated with immobilization atrophy (**Chapter 5**) constitutes an interesting difference in risk/benefit analysis. A sustained program of exercise is required to reduce the rate of postmenopausal bone loss; the rate quickly increases to the pre-exercise value when the steady state level of exercise is discontinued.

**Chapter 17** discusses the occurrence of menstrual cycle disturbances, their impact on bone during the premenopausal period and the subsequent effect in the postmenopausal period. These disturbances, whether clinical or subclinical, result in a premenopausal loss of trabecular bone, which is the bone type most susceptible to fractures in the postmenopausal period. The frequency of these disturbances is not increased by gradual intensification of exercise, provided that it is not accompanied by weight loss. Inadequate energy intake appears to be the most important nutritional cause of menstrual cycle disturbances.

The aforesaid chapters indicate that a calcium and/or vitamin D intake substantially below the mean for the adult population of several countries is associated with an accelerated rate of aging bone loss. Increased consumption of these nutrients therefore can be predicted to fractionally reduce the incidence of osteoporotic bone fractures in that segment of the aging population with such intakes. For the bulk of this population, however, prevention must be sought in non-nutritional measures.

H. H. Draper

# Contents

Chapter 1

# The Gain and Loss of Bone in the Human Life Cycle

John F. Aloia

## 1. Introduction

Osteoporosis may be defined as a diminished quantity and quality of bone that increases the risk for fracture. The fractures given most attention involve the vertebrae, femur and radius, although other fractures are also related to osteoporosis. Fractures of the distal forearm (Colles') and vertebrae, which contain large amounts of trabecular bone, increase after menopause. The increase in fracture incidence reaches a plateau at 65 years for the wrist, but the incidence continues to rise with increasing age for the vertebrae. The incidence of fractures with a more proportionate mix of compact and trabecular bone increases slowly, and then exponentially in the elderly, resulting in increased fractures in the femur, proximal humerus, proximal tibia and pelvis (Figure 1).

Over 1.2 million osteoporotic fractures occur in the U.S. alone each year. By age 65, one-third of women have vertebral fractures and in old age one of three women and one of six men have hip fractures. The direct medical cost of treating osteoporosis in the United States in 1986 was estimated at $5.2 billion and more recently at $10 billion. As a result of the rapidly accelerating size of the aged population, osteoporosis will have an increasing impact on the public health.

---

John F. Aloia • Department of Medicine, Winthrop-University Hospital, 259 First Street, Mineola, New York 11501.

*Advances in Nutritional Research,* Vol. 9
Edited by Harold H. Draper
Plenum Press, New York, 1994

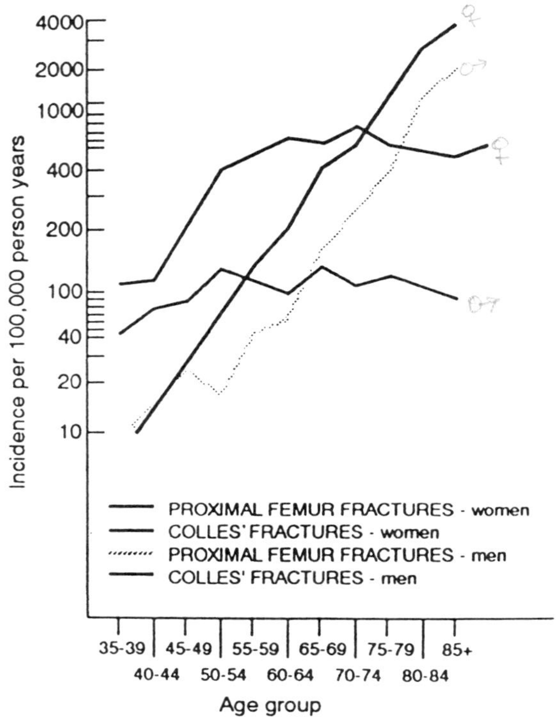


Figure 1. The increase in fracture incidence with age. Reproduced from Avioli (1987) with permission.

Since 80% of the strength of a bone is explained by its mass or density, having an optimal bone mass is one of the major goals in the prevention of osteoporotic fractures. Bone mass at any given time in the life cycle represents the interaction between (a) the genetic component of maximal possible bone mass for an individual and (b) all the environmental factors that influence bone gain and bone loss. The readily identifiable stages of bone gain and loss include growth, adolescence and consolidation, culminating in peak bone mass, followed by a period of slow involutional bone loss in men and women and a more rapid rate of loss in women beginning prior to menopause. Before considering models from the literature that describe bone gain and loss in detail, it may be worthwhile to briefly review some fundamentals of skeletal anatomy and physiology.

## 2. Skeletal Anatomy and Physiology

The skeleton may be thought of as consisting of two types of bone tissue. Compact bone is dense and is found in the shafts of long bones and as a shell around vertebral bodies; cancellous bone is arranged in a latticework, is found in the vertebrae, at the end of the long bones, and in the pelvis and other flat bones. Cancellous bone comprises only 20% of total bone mass but has a greater surface area, is in greater contact with the marrow and is therefore more responsive to hormonal influences and to the physiologic demands for mineral homeostasis than is compact bone. Compact bone is organized as a system of longitudinally oriented osteons (Haversian systems) consisting of concentrically arranged layers of tissue with a central vascular channel. In cancellous bone, mineralized spicules (trabeculae) are arranged along lines of mechanical force and are intersected both obliquely and transversely by other spicules which may be thought of as struts. This architectural arrangement strengthens cancellous bone.

The skeleton has two major functions: structural and chemical. The structural function subserves locomotion and protection of the internal organs, whereas the chemical function subserves the organism's needs for mineral, acid-base and endocrine homeostasis. The latter functions are more ancient in evolutionary development and can override structural demands. Structural demands are met by bone growth and bone modeling, whereas chemical demands are served by bone remodeling. Linear growth occurs through mineralization of the endochondral growth plates and by radial growth. Radial growth occurs by periosteal apposition that exceeds endosteal resorption. The growth plate closes at about age 20 years, yet radial growth continues for a decade to a decade and a half.

Bone modeling has been considered to occur in response to a "mechano-stat." This concept is similar to the concept of endocrine negative feedback loops, with mechanical stress as the stimulus to skeletal mass. Thus, when mechanical stress is placed on a skeletal region, there is an increase in bone mass. Alternatively, when mechanical stress is reduced, there is a subsequent reduction in bone mass. The stimuli to bone modeling are regional, i.e., increased mechanical stress results in a change in the shape of a bone so that it is better able to withstand the increased stress.

Bone remodeling occurs in discrete foci called bone remodeling units which complete remodeling in a prescribed time period. Overall bone remodeling represents the sum of all remodeling units (Riggs *et al.*, 1991a) (Figure 2).

At the beginning of each cycle, lining cells that cover the bone surface retract to expose the underlying bone. The beginning of this process is called activation. Osteoclast precursors are recruited to the area from marrow

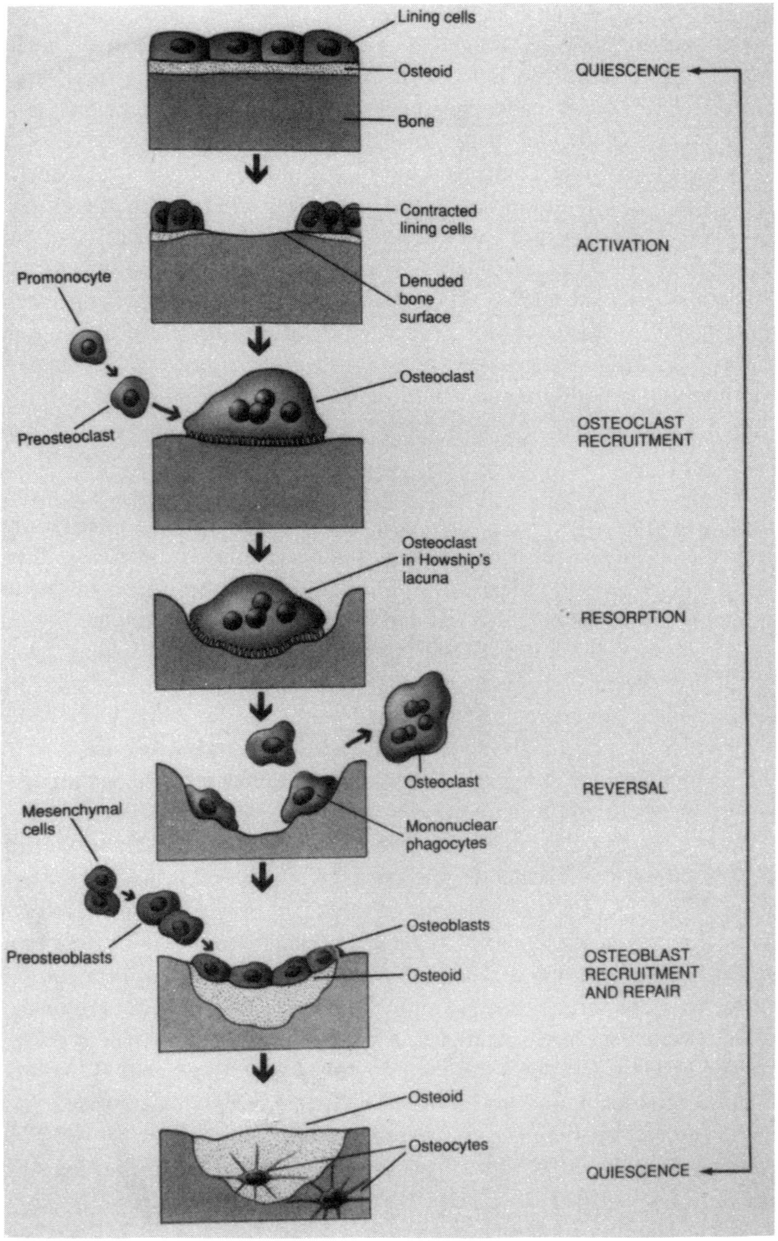


Figure 2. The bone remodeling cycle. Reproduced from Riggs (1991a) with permission.

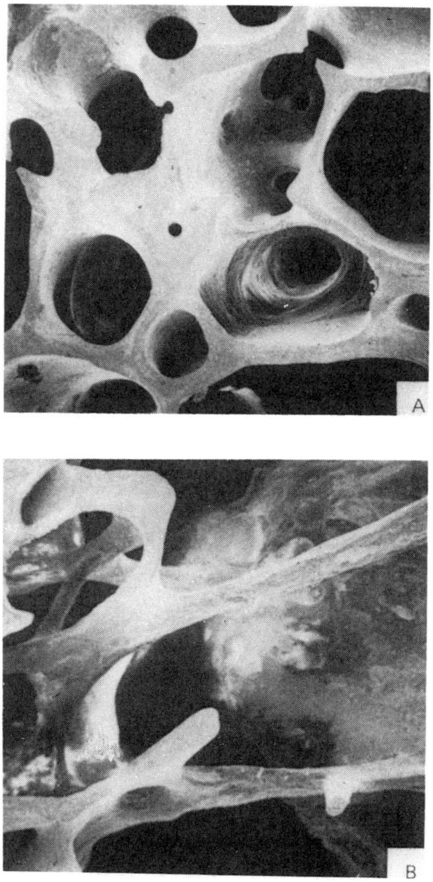


Figure 3. Normal bone (A) with osteoporotic bone showing perforation and loss of trabeculae (B). Reproduced from Dempster *et al.* (1986) with permission.

and fuse to form multinucleated osteoclasts. Over the next few weeks, the osteoclasts produce a resorption tunnel in cortical bone; in cancellous bone they form a lacuna on the trabecular surface. A reversal phase lasting 3–4 months follows, during which osteoclasts disappear and osteoblasts fill in the resorption cavity with bone matrix, which becomes mineralized. This process serves to accomplish goals such as repair of a stress microfracture. The resorption cavities created can be underfilled or overfilled. Remodeling balance refers to how completely these cavities are filled, i.e., complete = zero balance;

underfilled = negative balance, and overfilled = positive balance. The extent of remodeling imbalance for the entire skeleton is determined by the frequency of activation of new remodeling units. Thus, if there are many new units in negative balance the effect on skeletal mass can be profound. During growth there is positive balance, during young adulthood balance is close to zero, and with aging there is remodeling imbalance, resorption exceeding formation. In trabecular bone there are normal or deeper resorption cavities that are incompletely filled with new bone by the osteoblasts, leading to a thinning of bony trabeculae. In high bone turnover states, as occurs with estrogen withdrawal, there are an increased number of osteoclasts with deeper resorption cavities that actually perforate bony trabeculae and result in loss of trabecular connectivity and even of entire trabeculae (Dempster *et al.*, 1986) (Figure 3). The loss of whole trabeculae produces a permanent architectural change in the skeleton, limiting the effect of pharmacologic treatments of osteoporosis to the laying down of new bone on the remaining bone surfaces.

Bone formation rates decrease dramatically with age, the percentage of bone surface with no activity increases, and osteoblasts disappear (Table 1). The time to complete a remodeling cycle lengthens. Cortical bone becomes more porous by incomplete filling of resorption cavities (increased intracortical porosity). Endosteal bone undergoes net loss and periosteal apposition persists (Garn, 1972; Rubin, 1991; Burkhardt *et al.*, 1987; Recker *et al.*, 1988; Thompson, 1980).

Table 1. Postulated Reasons for Reduced Involutional Bone Formation

1. An age-related decrease in the population of osteoblastic progenitor cells.
2. An age-related decrease in the metabolic activity of osteoblasts.
3. Disruption in final transduction:
   a. Aging changes in calciotrophic hormones.
   b. Age-related changes in local mediators.
   c. Changes in chemoattractant matrix proteins.
   d. Changes in cell responsiveness to a, b, or c.
4. Decreased blood flow to bone:
   a. Decreased delivery of progenitor cells.
   b. Decreased transport of calciotrophic hormones.
5. Changes in the composition or structure of bone matrix, limiting signal transduction.
6. Changes in bone crystallinity that limit the conduction and distribution of strain-induced signals.

### 2.1. Tissue Quality

As a result of the changes in bone remodeling with aging, the strength of bone is compromised by architectural changes (Basle *et al.*, 1990; Beck *et al.*, 1992; Birkenhager-Frenkel *et al.*, 1988; Chalmers and Weaver, 1966; Chappard *et al.*, 1991; Cohn *et al.*, 1977a; Compston *et al.*, 1987; Evans, 1976; Eyre *et al.*, 1988; Field *et al.*, 1990; Foldes *et al.*, 1991; Hoiseth *et al.*, 1990; Kragstrup *et al.*, 1983; Marcus *et al.*, 1983; Mosekilde *et al.*, 1987; Mosekilde, 1988; Parfitt *et al.*, 1983; Pesch *et al.*, 1990; Davis *et al.*, 1991; Twomey *et al.*, 1983; Whyte *et al.*, 1982). There is a reduction in interconnectivity, trabecular struts are lost, the medullary cavity enlarges, vascular canals increase in size, cortical endosteal bone is replaced, and cortical bone becomes more porous. There may be some architectural compensation for these aging changes by concurrent subperiosteal bone apposition in response to physical activity. Bone fatigue with the accumulation of microfractures also occurs with age, comparable to the stress or fatigue defects occurring in airplanes. Their accumulation may eventually result in larger defects and fracture. With aging, microfracture repair through the process of remodeling is impaired.

Because older bone is more highly mineralized than younger bone, the density of the bone that is not in the area of porosities increases. The distribution of osteons is altered so that there is more bone without osteons, partial osteons, and overlapping osteons. Osteocytic lacunae become calcified as do the canaliculi. Thus, some areas of aged bone are undermineralized and others are hypermineralized. It is believed that hypermineralized bone may be compared to microscopic osteopetrosis with increased susceptibility to microfracture.

Either crystalline size increases or lattice imperfections increase with age, resulting in an increased crystallinity of bone. It is not clear whether this impacts adversely on bone quality by, for instance, altering electrical conductivity. Similarly, age-related changes in collagen and in other mineral-related proteins in bone matrix have been documented and related to alterations in structure in the process of mineralization, to bone turnover rates, and to the biomechanical properties of bone.

## 3. Changes in Bone Mass and Bone Density with Aging

Although bone quality is important in disease states, 80% of the strength of bone may be explained by its mass. Models for involutional bone loss have been developed using a variety of approaches and techniques. Thus, studies may be made on anthropologic collections, on histologic samples, or on the total skeleton or specific skeletal regions *in vivo*, measuring integral bone or attempting to separate compact from cancellous bone. Models developed from cross-sectional population studies suffer from inherent secular trends. Older individuals frequently have been subjected to environmental factors different from those for younger individuals. Thus, all models proposed from cross-sectional studies must be confirmed by prospective studies. Furthermore, because of ethnic and environmental differences, population standards must be drawn from the same population to which they are applied.

### 3.1. Models for Bone Gain and Loss

There is voluminous literature on models for bone growth and involutional bone mass loss. When reviewing these models, one finds many contradictions, some of which may be resolved by asking pertinent questions such as the following. What technique was used for measurement? Did the technique measure compact, cancellous or integral bone? What region of bone was measured and what was the proportion of compact and cancellous bone in the area measured? How was the study population recruited and what exclusion criteria were applied? What was the sample size? What was the ethnic background of the population? What was the sex of the population? Was the study cross-sectional or longitudinal? If longitudinal, how many measurements were made, at what intervals, for what duration? If cross-sectional, were there cohort effects that influenced the model? In women, what age span was studied? Was the number of subjects uniform through the age span? Were the data corrected for body size? Was the actual age of menopause used in the model? Were the conclusions confined to the population studied?

Some caveats in the interpretation of bone mineral data are in order. The interpretation of data on involutional bone loss depends on the mathematical approach to curve fitting. To examine whether there is a difference between premenopausal and postmenopausal women, a two-component curve (two linear regressions) may be most appropriate. However, to describe all phases of bone loss it is necessary to employ more complex curve fitting techniques. The size of the population is critical in determining whether changes will be observed. For example, using computer modeling, Sambrook et al. (1987) found that a sample size greater than 300 was needed to distinguish between linear and nonlinear patterns of loss if measurements were made throughout the life span. If no change is assumed in lumbar density prior to menopause, only 100 subjects are needed to detect nonlinear loss. However, if some loss occurs before menopause, over 1000 subjects are needed. Thus, sample size affects the conclusion as to whether bone loss is linear throughout life. Because of the possibility of cohort effects, any cross-sectional model should be confirmed using longitudinal measurements. Generally, measurement of longitudinal rates of change require even larger sample sizes because of errors in instrumentation as well as biologic variability. There is a great paucity of population studies that meet the requirements for proper population sampling and sample size necessary to draw firm conclusions. There are even fewer longitudinal studies that are necessary to validate models based on cross-sectional studies.

Physiologic questions that must be asked in development of any model include the following: How much bone mass is gained, at which skeletal sites, for how long a period of time? Is there an adolescent increment in bone mass? When does it start in males and females; is it different in cancellous and compact bone, in different bones, and at different sites in the same bones? When is peak bone mass attained in males and females? When does adult bone loss begin in males and females, what is its magnitude, does it vary by skeletal site, does it differ in cancellous and compact bone? What curve best describes menopausal bone loss? Does it differ by skeletal site or compact vs. cancellous bone? Does bone loss stop in very old age? Are there ethnic differences in bone gain or loss? All these questions must be kept in mind in the choice of an appropriate model for the study of bone gain and loss.

## 3.2. Measurement of Bone Gain and Loss by Radiographic Morphometry

Some of the most extensive work using the older technique of radiographic morphometry has been done by Garn (1972). This technique examines bone envelopes and surfaces. Bone is added at the subperiosteal surface throughout life, lost at the endosteal surface until late adolescence, gained through mid-adulthood and then lost again. Garn examined the second metacarpal at midshaft in 12,290 participants in his studies. He observed bone expansion between 1 and 25 years of age to be more than threefold in females and fourfold in males. There is a large increase in the first year of life as a result of subperiosteal apposition. There is a 20-25% expansion in the outer bone area between 1 and 2 years in both sexes. Then the rate of accretion slows before reaching a maximum for 3–4 years in adolescence. The adolescent spurt occurs at 10–14 years in girls and at 12–16 years in boys. In adolescence there is also periosteal apposition beginning at 13 in girls and 15 in boys. Boys continue to gain during the third decade.

Endosteal resorption begins late in the fourth decade. Studies of 459 males and 866 females showed that loss begins at about 30 years. The area of bone remaining in older females may be less than that in adolescence. Garn also noted that subperiosteal apposition at the femoral midshaft is greater than at other sites. Subperiosteal apposition in the femur, ribs and vertebrae continues throughout life.

Garn also examined racial differences in bone morphometry. African Americans, compared to whites, exhibited greater bone size and compact bone area, and have a greater rate of subperiosteal apposition with more endosteal apposition than resorption. Consequently, the values for blacks are greater at all ages. In addition, bone loss is less in blacks than in whites.

In summary, in the model developed by Garn, the outer subperiosteal surface undergoes lifelong apposition that is greater in males than in females. There is an adolescent spurt and then a slower rate of gain until death. The inner endosteal surface undergoes resorption from childhood to adolescence, apposition from mid-adolescence through the fourth decade, and then a second resorptive phase which is greater in the female. Cortical area increases from infancy through adulthood and then greatly decreases as the result of losses in the fifth and sixth decades. This model, developed from data acquired from radiographic morphometry, has the advantage of a large sample size. However, the site measured reflects compact bone gain and loss and the results may not apply to cancellous bone, the axial skeleton or sites subject to fracture.

A model for the development of vertebral osteoporosis (Riggs *et al.*, 1981, 1982, 1986, 1991b) suggests that there are two types of osteoporosis. Type I involves excess cancellous bone loss from the spine (Table 2) and Type II involves senescent loss of both compact and cancellous bone. On the other hand, Foldes *et al.* (1991) noted that compact bone loss may be slower in the axial than the appendicular skeleton, and that the deficit in osteoporosis may simply be greater in the axial than the appendicular skeleton (rather than the result of excessive cancellous loss). They point out that compact bone contributes relatively less volume and more surface area in the axial than in the appendicular skeleton.

## 3.3. Other Techniques that Measure Predominantly Compact Bone

Single photon absorptiometry was developed in the early 1960s and applied to population studies of bone density. Shortly thereafter, the technique of *in vivo* neutron activation analysis (IVNAA) and whole body counting was applied in human studies, permitting the development of a model that applies to the skeleton as a whole (Cohn *et al.*, 1976a, 1981, 1984; Aloia *et al.*, 1982, 1983, 1985a, 1991). However, most single photon absorptiometry measurements were made at radial sites, which are comprised predominantly of compact bone, whereas IVNAA reflects the total skeleton, which is 80% compact bone. As pointed out many years ago, measurement of total body calcium must be corrected for body size. Using these techniques, it was found that in adults: men have higher bone mass (or density) than women; there is a slower, linear loss of bone in men; there is a rapid postmenopausal bone loss in women that may be preceded by a slower loss rate similar to that observed in men; bone mass in black men and women is greater than in white men and women throughout adulthood. Longitudinal studies performed using these two techniques have confirmed the above model.

More recently, dual energy absorptiometry utilizing either a radioactive (DPA) or X-ray (DXA) source has been used for the measurement of total body bone mineral (as well as lean and fat mass). Gallagher *et al.* (1987) reported the results of a cross-sectional study in 392 white women aged 20-80 years. They found a small decline in total body bone mineral density (TBBMD) between the ages of 25 and 50 years, and a greater decline after menopause, similar to the model developed using total body calcium data. The largest decline was seen in the first five years after menopause, when the total decline was ten times greater than in premenopausal women. Similar findings were observed with respect to spinal bone mineral density (BMDs). Overall, the loss of TBBMD and BMDs was 2.5–3.0 times greater in the 25 years after menopause than in the 25 years before menopause. Other studies, using a smaller number of subjects, have failed to detect the slow premenopausal skeletal loss observed in studies using a larger sample size.

## 3.4. Studies of Cancellous Bone Loss

Much early information on cancellous bone loss was derived from histologic studies, primarily on the iliac crest. Most of these studies suffer from a small sample size that was not drawn from a healthy population. Nevertheless, many of these studies are of interest (Riggs *et al.*, 1981; Hansson and Roos, 1980; Krolner and Pors Nielsen, 1980; Hurxthal and Vose, 1969; Haasner *et al.*, 1967; Heuck, 1970; Banzer *et al.*, 1976; Mueller *et al.*, 1966; Weaver and Chalmers, 1966; Arnold, 1973; Dequeker *et al.*, 1971; Ahuja, 1969; Havivi *et al.*, 1971). A good review of this information has been published by Mazess (1982). A summary from that review is reproduced in Table 2.

Mazess (1982) has pointed out that 80% of the variance in compressive strength of cancellous bone and 90% of the variance in compact bone is explained by the mass of bone. The difference in variance is undoubtedly related to architectural differences in the two types of bone. There are geometric changes in compact bone with aging (long bones increase in diameter and area through endosteal resorption and periosteal apposition). Microfractures tend to accumulate with age more rapidly in cancellous bone and there is an increase in cancellous bone fat content that may alter energy metabolism. These changes, as well as the architectural changes that occur in cancellous bone, lead to a greater reduction in cancellous than compact bone strength than would be revealed only by measurement of mass. Mazess concluded that iliac crest cancellous bone decreases from early adulthood through old age in both sexes.

Table 2. Age Changes (Age of Onset and Rate per Decade) of Trabecular Bone by Noninvasive (Radiologic) Methods *in vivo* and by Analysis of Apparent Density in Postmortem Series

| | Method | Site | Men | | | Women | | |
|---|---|---|---|---|---|---|---|---|
| | | | No. | Onset | Rate | No. | Onset | Rate |
| Noninvasive | | | | | | | | |
| Riggs *et al.* (1981) | DPA | $L_1$-$L_4$ | 82 | 20 | 2% | 105 | 20 | 6% |
| Hansson and Roos (1980) | DPA | $L_3$ | -- | -- | -- | 214 | 20 | 10% |
| Krolner and Pors Neilson (1980) | DPA | $L_2$-$L_4$ | -- | -- | -- | 41 | 30 | 10% |
| Hurxthal and Vose (1969) | RP | $L_3$ | 167 | 20 | 9% | 237 | 20 | 10% |
| Haasner *et al.* (1967) | RP | L | 288 | 45 | 6% | 305 | 38 | 8% |
| Heuck (1970) | RP | Femur | 203 | 20 | 6% | 193 | 20 | 7% |
| Banzer *et al.* (1976) | SPA | CALC | 93 | 20 | 2% | 99 | 20 | 2% |
| Anatomical | | | | | | | | |
| Mueller *et al.* (1966) | DENS | L & IC | ≈ 50 | 50 | 13% | ≈50 | 50 | 13% |
| Weaver and Chalmers (1966) | DENS | $L_3$ | 67 | 40 | <5% | 66 | 40 | 9% |
| Arnold (1973) | DENS | $L_2$-$L_3$ | ≈300 | 25 | 10% | ≈300 | 25 | 10% |
| Dequeker *et al.* (1971) | DENS | IC | 70 | 20 | 2% | 59 | 20 | 8% |
| Ahuja (1969) | DENS | L & IC | 149 | 30 | 6% | 51 | 30 | 10% |
| Havivi *et al.* (1971) | DENS | $L_3$ | 150 | 30 | 7% | 65 | 30 | 8% |

(DPA - dual-photon absorptiometry; SPA - single-photon absorptiometry; RP - radiographic photodensitometry; DENS - apparent density; L - lumbar vertebrae; CALC - calcaneous; IC - iliac crest).

The development of computed tomography has made possible measurement of the central portion of the vertebral body, which is comprised of cancellous bone. The findings from these studies, using *in vivo* cross-sectional analysis of data for large populations of healthy women, appear to contradict the results of histologic studies. Rosenthal *et al.* (1989) reported on 57 women between the ages of 18 and 50 and found no decline in cancellous bone density between the ages of 18 and 44 years. Block *et al.* (1989), in a study on 538 healthy women aged 20–80 years, found a 60% loss from the spine over the life span. When a single curve was applied to the data, the best fit was obtained using a cubic form. These investigators also found only a small premenopausal loss, with an acceleration of loss around the menopause. When displayed by decades their data indicated that loss began at about age 40 years, but when displayed in 5-year intervals it appeared to begin later in the decade.

Nordin *et al.* (1990, 1992) reported the results of a study on 16 premenopausal women and 243 postmenopausal women. They found a self-limiting fall in cancellous bone density of the spine amounting to 35% in 25 years, with most of the loss occurring in the first 5 years postmenopause. Kalender *et al.* (1989) measured both compact and cancellous bone loss rates in 135 males and 139 females. They found a slightly higher rate of loss in women. They also noted that loss of bone mineral was higher in cancellous than in compact bone. A cubic model and a menopause breakpoint model fit the data for cancellous bone density loss in women better than a linear model.

### 3.5. Bone Growth Using Newer Techniques in Prepubertal and Adolescent White Children

Recently, DPA has been used to measure bone gain in children. DeSchepper *et al.* (1991) found no differences between boys and girls in a sample of 136 growing children aged 1–18 years. They observed prepubertal linear growth in lumbar spine density with a pubertal spurt (Figure 4).

Lloyd *et al.* (1992) measured total body bone mineral (TBBM) in 112 premenarchal white girls. They had achieved 90% of their predicted adult height, 68% of adult weight, 83% of adult total body bone mineral density (TBBMD) and 53% of total body bone mineral content (TBBMC). The authors cited data from the literature indicating there is no increase in BMD in women after age 20 years (Sowers *et al.*, 1991; Rosenthal *et al.*, 1989; Buchanan *et al.*, 1988; Dhuper *et al.*, 1990). This study gives some picture of the tremendous change in growth and increase in TBBMD that follow in adolescence, since the rate of increase in mineral accretion in adolescence will greatly exceed the increase in bone area. However, since this investigation did not include adults, it remains possible that there is bone gain after the completion of puberty.

DePriester *et al.* (1991) reported single photon absorptiometry measurements made at the 1/3 radial site in 420 children aged 4–10 years. They addressed body size issues in the study of growth by adjusting for height and weight. In attempting to adjust for bone width, they noted that the slope of bone BMC vs. BW was steeper in girls than in boys. The two curves intersected at age 8.0 in boys and age 9.2 years in girls, implying to these investigators that the boys had greater accretion in early childhood than girls. They also cite Garn's work suggesting that cortical width covers greater areas in boys, so that an increase in bone volume rather than BMD may underlie the difference in BMC between the sexes. Radial bone width diverges in the sexes prior to puberty without a divergence of BMC.

Trouerbach *et al.* (1991) point out that growth is three-dimensional. When they performed roentgen microdensitometry (aluminum wedge) on the left second digit of 1190 children aged 6.8–16.7 years, using both PA and lateral values, they found that girls had higher BMD values than boys at both diaphyseal sites and metaphyseal sites, and that there was a relationship between these values and skeletal age. The increase in BMD with increasing skeletal age was gradual in girls but steep in boys from age 8–10 years, corresponding to their relative increases in height and weight. Ponder *et al.* (1990), using DPA of the lumbar L2–L4 vertebrae of 184 boys and girls aged 5–11.9 years, found that a quadratic curve provided the best fit for data on the increase in bone mineral density, with body weight being an independent variable.

## 3.6. The Pubertal Spurt and Peak Bone Mass in Whites

Data on pubertal bone density changes are affected by body size, pubertal development, and by correction of regional density measurements for bone size. The range in bone density during adolescence is wide. In white girls the 95% confidence interval generally comprises 20% above and below average. Thus, a girl in the lowest percentile in adolescence has the same value as a woman near the top of the curve at age 70 years.

Bell *et al.* (1991) noted that the BMD of the lumbar spine is greater in girls than in boys between 7 and 12 years of age, whereas the trochanteric and femoral neck values are greater in boys. Katzman (1991) corrected for growth in the spine by introducing *BMAD,* the BMC corrected to a derived bone reference volume. They found that there was a plateau in bone density by age 16 years and that 50% of the change in spine mineral and 99% of the change in whole body mineral was due to bone expansion.

→

Figure 4. The change in spinal bone density in boys (upper) and girls (lower). Reproduced from De Schepper *et al.* (1991) with permission.

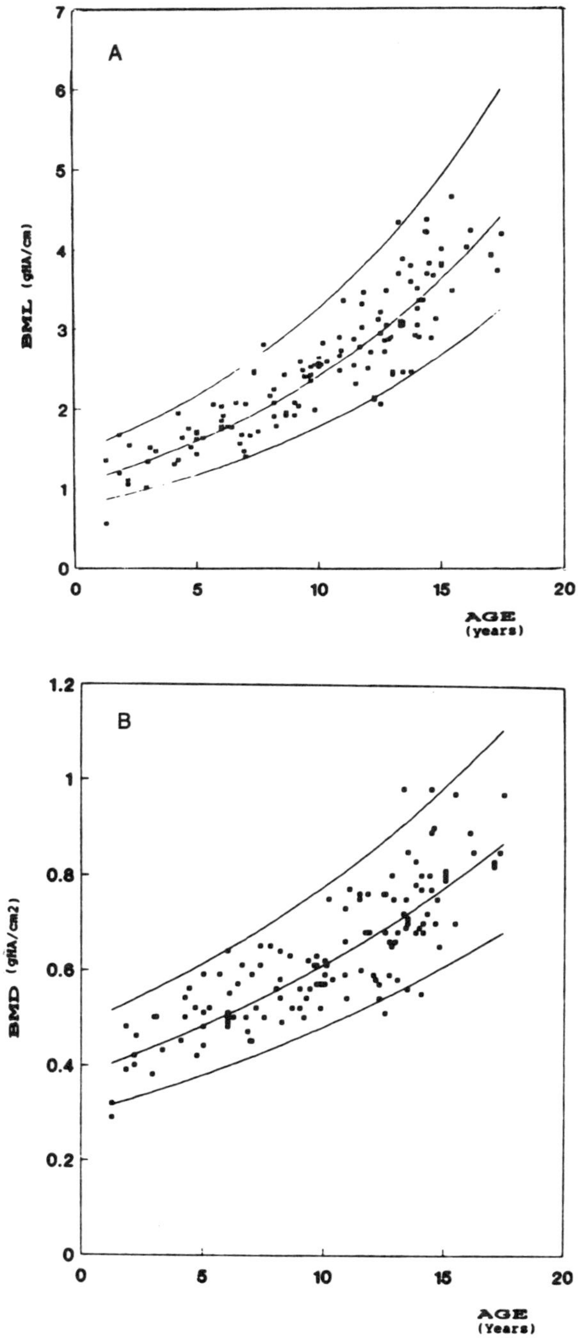
A
BML (gHA/cm)
AGE
(years)
B
BMD (gHA/cm2)
AGE
(Years)

Using DPA and CT to measure spinal density, others (Glastre *et al.,* 1990; Wahner *et al.,* 1988; Gilsanz *et al.*, 1988), have clearly shown that the major increase in BMD is related to stage of puberty and that differences between boys and girls are a result of girls entering and ending puberty at a younger age than boys. Thus, at age 12 the BMD of the spine is higher in girls because they are at a more advanced pubertal stage (Glastre *et al.,* 1990).

Bonjour *et al.* (1991) made measurements at L2–4, the femoral neck and the femoral shaft on 2078 white boys and girls aged 9–18 years and corrected the data for pubertal stage. There was a marked delay in increase in L2–L4 vs. age in boys compared to girls which vanished when the data were corrected for pubertal stage. At the end of the rapid growth spurt, males had higher mean values for the BMC of L2–L4, BMD of the femoral neck and BMD of the femoral shaft, but there was no sex difference for BMD of L2–L4. The girls exhibited a reduction in growth after 15 years of age and the changes in the BMD of L2–L4 and the femoral neck occurred between the 2d and 4th years after menarche (Figure 5, Figure 6). The BMD values at age 15 were similar to those reported for women between ages 20 and 35 years. This finding suggests that peak bone mass in the spine and radius is attained by age 15 years in girls. Other studies have also indicated that there is a peak by the end of the second decade. For example, in a study of 941 normal females using single photon absorptiometry of the radius, Johnston *et al.* (1975, 1979) found an increase in bone mass to age 20 years, then a plateau and then a decrease at age 40 years. Pun *et al.* (1991), in a study of 99 females, found a peak for total bone mineral in the 2d and 3d decade. In 120 women and 121 men aged 15–29 years, Rico *et al.* (1992), found that peak total bone mass occurred earlier in women than in men and at a lower value (Figure 7). Bonjour *et al.* (1991) also observed that males, consistent with their longer growth period, continued to increase in BMD between 15 and 18 years of age.

On the other hand, in a study of 351 Finnish women using DXA, peak BMD of the spine was observed at 31–35 years of age and in the femoral neck and Ward's triangle at 20–25 years (Laitinen *et al.,* 1991). The study of this population began at age 20 years so a peak before this age could not be detected. In another study on 165 women, lumbar spine density was found to peak in the 4th decade (Hall *et al.,* 1990), whereas Stevenson *et al.* (1989) found peak bone density in the spine and femur occurred at the end of linear skeletal growth. Measurements using other techniques and sites may give different values for peak bone density (Trouerbach *et al.,* 1991).

→

Figure 5. Bone mineral density of the lumbar spine and femur in boys and girls age 1-18. Reproduced from Bonjour *et al.* (1991) with permission.

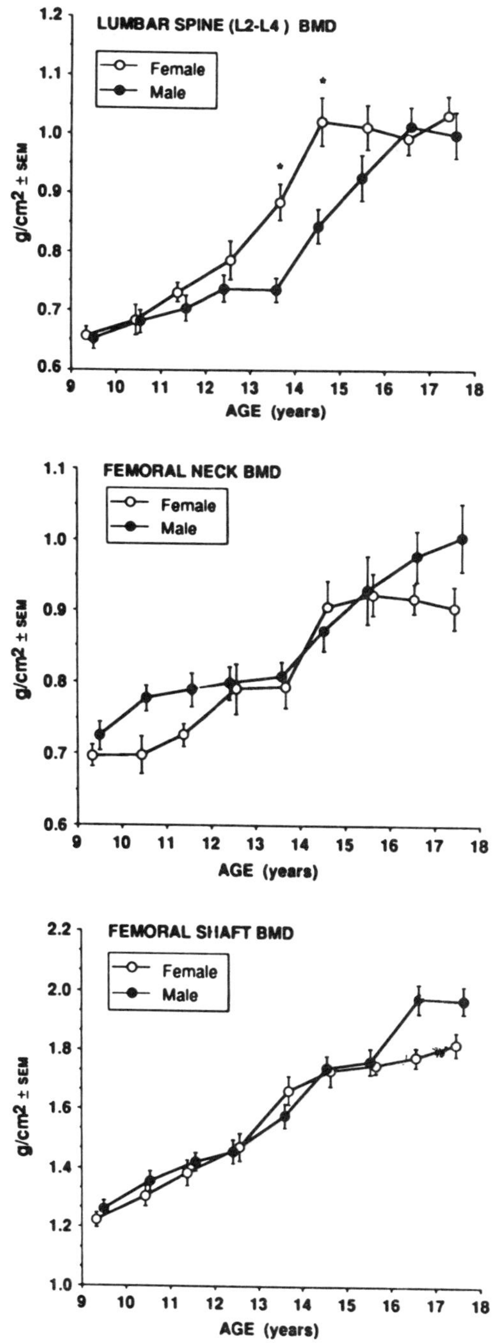
LUMBAR SPINE (L2-L4) BMD
Female
Male
g/cm2 ± SEM
AGE (years)
FEMORAL NECK BMD
Female
Male
g/cm2 ± SEM
AGE (years)
FEMORAL SHAFT BMD
Female
Male
g/cm2 ± SEM
AGE (years)

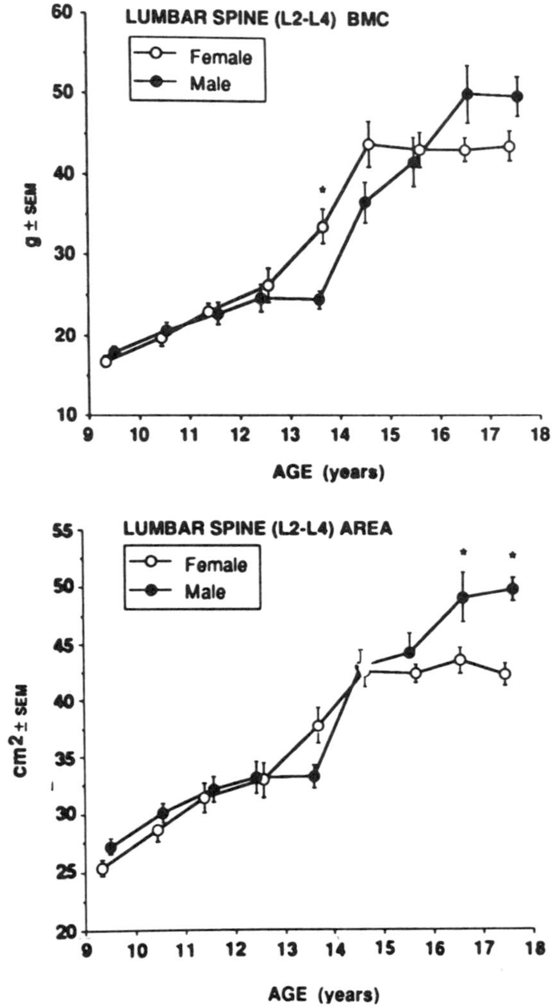


Figure 6. Lumbar spine BMC and lumbar spine area showing continued increases from age 15-18 years only in males. Reproduced from Bonjour *et al.* (1991) with permission.

DeSchepper *et al.* (1991), in a study on children, found a deceleration of the growth spurt in stage 4 of puberty. However, as their study included only children aged 1-18 years, their belief that the peak may have been as late as 30 years is only speculative.

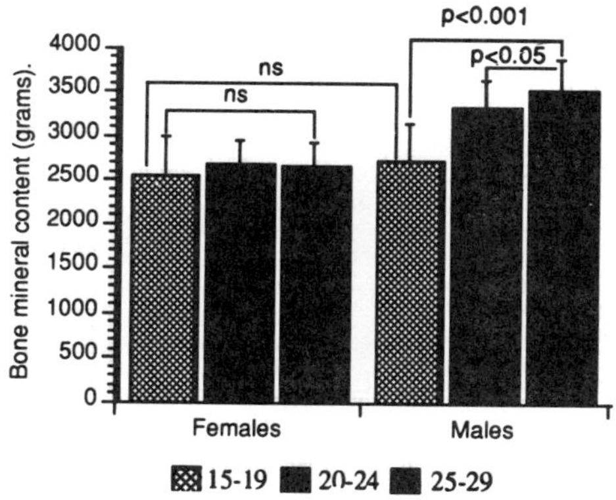


Figure 7. Peak bone mass occurs earlier in women and has a low value. Reproduced from Rico *et al.* (1992) with permission.

## 3.7. Adult Bone Gain and Loss

Early studies (Cohn *et al.*, 1981), using neutron activation analysis and single photon absorptiometry of the radius, suggested that there is a slow rate of loss beginning before menopause. This conclusion was derived by fitting a two-component linear regression line to the data. This model provided for a slow adult loss followed by a rapid postmenopausal loss and then resumption of the slower loss. An important question to be posed is whether there is a slow rate of loss (or gain) in women in the childbearing years. In Garn's model (1972) there was a period of skeletal consolidation. The practical issue here is whether efforts need to be made to maintain or improve bone mass in the adult premenopausal years. Unfortunately, the data on this question are conflicting, primarily because of differences in statistical treatment, sample size, and skeletal site tested. For instance, Nordin *et al.* (1990), who measured forearm bone mineral content in 485 normal women, found a slow linear premenopausal loss and an exponential postmenopausal loss. Nilas *et al.* (1988), using a polynomial fit, found a linear loss from the spine before menopause, and an exponential loss thereafter. Riggs *et al.* (1981, 1982, 1986) and Aloia *et al.* (1985b) obtained the most convincing evidence that there is a substantial premenopausal reduction in the bone density of the spine and femur.

In a study on 105 women aged 30–89 years, Riggs *et al.* (1981) observed a loss from the spine beginning in young adulthood and a loss from the radius beginning around menopause. They calculated a lifelong loss of 47% from the vertebrae, 30% from the midradius and 39% from the distal radius. This work was followed by a longitudinal study (n = 139, ages 20–88 years) of the midradius and lumbar spine (Riggs *et al.*, 1986). No change was found in the radius before menopause and a decline of 1.01% per year was observed thereafter. No difference was found in the rates of loss from the lumbar spine between premenopausal and postmenopausal women (−1.32% per year before and −0.97% per year after menopause). The authors proposed that half of vertebral bone loss occurs prior to menopause. The results of a longitudinal study (Aloia *et al.*, 1985a) also indicated that there is substantial premenopausal loss from the spine and femur. In an earlier cross-sectional study, Mazess *et al.* (1987) found that half of the 20–25% loss from the spine and femur occurred prior to menopause and that there was no loss over the age span 26–40 years. These findings assume a rapid perimenopausal loss. In a later study of 300 white women aged 20–39 years (a more homogeneous population), Mazess and Barden (1991) found no loss of bone density from the spine, femur, radius or humerus when patients were grouped by age increments of 5 years. In a 2-year longitudinal study no change was found in the spine and 1/3 radial site. These longitudinal studies suggesting that there is bone loss in 20–40 year old females were based on only about 19 subjects in the case of Riggs *et al.* (1986) and a similar number in that of Aloia *et al.* (1985a). Aloia *et al.* grouped their data by age (20–45 years) and found a doubling in the rate of loss in the older years. Recently, Recker *et al.* (1992) reported on 156 healthy women of college age followed for 5 years. They found gains in the third decade of life as follows: 4.8% for the radius, 5.8% for the spine, and 12.5% for total body bone mineral. They found positive correlations with calcium intake, physical activity and oral contraceptive use. Based on information obtained from prospective studies, it appears likely that there is a stage of skeletal consolidation after adolescence that is followed by a phase of slow premenopausal bone loss.

### 3.8. Perimenopausal and Early Postmenopausal Bone Loss

An accelerated phase of perimenopausal and postmenopausal bone loss has been found at both compact and cancellous bone sites for the total skeleton, the radius and the lumbar spine (Elliott *et al.*, 1990; Nilas *et al.*, 1988). Using DPA of the spine, the menopausal acceleration of bone loss was confirmed in a longitudinal study of 130 women (Lindquist *et al.*, 1983). Krolner *et al.* (1982) found that in the years approaching menopause, bone loss accelerates to a level of almost 6% per year. Sample size is critical in cross-sectional studies designed to detect accelerated bone loss over a short period of time. This has been demonstrated for the lumbar spine by Hui *et al.* (1987). Although van Berkum *et al.* (1988) could not find much postmenopausal bone loss in 171 Dutch women, Reginster *et al.* (1990), in a study of 695 postmenopausal women, found that a maximum loss occurred 5 years after menopause. Twenty years after menopause 50% of women had values below the 90th percentile of osteoporotic fracture patients. Nordin *et al.* (1992), using computed tomography of the spine, noted that most of the 35% of bone loss in the 25 years after menopause occurred within the first 5 years. In a large study of Japanese women with irregular menses, it was shown that bone loss from the spine began before menopause (Norimatsu *et al.*, 1989).

There are fewer published data on the femur; DPA measurement of the femur has been available for a shorter period of time than for the spine; consequently few studies of long duration have been conducted. Moreover, most such studies have used samples of insufficient size to detect premenopausal or postmenopausal accentuation of bone loss. The data available suggest that there is a linear loss with age from the femoral neck, Ward's triangle and intertrochanteric areas. Using total body and regional DXA, Pun *et al.* (1991) observed an acceleration of bone loss from Ward's triangle after menopause. On the basis of a study of a large heterogeneous population, Mazess *et al.* (1987) concluded that there is no loss from the femoral neck between the ages of 20 and 35 years but there is a loss of 25% between 40 and 70 years. In a study of 392 women, Gallagher *et al.* (1987) found a postmenopausal decline in the mass of the femur. More longitudinal studies will probably demonstrate an estrogen-dependent accentuation of bone loss from the femur in postmenopausal women.

It is pertinent to recall that during rapid perimenopausal bone loss there is not only a change in bone mass but also in cancellous architecture that has long-term consequences for the prevention of osteoporotic fractures. These changes, reviewed by Parfitt *et al.* (1983), include focal perforation of trabecular plates caused by increased resorption depth, generalized or focal reductions in trabecular thickness, preferential remodeling activity on inner trabeculae, and a stoichiometric relationship between resorption depth and trabecular thickness. The result is loss of trabeculae and connectivity. Since it appears that lost trabeculae cannot be reformed, this represents a loss of bone quality that cannot be changed by increasing the mass of remaining trabeculae. This is one reason why hormonal replacement therapy (or an alternative) is critical in the prevention of osteoporosis in women who are already on the low side of "average" in bone mass at menopause.

## 3.9. The Postmenopause

Most studies of aging bone loss suggest that there is a slowing of bone loss after the rapid perimenopausal loss of bone (Nilas *et al.,* 1988; Reginster *et al.,* 1990). Thus, when curve-fitting data for bone loss in cross-sectional studies in postmenopausal women, an exponential form (showing a reduction of bone loss with increasing age) usually fits best. This was found for the radius and the lumbar spine using computed tomography or dual energy absorptiometry (Nordin *et al.*, 1990). Again, there are fewer data available for the femur. Most studies indicate a lifetime linear loss from age 40 or younger with a more rapid loss from Ward's triangle and a slower loss from the trochanteric region (Vega *et al.*, 1991; Mazess *et al.*, 1987; Aloia *et al.*, 1985a; Riggs, 1982). Mazess *et al.* (1987) reported that, by age 70, women had lost 20% of mineral density from the spine, 25% from the femoral neck and 40% from Ward's triangle. Riggs *et al.* (1981, 1982, 1986), using a different absorptiometer, reported a larger lifetime loss from the spine of 42%, a loss of 58% from the femoral neck and 53% from the intertrochanteric area. Using the same instrument (dual energy photon absorptiometry) as used by Mazess, Aloia *et al.* (1985a) found a loss from age 20–80 of 19% for the spine, 26% from the femoral neck, 37% from Ward's triangle and 14% from the intertrochanteric area. The values are remarkably similar for white women in different geographic regions and with different ethnic backgrounds. The various dual energy absorptiometry instruments provide different values. It is critical that their manufacturers cross-calibrate to the same standard, something which they have agreed to do in principle.

### 3.10. Loss and Gain of Bone in the Aged

Hui *et al.* (1982) found that a quadratic function best fit their data for bone loss. From their model, it would be predicted that bone gain resumes at age 86 years (presumably because of continued subperiosteal apposition while endosteal resorption declines). On the other hand, Recker *et al.* (1988), using histomorphometry, have shown that the mineral apposition rate declines with aging. There are too few data to draw a conclusion regarding this aspect of the bone loss model.

### 3.11. Bone Gain and Loss in Men

Although bone gain and loss in men have been referred to from time to time, this review has been concentrated on women because osteoporosis is more common in women than in men. Boys are behind girls in bone density when considered by age but not by pubertal stage. By the end of the rapid growth spurt, males have higher values for lumbar BMC, femoral neck BMD and femoral shaft BMD but not for L2–4 BMD (Bonjour *et al.,* 1991). Males continue to increase their bone density to age 15–18 years, later than girls, an age corresponding to puberty. The greater sex deficit in the lumbar spine in later years must be due to gonadal deficiency in women.

A unique approach to sex differences was used by Kelly *et al.* (1990), who studied 29 pairs of dizygotic twins of different sex (aged 21–55 years) in which the female twin was premenopausal. The males had a higher value for bone mass of the radius but there was no difference in the femur and the females actually had higher lumbar spine values. This study design removes the influence of heredity, which may explain as much as 60–80% of bone mass. Riggs *et al.* (1982) found that lifetime loss from the spine in men was 1/4 that observed in women and that loss from the femoral neck and intertrochanteric region was 2/3 of that in women.

In a study of lumbar and proximal femoral bone density, Mazess *et al.* (1987) found that loss from the spine and trochanter was small between 26 and 70 years of age, whereas there was notable loss from the femoral neck (21%) and Ward's triangle (34%). The rate of loss was less than 30–40% of that seen in the femoral neck and Ward's triangle of women. The loss of 20% in the femoral neck of women still leaves men with a value for BMD of 0.15 g/cm$^2$ above that for age-matched women. A loss of 0.1 g/cm$^2$ doubles the risk for spine fracture and triples the risk for wrist fracture. The difference in spine BMD in a 70-year-old male and female is 0.3 g/cm$^2$, representing an 8–9-fold increase in risk in women. Loss of bone density from the distal radius (and os calcis) in men is 1/3 that observed in women.

## 3.12. Ethnic Heterogeneity

A similarity of values for bone density and patterns of bone loss has been well established (with occasional differences) for white men and women of European origin, whether they live in North America, Europe, Australia or New Zealand (Figure 8).

Garn *et al.* (1973) and Trotter *et al.* (1960) observed higher bone mineral density in American blacks than in whites, and this was confirmed with the measurement of total body calcium and bone density of the radius of healthy adults (Cohn *et al.,* 1977b). Bell *et al.* (1991) found that BMD of the lumbar spine, femoral neck and intertrochanteric region was greater in black children

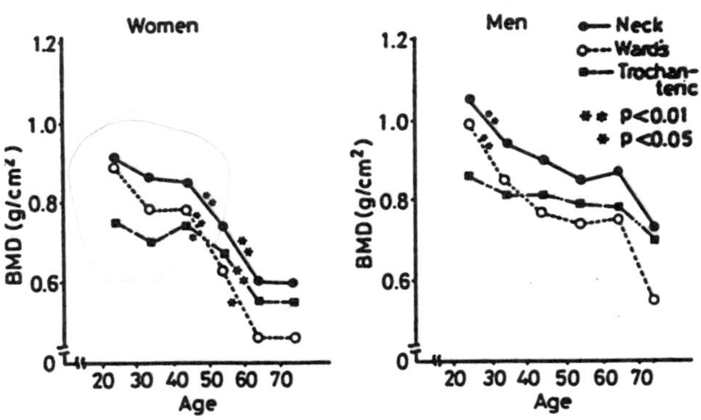


Figure 8. Bone mineral density of the femur from a European population. Reproduced from Norimatsu *et al.* (1989) with permission.

aged 7–12 years than in white children. Similar findings were reported by Liel *et al.* (1988). McCormick *et al.* (1991) combined data from studies on the lumbar spine of two age groups (ages 5–10.99 and 11–18.95 years). They found higher bone density values for blacks beginning by age 5 years. The mechanism is believed to involve slower bone resorption in blacks than in whites. In a study by Gilsanz *et al.* (1991), cancellous bone of the vertebrae was measured using computed tomography in 2–20 year olds matched for age and sexual development. Central vertebral bone density was the same for black and white males before puberty, but the pubertal increase in density in late puberty was much greater in black males (34% compared to 11%). These findings differ from those obtained using energy absorptiometry. Changes in compact bone are not detected using computed tomography. DPA is a measure of integral bone and is influenced by the size of the vertebral body. During growth the volume of the vertebrae increases more rapidly than the area and therefore the "overall density" increases more than the true density. Meier *et al.* (1992) recently showed, using longitudinal measurements, that premenopausal bone loss from the radius and lumbar spine occurred at a similar rate in blacks and whites. They also observed similar postmenopausal rates of loss from the lumbar spine and a lesser loss from the radial site in blacks in the early menopause.

Studies on Asians have also revealed differences in bone density values compared to those obtained on Americans and Western Europeans. In the study of Kin *et al.* (1991) on 1,048 Japanese women and 248 men, using DXA of the lumbar spine, peak bone mass in men was observed at age 20–29 and bone loss was seen in women prior to menopause, a pattern similar to the findings of several studies on U.S. white women. Seto *et al.* (1990) reported normative values for Japanese in a smaller number of subjects, and found lumbar bone density loss beginning in the mid-thirties. They pointed out that the "fracture threshold" is much lower for Japanese women than American women. Longitudinal studies of radial BMD in Japanese women confirmed that a pattern of substantial loss occurs only after menopause (Hagino *et al.*, 1992). Despite a lower BMD in Japanese than in American white women and men, there is a lower fracture incidence in the Japanese (Figure 9).

Tsai *et al.* (1991) reported DPA measurements of the lumbar spine and hips of 116 healthy Taiwanese women. The pattern of loss for the spine was similar to that observed in the United States, but the values for the hip were 10–15% lower than those for U.S. whites and were similar to values observed in the Japanese. Yano *et al.* (1984) found that the forearm BMC of Japanese immigrants in Hawaii was lower than that of U.S. mainland Japanese immigrants, suggesting an influence of environmental factors. Sugimoto *et al.* (1992) compared the spine and femur BMD of Japanese, Koreans, and Taiwanese using DPA. There were differences in the values and perhaps even in the pattern of bone loss (Figure 10).

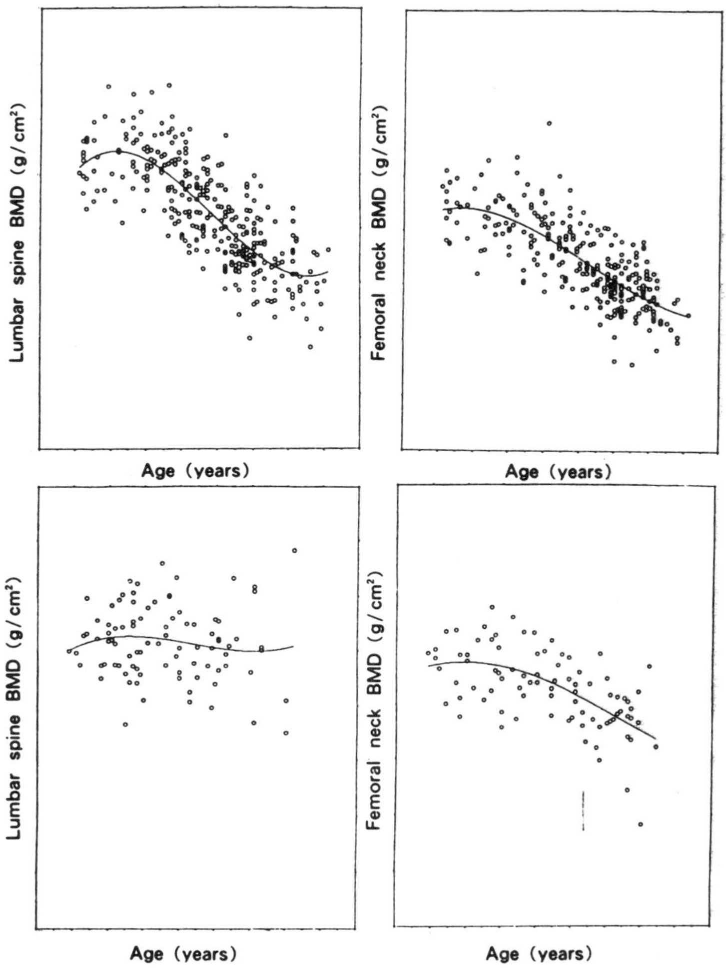


Figure 9. Bone mineral density of the lumbar spine (left) and femoral neck in Japanese women (upper) and men (lower). Reproduced from Sugimoto *et al.* (1992) with permission.

It is evident that there are ethnic and cultural differences that influence normative bone density data. This should be expected, since osteopenia is multifactorial in its origin, involving not only a strong heritable influence but also environmental factors such as physical activity, calcium intake, mechanical loading, cigarette smoking, alcohol intake, and hormonal influences. There is an inference in the literature that the spine may be influenced more than the femur by nutritional and hormonal factors, whereas mechanical loading may have more influence on the femur.

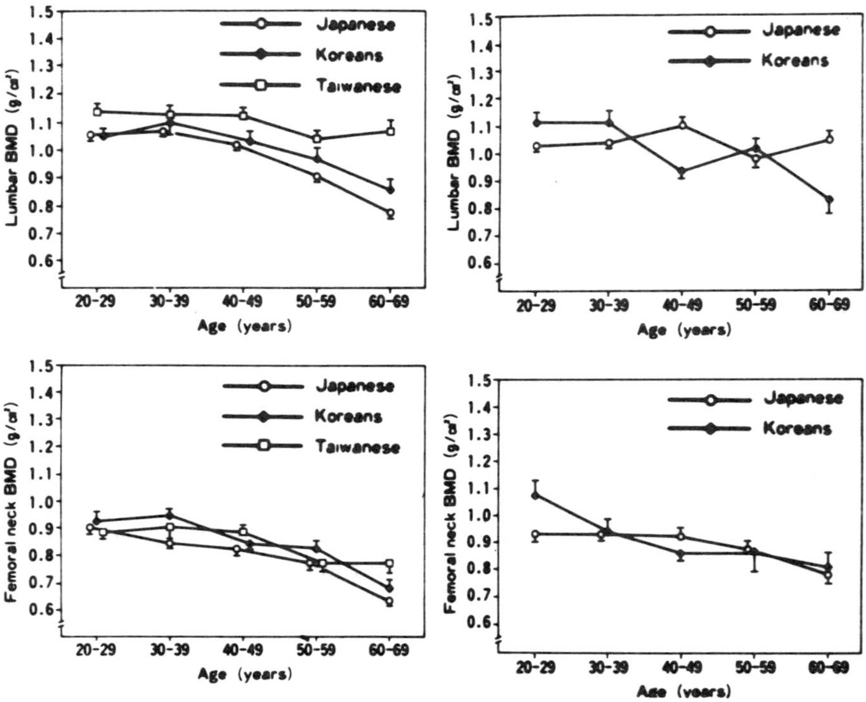


Figure 10. A comparison using dual photon absorptiometry of BMD in Japanese, Koreans and Taiwanese. Reproduced from Sugimoto (1992) with permission.

## 4. Conclusion

In reading the following chapters against this backdrop of information on bone gain and loss, it will be useful to recall that bone strength is related not only to bone mass but also to bone quality. The architectural changes that occur in the perimenopausal period, involving loss of whole trabeculae and of connectivity are probably irreversible, making prevention of osteoporosis much more desirable than treatment. Propensity to fall is an important consideration in osteoporotic fractures, particularly in hip fracture in the elderly. There are ethnic differences in bone density, blacks having higher values than whites and whites than Asians. Despite their lower bone density Japanese women have a lower incidence of fractures.

There is an extraordinary doubling of the skeletal mass during the first year of life. There is quadratic growth in boys and girls prior to the onset of puberty. Prepubertal boy/girl comparisons are complicated by changes in the size of bones and differences in height and weight and skeletal age. Boys may have greater accretion in early childhood than girls, but the results of most studies, when corrected for height and weight, indicate that prepubertal girls and boys have equal bone density. The rapid growth spurt that follows in adolescence produces male/female differences that are related to pubertal stage and body size. Girls start and end puberty earlier than boys. Prior to puberty girls have attained 83% of their total body bone mineral density but only 53% of their total body bone mineral content. Since peak bone density is attained by the end of adolescence, the tremendous adolescent accretion of mineral is apparent. Although males may have higher BMD of the femoral neck, there are no differences in BMD of the lumbar spine at the end of adolescence. The proposition that further skeletal accretion occurs between the ages of 20 and 40 years has received little support from cross-sectional studies of total body bone mineral and density of the spine, radius, and femur in women, but total body bone mineral may be increased slightly in males 25–29 years old. One prospective study indicates that there is bone gain in women during the third decade.

There is controversy as to how much premenopausal bone loss occurs. However, it appears that there is a slow loss of bone in adulthood that accelerates about 5 years before menopause and continues for 5–6 years after cessation of menses. There is evidence for menopausal accentuation at every skeletal site tested, including cancellous and compact bone tissue. Men undergo a slow linear loss of bone from middle adulthood until old age. The question whether bone loss ends in very old age is not entirely resolved.

Many components of this model require large scale longitudinal studies for confirmation. In evaluating the evidence for an influence of various factors on bone mass, remaining questions in the development of a definitive model of bone gain and loss must be considered. It is a rational inference that the stages of the life cycle when there is the most bone gain and loss (childhood, adolescence, menopause) are the stages when intervention to maximize bone mass will be most successful. However, it remains possible that increments in bone mass also may be successfully achieved during those periods when there are slower rates of skeletal change.

## References

Ahuja, M., 1969, Normal variation in the density of selected human bones in North India, *J. Bone Joint Surg.* **51B**:719.

Aloia, J.F., Ross, P., Vaswani, A., Zanzi, I., and Cohn, S., 1982, Rate of bone loss in postmenopausal and osteoporotic women, *Am. J. Physiol.* **242:**E82.

Aloia, J.F., Vaswani, A., Yeh, J.K., Ross, P., Ellis, K., and Cohn, S.H., 1983, Determinants of bone mass in postmenopausal women, *Arch. Intern. Med.* **143:**1700.

Aloia, J.F., Vaswani, A., Ellis, K., Yuen, K., and Cohn, S.H., 1985a, A model for involutional bone loss, *J. Lab. Clin. Med.* **106**:630.

Aloia, J.F., Cohn, S.H., Vaswani, A., Yeh, J.K., and Ellis, K., 1985b, Risk factors for postmenopausal osteoporosis, *Am. J. Med.* **78**:95.

Aloia, J.F., McGowan, D., Vaswani, A., Ross, P., and Cohn, S.H., 1991, The relationship of menopause to skeletal and muscle mass, *Am. J. Clin. Nutr.* **53**:1378.

Arnold, J.S., 1973, Amount and quality of trabecular bone in osteoporotic vertebral fractures, *Clin. Endocrinol. Metab.* **2**:221.

Avioli, L. V., 1987, *The Osteoporotic Syndrome: Detection, Prevention and Treatment*, 2d ed., W. B. Saunders, Philadelphia.

Banzer, D.H., Schneider, U., Risch, W.D., and Botsch, H., 1976, Roentgen signs of vertebral demineralization and mineral content of peripheral cancellous bone. *Am. J. Roentgen.* **126**:1306.

Basle, M.F., Mauras, Y., and Audran, M., 1990, Concentration of bone elements in osteoporosis, *J. Bone Min. Res.* **5**:41.

Beck, T.J., Ruff, C.B., Scott, W.W., Jr., Plato, C., Tobin, J.D., and Quan, C.A., 1992, Sex differences in geometry of the femoral neck with aging: A structural analysis of bone mineral data, *Calcif. Tissue Int.* **50**:24.

Bell, N.H., Shary, J., Stevens, J., Garza, M., Gordon, L., and Edwards, J., 1991, Demonstration that bone mass is greater in black than in white children, *J. Bone Min. Res.* **6**:719.

Birkenhager-Frenkel, D.H., Courpron, F., Hupscher, E.A., Clermonts, E., Coutino, M.F., Schmitz, P.I.M., and Meunier, P.J., 1988, Age-related changes in cancellous bone structure: A two-dimensional study in the transiliac and iliac crest biopsy sites, *Bone and Mineral* **4**:197.

Block, J.E., Smith, R., Steiger, P., Glueer, C.C., Ettinger, B., and Genant, H.K., 1989, Models of spinal trabecular bone loss as determined by quantitative computed tomography, *J. Bone Min. Res.* **4**:249.

Bonjour, J.P., Theintz, G., and Buchs, B., 1991, Critical years and stages of puberty for spinal and femoral bone mass accumulation during adolescence, *J. Clin. Endocrinol. Metab.* **73**:555.

Buchanan, J.R., Myers, C., Lloyd, T., and Greer, R.B., 1988, Early vertebral trabecular bone loss in normal premenopausal women, *J. Bone. Min. Res.* **3**:583.

Burkhardt, R., Kettner, G., Bohm, W., Schmidmeier, M., Schlag, R., Frisch, B., Mallmann, B., Eisenmenger, W., and Gilg, T.H., 1987, Changes in trabecular bone, hematopoiesis and bone marrow vessels in aplastic anemia, primary osteoporosis, and old age: a comparative histomorphometric study, *Bone* **8**:157.

Chalmers, J. and Weaver, J.K., 1966, Cancellous bone: its strength and changes with aging and an evaluation of some methods for measuring its mineral content, *J. Bone Joint Surg.* **48A**:299.

Chappard, D., Alexandre, CH., Robert, J.M., and Riffat, G., 1991, Relationships between bone and skin atrophies during aging, *Acta Anat.* **41**:239.

Cohn, S.H., Vaswani, A.N., Aloia, J.F., Roginsky, M.S., Zanzi, I., and Ellis, K.J., 1976a, Changes in body chemical composition with age measured by total body neutron activation, *Metabolism* **25**:85.

Cohn, S.H., Vaswani, A.N., Zanzi, I., and Ellis, K., 1976b, The effect of aging on bone mass in adult women, *Am. J. Physiol.* **230**:143.

Cohn, S.H., Abesamis C., Zanzi, I., Aloia, J.F., Yasumura, S., and Ellis, K.J., 1977a, Body elemental composition: comparison between black and white adults, *Am. J. Physiol.* **232:**E419.

Cohn, S.H., Abesamis C., Yasumura, S., Aloia, J.., Zanzi, I., and Ellis, K.J., 1977b, Comparative skeletal mass and radial bone mineral content in black and white women, *Metabolism* **26**:171.

Cohn, S.H., Aloia, J.F., Vaswani, A.N., Zanzi, I., Varetsky, D., and Ellis, K., 1981, Age and sex related changes in bone mass measured by neutron activation. Abstract at the International Symposium on Osteoporosis, Jerusalem, Israel, 5/31-6/4/81 (unpubl.).

Cohn, S.H., Aloia, J.F., Vaswani, A.N.,Yuen, K., Yasumura, S., and Ellis, K.J. 1984, "Model for determining women at risk for developing osteoporosis: by total body neutron activation, photon absorptiometry of spine and radius" (abstract), in: Proceedings of the Copenhagen International Symposium on Osteoporosis, June 3-8, 1984, Wiley, New York.

Compston, J.E., Mellish, R.W., and Garrahan, N.J., 1987, Age-related changes in iliac crest trabecular microanatomic bone structure in man, *Bone* **8:**289.

Davis, J.W., Ross, P.D., Vogel, J.M., and Wasnich, R., 1991, Age-related changes in bone mass among Japanese-American men, *Bone and Mineral* **15:**227.

Dempster, D.V., Shane, E., Horbert, W., and Lindsay, R., 1986, A simple method for correlative light and scanning electron microscopy of human iliac crest bone biopsies: qualitative observations in normal and osteoporotic subjects, *J. Bone Min. Res.* **1:**15.

DePriester, J.A., Cole, T.J., and Bishop, N.J., 1991, Bone growth and mineralization in children aged 4 to 10 years, *Bone and Mineral* **12:**57.

Dequeker, J., Remans, J., Franssen, R., and Waes, J., 1971, Ageing patterns of trabecular and cortical bone and their relationship, *Calcif. Tissue Res.* **7:**23.

DeSchepper, J., Derde, M.P., Van den Broeck, M., Piepsz, A., and Jonckheer, M.H., 1991, Normative data for lumbar spine bone mineral content in children: influence of age, height, weight, and pubertal stage, *J. Nuclear Med.* **32**:216.

Dhuper, S., Warren, M.P., Brooks-Gunn, J., and Fox, R., 1990, Effects of hormonal status on bone density in adolescent girls, *J. Clin. Endocrinol. Metab.* **71:**1083.

Elliott, J.R., Gilchrist, N.L., Wells, J.E., Turner, J.G., Ayling, E., Gillespie, W.J., Sainsbury, R., Hornblow, A., and Donald, R.A., 1990, Effects of age and sex on bone density at the hip and spine in a normal Caucasian New Zealand population, *NZ Med. J.* **103:**33.

Evans, F.G., 1976, Mechanical properties and histology of cortical bone from younger and older men, *Anat. Rec.* **185:**1.

Eyre, D.R., Dickson, I.R., and Van Ness, K., 1988, Collagen cross-linking in human bone and articular cartilage, *Biochem. J.* **252:**495.

Field, R.E., Dixon, A.K., Lawrence, J.P., and Rushton, N., 1990, Bone density distribution within the femoral head and neck, *Skeletal Radiol.* **19:**319.

Foldes, J., Parfitt, A.M., Shih, M.S., Rao, D.S., and Kleerekoper, M., 1991, Structural and geometric changes in iliac bone: relationship to normal aging and osteoporosis. *J. Bone Min. Res.* **6:**759.

Gallagher, J.C., Goldgar, D, and Moy, A., 1987, Total bone calcium in normal women: Effect of age and menopause status, *J. Bone Min. Res.* **2:**491.

Garn, S.M., 1972, The course of bone gain and the phases of bone loss, *Orthop. Clin. N. Am.* **3:**503.

Garn, S.M., Clark, D.C., and Trowbridge, F.L., 1973, Tendency toward greater stature in American black children, *Am. J. Dis. Child.* **126:**164.

Gilsanz, V., Gibbens, D.T, and Roe, T.F., 1988, Vertebral bone density in children: effect of puberty, *Radiology* **166:**847.
Gilsanz, V., Roe, T.F., Mora, S., Costin, G., and Goodman, W., 1991, Changes in vertebral bone density in black girls and white girls during childhood and puberty, *New Engl. J. Med.* **535:**1597.
Glastre, C., Braillon, P., David, L., Cochat, P., Meunier, P.J., and Delmas, P.D., 1990, Measurement of bone mineral content of the lumbar spine by dual energy x-ray absorptiometry in normal children: Correlations with growth parameters, *J. Clin. Endocrinol. Metab.* **70:**1330.
Haasner, E., Krokowski, E., and Bach, K., 1967, Normalwerte des Hydroxylapatitgehaltes im Skelet in Abhangigkeit von Lakalisation, Lebensalter und Geschlecht, *Klin. Wochenschr.* **45**:575.
Hagino, H., Yamamoto, K., Teshima, R., Kishimoto, H., and Kagawa, T.,1992, Radial bone mineral changes in pre- and postmenopausal healthy Japanese women: Cross-sectional and longitudinal studies, *J. Bone Min. Res.* **7:**147.
Hall, M.L., Heavens, J., Cullum, I.D., and Ell, P.J., 1990, The range of bone density in normal British women, *Br. J. Radiol.* **6:**366.
Hansson, T., and Roos, B., 1980, The amount of bone mineral in the lumbar spine in women 35 to 80 years of age, Presented at Tenth European Symposium on Osteoarthrology, Malmo, Sweden.
Havivi, E., Reshef, A., Schwartz, A., Guggenheim, K., Bernstein, D.S., Hegsted, D.M., and Stare, F.J., 1971, Comparisons of metacarpal bone loss with physical and chemical characteristics of vertebrae and ribs, *Israel J. Med. Sci.* **7:**1055.
Heuck, F.H.W., 1970, "Quantitative measurements of mineral content in bone diseases," in: *Symposium Ossium* (A. M. Jelliffe and B. Strickland, eds.), E.S. Livingstone, Edinburgh.
Hoiseth, A., Alho, A., and Husby, T., 1990, Femoral cortical/cancellous bone related to age, *Acta Radiol.* **31:**626.
Hui, S.L., Wiske, P.S., Norton, J.A., and Johnston, C.C., 1982, A prospective study of change in bone mass with age in postmenopausal women, *J. Chron. Dis.* **35:**715.
Hui, S.L., Slemenda, C.W., Johnston, C.C., and Appledorn, C.R., 1987, Effects of age and menopause on vertebral bone density, *Bone and Mineral* **2:**141.
Hurxthal, L.M., and Vose, G.P., 1969, The relationship of dietary calcium intake to radiographic bone density in normal and osteoporotic persons, *Calcif. Tissue Res.* **4:**245.
Johnston, C.C., Jr., Smith, D.M., and Khairi, M.R.A., 1975, "Prospective and cross-sectional study of radial bone loss in post-menopausal women," in: *Calcified Tissues,* Proceedings of XIth European Symposium on Calcified Tissues (S. Pors Nielsen, ed.), Springer-Verlag, New York.
Johnston, C.C., Jr., Norton, J.A., Jr., Khairi, R.A., and Longcope, C., 1979, "Age-related bone loss," in: *Osteoporosis II.* ( U. Barzell, ed.), pp. 59-72, Grune & Stratton, New York.
Kalender, W.A., Felsenberg, D., Louis, O., Lopez, P., Klotz, E., Osteau, M., and Fraga, J., 1989, Reference values for trabecular and cortical vertebral bone density in single and dual-energy quantitative computed tomography, *Europ. J. Radiol.* **9:**75.
Katzman, D.K., Bachrach, L.K., Carter, D.R., and Marcus, R., 1991, Clinical and anthropometric correlates of bone mineral acquisition in healthy adolescent girls, *J. Clin. Endocrinol. Metab.* **73:**1332.

Kelly, P.J., Twomey, L., Sambrook, P.N., and Eisman, J.A., 1990, Sex differences in peak adult bone mineral density, *J. Bone Min. Res.* **5:**1169.

Kin, K., Kushida, K., Yamazaki, K., Okamoto, S., and Inoue, T., 1991, Bone mineral density of the spine in normal Japanese subjects using dual-energy x-ray absorptiometry: effect of obesity and menopausal status, *Calcif. Tissue Int.* **49:**101.

Kragstrup, J., Melsen, F., and Mosekilde, L., 1983, Thickness of lamellae in normal human iliac trabecular bone, *Metab. Bone Dis. Rel. Res.* **4:**291.

Krolner, B., and Nielsen, S., 1982, Bone mineral content of the lumbar spine in normal and osteoporotic women: cross-sectional and longitudinal studies, *Clin. Sci.* **62:**329.

Krolner, B., and Pors Nielsen, S., 1980, Measurement of bone mineral content (BMC) of the lumbar spine. I: Theory and application of a new two-dimensional dual-photon attenuation method, *Scand. J. Clin. Lab. Invest.* **40:**653.

Laitinen, K., Valimaki, M., and Keto, P., 1991, Bone mineral density measured by dual-energy x-ray absorptiometry in healthy Finnish women, *Calcif. Tissue Int.* **48:**224.

Liel, Y., Edwards, J, Shary, J., Spicer, D.M., Gordon, L., and Bell, N.H., 1988, The effect of race and body habitus on bone mineral density of the radius, hip, and spine in premenopausal women, *J. Clin. Endocrinol. Metab.* **66:**1247.

Lindquist, O., Bengtsson, C., Hansson, T., and Jonsson, R., 1983, Changes in bone mineral content of the axial skeleton in relation to aging and the menopause, *Scand. J. Clin. Lab. Invest.* **43:**333.

Lloyd, T., Rollings, N., Andon, M.B., Demers, L.M., Eggli, D.F., Kieselhorst, K., Kulin, H., Landis, J.R., Maratel, J.K., Orr, G., and Smith, P., 1992, Determinants of bone density in young women: I. Relations among pubertal development, total body bone mass, and total body bone density in premenarchal females, *J. Clin. Endocrinol. Metab.* **75:**383.

Marcus, R., Kosek, J., Pfefferbaum, A., and Horning, S., 1983, Age-related loss of trabecular bone in premenopausal women: a biopsy study, *Calcif. Tissue Int.* **35:**406.

Mazess, R.B., 1982, On aging bone loss, *Clin. Orthop. Rel. Res.* **165:**239.

Mazess, R.B., and Barden, H.S., 1991, Bone density in premenopausal women: effects of age, dietary intake, physical activity, smoking, and birth-control pills, *Am. J. Clin. Nutr.* **53:**132.

Mazess, R.B., Barden, H.S., Ettinger, M., Johnston, C., Dawson-Hughes, B., Baran, D., Powell, M., and Notelovitz, M., 1987, Spine and femur density using dual-photon absorptiometry in US white women, *Bone and Mineral* **2:**211.

McCormick, D.P., Ponder, S.W., Fawcett, H.D., and Palmer, J.L., 1991, Spinal bone mineral density in 335 normal and obese children and adolescents: evidence for ethnic and sex differences, *J. Bone Min. Res.* **6:**507.

Meier, D., Luckey, M., Wallenstein, S., and Lapinski, R., 1992, Significant premenopausal bone loss in white and black women: a longitudinal study, *J. Bone Min. Res.* **7:**S135 (abstract).

Mosekilde, L., 1988, Age-related changes in vertebral trabecular bone architecture—assessed by a new method, *Bone* **9:**247.

Mosekilde, L., Mosekilde, L., and Danielsen, C.C., 1987, Biomechanical competence of vertebral trabecular bone in relation to ash density and age in normal individuals, *Bone* **8:**79.

Mueller, K.H., Trias, A. and Ray, R.D., 1966, Bone density and composition: Age-related and pathological changes in water and mineral content, *J. Bone Joint Surg.* **48A:**140.

Nilas, L., Gotrfredsen, A., Hadberg, A., and Christiansen, C., 1988, Age-related bone loss in women evaluated by the single and dual photon technique, *Bone and Mineral* **4**:95.

Nordin, B.E.C., Need, A.G., and Chatterton, B.E., 1990, The relative contributions of age and years since menopause to postmenopausal bone loss, *J. Clin. Endocrinol. Metab.* **70**:83.

Nordin, B.E.C., Need, A.G., Bridges, A., and Horowitz, M., 1992, Relative contributions of years since menopause, age, and weight to vertebral density in postmenopausal women, *J. Clin. Endocrinol. Metab.* **74**:20.

Norimatsu, H., Mori, S., Uesato, T., Yoshikawa, T., and Kasuyama, N., 1989, Bone mineral density of the spine and proximal femur in normal and osteoporotic subjects in Japan, *Bone and Mineral* **5**:213.

Parfitt, A.M., Mathews, C.H.E., Villanueva, A.R., Kleerekoper, M., Frame, B., and Rao, D.S., 1983, Relationships between surface, volume, and thickness of iliac trabecular bone in aging and in osteoporosis, *J. Clin. Invest.* **72**:1396.

Pesch, H.J., Becker, T., and Bischoff, W., 1990, 'Physiological osteoporosis' and 'osteoblast insufficiency' in old age, *Arch. Orthop. Trauma Surg.* **110**:1.

Ponder, S.W., McCormick, D.P., Fawcett, H.D., Palmer, J.L., McKernan, M.G., and Brouhard, B.H., 1990, Spinal bone mineral density in children aged 5.00 through 11.99 years, *Am. J. Dis. Child.* **144**:1346.

Pun, K.K., Wong, F.H.W., and Loh, T., 1991, Rapid postmenopausal loss of total body and regional bone mass in normal southern Chinese females in Hong Kong, *Osteoporosis Int.* **1**:87.

Recker, R.R., Kimmel, D.B., and Parfitt, A.M., 1988, Static and tetracycline-based bone histomorphometric data from 34 normal postmenopausal females, *J. Bone Min. Res.* **3**:133.

Recker, R.R, Davies, K.M., Hinders, S.M., Heaney, R.P., Stegman, M.R., and Kimmel, D.B., 1992, Bone gain in young adult women, *JAMA* **368**:2403.

Reginster, J.Y., Deroisy, R., Albert, A., Sarlet, N., Collette, J., and Franchimont, P., 1990, Dual photon absorptiometry of lumbar spine in West European (Belgian) postmenopausal females: normal range and fracture threshold, *Clin. Rheumatol.* **9**:220.

Rico, H., Revilla, M., Hernandez, E.R., Villa, L.F., and Alvarez del Buergo, M., 1992, Sex differences in the acquisition of total bone mineral mass peak assessed through dual-energy x-ray absorptiometry, *Calcif. Tissue Int.* **51**:251.

Riggs, B.L., 1991a, *Physician's Resource Manual on Osteoporosis*, 2d ed., National Osteoporosis Foundation, Washington.

Riggs, B.L., 1991b, Overview of osteoporosis, *West. J. Med.* **154**:63.

Riggs, B.L., Wahner, H.W., Seeman, E., Offord, K.P., Dunn, W.L., Mazess, R.B., Johnson, K.A. and Melton, L.J., 1982, Changes in bone mineral density of the proximal femur and spine with aging, *J. Clin. Invest.* **70**:716.

Riggs, B.L., Wahner, H.W., Dunn, W.L., Mazess, R.B., Offord, K.P., and Melton, L.J., III, 1981, Differential changes in bone mineral density of the appendicular skeleton with aging, *J. Clin. Invest.* **67**:328.

Riggs, B.L., Wahner, H.W., Melton, L.J., III, Richelson, L.S., Judd, H.L. and Offod, K.P., 1986, Rates of bone loss in the appendicular and axial skeletons of women, *J. Clin. Invest.*, **77**:1487.

Rosenthal, D.I., Mayo-Smith, W., Hayes, C.W., Khurana, J.S., Biller, B., Neer, R.M., and Klibanski, A., 1989, Age and bone mass in premenopausal women, *J. Bone Min. Res.* **4**:533.

Rubin, C.D., 1991, Southwestern internal medicine conference: Age-related osteoporosis, *Am. J. Med. Sci.* **301**:281.

Sambrook, P.N., Eisman, J.A., Furler, S.M., and Pocock, N.A., 1987, Computer modeling and analysis of cross-sectional bone density studies with respect to age and the menopause, *J. Bone Min. Res.* **2**:109.

Seto, H., Kamei, T., Futatsuya, R., Banba, Y., Ihaa, F., Kakishita, M., and Nanbu, I., 1990, Bone mineral density of the lumbar spine by dual photon absorptiometry: Age-related regression in normal Japanese subjects and fracture threshold in osteoporosis, *Radiation Med.* **8**:61.

Sowers, M.F., Kshirsagar, A., Crutchfield, M., and Updike, S., 1991, Body composition, age, and femoral bone mass of young adult women, *Ann. Epidemiol.* **1**:245.

Stevenson, J.C., Lees, B., Devenport, M., Cust, M.P., and Ganger, K.F. 1989, Determinants of bone density in normal women: risk factors for future osteoporosis? *BMJ* **298**:924.

Sugimoto, T., Tsutsumi, M., Fujii, Y., Kawalaatsi, M., Negishi, H., Lee, M., Sai, K., Ukase, M., and Fujita, T., 1992, Comparison of bone mineral content among Japanese, Koreans, and Taiwanese assessed by dual-photon absorptiometry, *J. Bone Min. Res.* **7**:153.

Thompson, D.D., 1980, Age changes in bone mineralization, cortical thickness, and Haversian canal area, *Calcif. Tissue Int.,* **31**:5.

Trotter, M., Broman, G.E., and Peterson, R.R., 1960, Densities of bones of white and Negro skeletons, *J. Bone Joint Surg. (Am.)* **42**:50.

Trouerbach, W.T., de Man, S.A., Gommers, D., Zwamborn, A., and Grobbee, D.E., 1991, Determinants of bone mineral content in childhood, *Bone and Mineral* **13**:55.

Tsai, K.S., Huang, K.M., and Chieng, P.U., 1991, Bone mineral density of normal Chinese women in Taiwan, *Calcif. Tissue Int.* **48**:161.

Twomey, L., Taylor, J. and Furniss, B., 1983, Age changes in the bone density and structure of the lumbar vertebral column, *J. Anat.* **136**:15.

van Berkum, F.N.R., Pols, H.A.P., Kooij, P.P.M., and Birkenhager, J.C., 1988, Peripheral and axial bone mass in Dutch women: relationship to age and menopausal state, *Neth. J. Med.* **32**:226.

Vega, E., Mautalen, C, Gomez, H., Garrido, A., Melo, L., and Sahores, A.O., 1991, Bone mineral density in patients with cervical and trochanteric fractures of the proximal femur, *Osteoporosis Int.* **1**:81.

Wahner, H.W., Dunn, W.L., Brown, M.L., Morin, R.L., and Riggs, B.L., 1988, Comparison of dual-energy x-ray absorptiometry and dual photon absorptiometry for bone mineral measurements of the lumbar spine, *Mayo Clin. Proc.* **63**:1075.

Weaver, J.K., and Chalmers, J., 1966, Cancellous bone: its strength and changes with aging and an evaluation of some methods for measuring its mineral content. I. Age changes in cancellous bone, *J. Bone Joint Surg.* **48A**:289.

Whyte, M.P., Bergfeld, M.A., Murphy, W.A., Avioli, L.V., and Teitelbaum, S.L., 1982, Postmenopausal osteoporosis, a heterogeneous disorder as assessed by histomorphometric analysis of iliac crest bone from untreated patients, *Am. J. Med.* **73**:193.

Yano, K., Wasnich, R.D., Vogel, J.M., and Heilbrun, K.L., 1984, Bone mineral measurements among middle-aged and elderly Japanese residents in Hawaii, *Am. J. Epidemiol.* **119**:751.

Chapter 2

# Low Bone Mass in Past and Present Aboriginal Populations

Susan K. Pfeiffer and Richard A. Lazenby

## 1. Introduction

A slight and gradual loss of bone mass is characteristic of all aging primates, if they live long enough (Garn, 1970; Burr, 1980). Nevertheless, the observation of reduced bone mass among ancestral human skeletal remains is limited to relatively recent populations. Since the domestication of plants roughly 12,000 years ago, skeletal remains from disparate parts of the world have occasionally shown low bone mass. Perhaps earlier populations did not suffer age-related bone loss because they died at young ages (Pfeiffer, 1990), or perhaps their diet or lifestyle facilitated effective bone maintenance. Past human populations were more dependent on local natural resources and their own physical labor for subsistence, a cultural pattern maintained by only a few geographically isolated aboriginal groups today. These "anthropological populations" have been portrayed as natural paradigms whose dietary habits might be studied as representations of our species' natural "set point" for nutritional requirements, and against which we might evaluate modern regimens and their biological consequences (Eaton *et al.*, 1988; Eaton and Nelson, 1991).

---

Susan K. Pfeiffer • School of Human Biology, University of Guelph, Guelph, Ontario, Canada, N1G 2W1. Richard A. Lazenby • Anthropology Programme, University of Northern British Columbia, Prince George, BC, Canada V1N 4Z9

*Advances in Nutritional Research,* Vol. 9
Edited by Harold H. Draper
Plenum Press, New York, 1994

With regard to calcium and bone mass, it has been argued that our mammalian, primate and earlier hominid ancestors consumed a diet much higher in bioavailable calcium than did our more recent ancestors of the last ten millennia. One scenario traces a 200 million year evolutionary sequence from insectivory, through frugivory, to omnivory, the latter stage incorporating increasingly greater amounts of scavenged or hunted meat with the appearance of the genus *Homo* some 2 million years ago (Eaton and Nelson, 1991; Leonard and Robertson, 1992). The suggestion that these calcium-rich diets were compatible with achieving a high peak bone mass is consistent with data documenting skeletal robusticity in early forms of *Homo* (Ruff, 1988; McHenry, 1992).

The study of cortical and trabecular bone mass and symptomatic fracture in aboriginal populations can help us understand the phenomenon of low bone mass as it appears in modern, industrial populations. Aboriginal peoples can have less varied diets, less genetic diversity, comparatively low socioeconomic variability, and more homogeneous—albeit sexually dimorphic—mechanical histories (Ruff, 1992). Thus, causal links between low bone mass and environmental variables should be more clearly perceived. This might allow more specific hypotheses to be formulated prior to testing in contemporary clinical settings.

Nevertheless, there are obvious limitations to the analysis of diet and bone mass in skeletal populations, even beyond those imposed by their intrinsically cross-sectional research designs. The attribution of fundamental parameters such as age and sex to individual adult skeletons is imprecise. Also, we may have only general information regarding a particular group's dietary regimen. The methods of bone mass quantification applied in the clinical arena may not be easily transferable to dry, long-buried bone tissue. Indeed, the burial environment may significantly alter skeletal tissues at both the microscopic (Hanson and Buikstra, 1987) and elemental (Hancock *et al.*, 1989) levels. The prehistoric dietary habits represented may themselves have become extinct, leaving no modern analogs and thereby relegating information about osteopenic side-effects to simple curiosity. Finally, where living populations are concerned, there is the possibility of confounding age- or sex-specific secular change in bone mass associated with culture contact and acculturation.

In the following sections, we first consider the methods used to infer or measure low bone mass in anthropological populations. This is followed by a summary of the evidence from prehistoric populations and historic Arctic groups, both of which are described as having skeletal mass below expected values. Nutritionally-based hypotheses dominate the explanations of low bone mass in these populations; however, such models now seem at odds with epidemiological data from modern populations (Hegsted, 1986).

## 2. Methods of Investigation

These can be divided into three broad categories: gross examination of skeletal remains; non-invasive techniques applied to both living and dead individuals; and invasive sampling of skeletal tissue.

Methods of gross examination, which involve recording the incidence of symptomatic fractures normally attributed to low bone mass, are considered insufficient for a clinical diagnosis of osteoporosis (Melton and Wahner, 1989). Cited most often are Colles' fractures of the distal radius (Roberts and Wakely, 1992) and compression fractures of vertebrae (Pfeiffer and Fairgrieve, 1994). Proximal femoral and humeral fractures are rarely attributed to low bone mass in the anthropological literature, possibly due to their more ambiguous appearance in dry bone and to obfuscation by post-depositional breakage. Caution must be applied when interpreting any prehistoric fracture as an indication of deficient skeletal mass since acute trauma in the absence of osteopenia could also be responsible.

Non-invasive techniques are necessarily the method of choice for studying living anthropological populations; we are unaware of any investigation which has used, for example, iliac crest biopsy to quantify bone mass in a living aboriginal population. In addition, non-invasive methods are often the only recourse in the study of skeletons, for which permission to take tissue samples may be denied for curatorial reasons.

The earliest studies of bone mass in prehistoric populations employed radiographic methods (Dewey *et al.*, 1969a,b; Van Gerven and Armelagos, 1970), following the practice of clinicians and epidemiologists (Barnett and Nordin, 1960; Garn *et al.*, 1964). While now out of favor in diagnostic settings due to their comparative imprecision (Southern, 1990), radiographic methods can provide useful information regarding changes in cortical bone mass in both living and dead populations. The more highly automated and objective methods of single (Southern, 1990) and dual (Bennike and Bohr, 1990) photon absorptiometry, as well as computed tomography (Bridges, 1989), have been used to measure bone mineral content and cross-sectional geometry of skeletal remains, but these techniques can be limited by problems peculiar to analyses of buried bone. For example, soil matrix embedded within the medullary cavity can affect attenuation coefficients used in constructing bone mineral profiles.

Single photon absorptiometry (SPA) of living Arctic groups (Mazess and Mather, 1974, 1975) was used to confirm earlier suspicions of accelerated bone loss in these populations based on cruder techniques (dry weight and ash density) (Mazess and Jones, 1972). However, the efficacy of correcting for body size variation by dividing SPA-determined bone mineral content by bone width has been questioned (Ruff and Hayes, 1984), casting doubt about whether this ratio (the bone mineral index, BMI) is a robust demonstration of skeletal change with age.

Although they are destructive, invasive methods provide access to unique information pertinent to low bone mass. These include histomorphometric measures of Haversian remodeling parameters and intracortical porosity (Ericksen, 1980; Laughlin, 1979; Richman *et al.*, 1979); elemental and structural changes in bone apatite (Thompson *et al.,* 1983); and differential trabecular involution (Mielke *et al.*, 1972; Weinstein *et al.,* 1981). For example, Roberts and Wakely (1992) have illustrated the uniformitarian nature of trabecular remodeling and reparative processes using electron microscopy to examine Romano-British and Medieval skeletal material (Figure 1).

## 3. Contemporary Aboriginal Populations

There are no studies that have examined bone mineral status in pristine aboriginal populations, i.e., groups whose traditional dietary adaptations and lifestyles have not been influenced by developed states. Cross-sectional studies can, however, sample older age cohorts whose members have experienced less acculturation, and whose bone mass may reflect more traditional lifeways. Investigations during the 1970s of Alaskan Eskimo and Canadian Inuit populations, many in conjunction with the IBP studies of circumpolar peoples, are widely cited in this context and constitute the largest body of data concerning bone mineral status among living aboriginal populations. Radiographic (Pawson, 1974) and single photon absorptiometric (Harper *et al.,* 1984; Mazess and Mather, 1974, 1975) investigation of these groups describe a unique pattern of bone loss when compared to U.S. Caucasians (Mazess and Cameron, 1974). The Eskimo and Inuit samples examined were shown to have a lower peak bone mass and a more rapid rate of loss (Figure 2); the oldest cohorts (70+ years of age) had a bone mineral content (BMC) 15% to 30% lower than that of age- and sex-matched Caucasians.

Mazess and co-workers attributed the observed difference to the traditional high protein carnivorous Arctic diet, arguing that it promotes serum acidosis and/or secondary hyperparathyroidism leading to increased bone turnover. However, Yuen *et al.* (1984) note that some epidemiological studies have not found a positive relationship between protein consumption and bone loss, and a diet high in phosphate may not in itself adversely affect calcium balance (Charles, 1992).

Data on secular growth acceleration (Schaefer *et al.*, 1980) and skinfolds (Hildes and Schaefer, 1972) suggest that younger Eskimo and Inuit consume more Western foods than their parents. However, they still exhibit bone mass values that are five to 10% below age-matched Caucasian standards. This failure to attain comparable peak bone mass levels may be a result of nutritional acculturation (Draper, 1986). Western diets, biased toward processed foods, are rich in carbohydrates and fats. They de-emphasize traditional sources of calcium such as fish bone and mammalian trabecular bone. At the same time, food additives common to these diets maintain traditionally high serum phosphorus levels and their potentially negative impact on calcium status. Thus, it appears that, while dietary constituents have changed, the effect on skeletal mass has not.

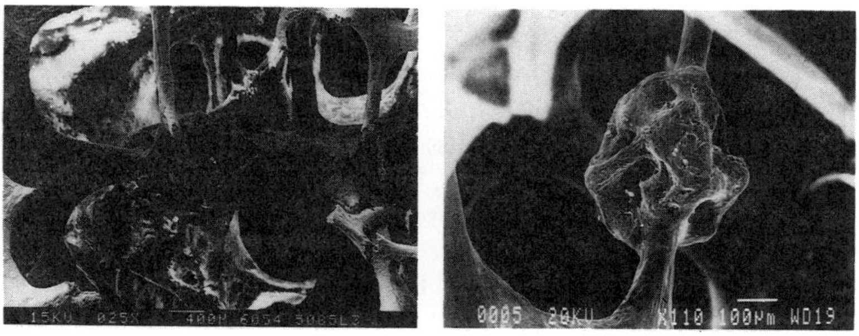

Figure 1. SEM micrographs of similarly-healing trabecular fractures from modern (right) and archeological (left) vertebrae. From Mosekilde (1990) and Roberts and Wakeley (1992).

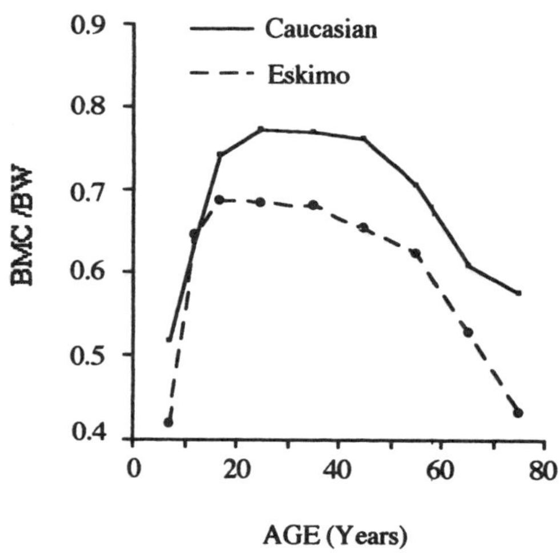


Figure 2. Eskimos lose radial bone mass at an earlier age, and at a greater rate, than U.S. whites. Data on St. Lawrence Island Eskimo and Caucasian females taken from Harper *et al.* (1984).

## 4. Bone Mass and Quality in Past Populations

Most available data concerning nutritional status, bone mass and bone quality in aboriginal populations have been obtained from archaeological skeletal samples from three geographic regions: (i) Sudanese Nubia, ca. 350 B. C. to A. D. 1450 (Armelagos *et al.*, 1972; Dewey *et al.*, 1969a,b; Hummert, 1983; Martin, 1983; Martin and Armelagos, 1979; Martin *et al.*, 1984, 1985; Mielke *et al.*, 1972; Van Gerven *et al.*, 1990), (ii) eastern and southwestern North America, ca. 2000 B. C. to A. D. 1500 (Bridges, 1989; Carlson, 1976; Cook, 1979; Ericksen, 1976, 1980; Huss-Ashmore *et al.*, 1982; Nelson, 1984; Perzigian, 1973a,b; Pfeiffer and King, 1983; Richman *et al.*, 1979; Saunders and Melbye, 1990; Southern, 1990; Stout and Teitlebaum, 1976), and (iii) the North American Arctic, including Greenland, ca. 700 B. C. to A.D. 1900 (Mazess, 1966; Mazess and Jones, 1972; Merbs, 1983; Thompson and Cowen, 1984; Thompson and Guiness-Hey, 1981; Thompson *et al.*, 1981; 1983). Low bone mass in past populations from Europe (Bennike and Bohr, 1990; Roberts and Wakely, 1992) and the Near East (Sambrook *et al.*, 1986; Smith *et al.*, 1983) has also been reported.

Modern clinical patterns of bone loss have been described in a number of prehistoric samples from both the New and the Old Worlds (Armelagos *et al.*, 1972; Carlson *et al.*, 1976; Van Gerven, 1973; Van Gerven and Armelagos, 1970). More commonly, anthropological studies of skeletal growth and aging take a diachronic and comparative perspective, exploring changing patterns within and between prehistoric populations from a specifically nutritional perspective (Bridges, 1989; Ericksen, 1976, 1980; Nelson, 1984; Richman *et al.*, 1979). Questions concerning the impact of nutrition on skeletal growth and senescence that have been addressed within the three major regions include: (i) the effect of chronic malnutrition associated with the origin and intensification of agriculture (North America; Nubia); and (ii) the effect of an ecologically-constrained dietary adaptation which is heavily reliant on animal protein (Arctic). Most of this research has focused on bone status in adult skeletons, although some work (Cook, 1979; Hummert, 1983; Keith, 1981; Saunders and Melbye, 1990; Van Gerven *et al.*, 1985) has examined the expression of nutritionally-induced low bone mass in juveniles.

### 4.1. Effects of Prehistoric Agriculture

The transition to agriculture drastically altered the human condition. It is believed that populations became more sedentary, growing in size and density; infectious disease incidence increased, and diets became less diverse (Roosevelt, 1984). Subsistence shifted away from a “broad-spectrum” adaptation, in which a wide variety of abundant plant, invertebrate, and aquatic resources were exploited, toward an emphasis on comparatively few prolific, storable and easily manipulated plants, and a reduction in the per capita consumption of meat (Cohen, 1989). Grains and legumes became dietary staples, e.g., maize, beans and squash in North America (Cassidy, 1984), and millet, sorghum and lentils in the Sudan (Van Gerven *et al.,* 1990). Such diets, while calorically rich, may be less than adequate sources of some vitamins (A, D) and minerals (zinc, iron) which are commonly obtained from animal foods. Moreover, cereal-based diets are high in fiber, phytate and phosphorus, constituents that can inhibit absorption of minerals such as zinc, iron, magnesium, and calcium (Charles, 1992; Peacock, 1991). Eaton and Nelson (1990), for example, report Ca:P ratios of 1.5 and 0.1 for uncultivated plants and grain cultivars, respectively. The low ratio for cultivated cereals reflects the differential allocation of calcium to the vegetative component, and phosphorus to the seed, the latter being the focus of consumption.

Anthropologists generally accept that the outcome of the transition away from hunting and gathering and toward farming was an increase in morbidity and mortality. This trend would have been exacerbated with agricultural intensification, and the subsequent loss of wild plant and animal dietary supplements. Numerous specific and non-specific indicators of anemias and other mineral and vitamin deficiencies associated with the rise and intensification of agriculture are recognized in the prehistoric record (Huss-Ashmore *et al.*, 1982; Van Gerven *et al.*, 1990).

Specific indicators of lower bone mass and bone quality in past human populations include measures of cortical thickness, percent cortical area, bone mineral content and histomorphometry. Ericksen (1976) compared radiographic data for mediolateral cortical thickness of the humerus and femur across three populations varying in terms of subsistence base: primarily meat (Alaskan Eskimo); mixed hunting and farming (Arikara); and cereal (primarily maize) agriculture (Pueblo). The most deficient skeletons belonged to the Puebloan sample (Table I). All groups were shown to lose bone mass with age, with females exhibiting greater magnitudes and rates of loss than males.

Both Ericksen (1980) and Richman et al., (1979) examined these same samples histologically. Of particular note is the finding of different frequencies of so-called "Type II" osteons (Richman *et al.,* 1979). These are Haversian systems in which a second and much smaller remodeling sequence has occurred within the existing Haversian canal; they were interpreted to be sites of rapid mobilization of skeletal calcium. The Eskimo sample shows the highest occurrence of these structures, Puebloans the lowest—a result which the authors find consistent with the presumed acidic nature of the Eskimo meat diet. Cassidy's comparisons of other populations of hunter-gatherers and maize horticulturalists (1984) are consistent with Ericksen's (Table I).

Pfeiffer and King (1983) observed low mean bone mass relative to modern European standards for both cortical and trabecular locations in two large late prehistoric (A.D. 1450-1600) horticultural samples from eastern North America. They cite possible dietary insufficiencies associated with a maize diet, but

Table I. Adult Bone Mass Indices and Subsistence[a]

| | | Male | | Female | |
|---|---|---|---|---|---|
| | | HG | MZ | HG | MZ |
| Ericksen[b] | Younger | 1.67 | 1.48 | 1.65 | 1.41 |
| | Older | 1.71 | 1.41 | 1.63 | 1.36 |
| Cassidy[c] | Younger | 51 | 37 | 51 | 37 |
| | Older | 47 | 36 | 36 | 33 |

a. HB = hunting and gathering; MZ = maize horticulture
b. Femur mediolateral index data, corrected for bone size, for Eskimo (HG) and Pueblo (MZ) samples; younger, 18-25; older, 30-50 years.
c. Tibial cortical index data, not corrected for bone size, for Indian Knoll (HG) and Hardin Village (MZ); younger, 17-29; older, 30-49 years.

acknowledge potentially exacerbating effects of endemic infectious diseases (e.g., tuberculosis). Ruff *et al.* (1984) reported declines in cortical bone quality in the humerus and femur with the transition to maize horticulture on the Georgia coast, but in a more recent extended analysis (Ruff and Larsen, 1990) the trend appears to reverse with European contact. In both cases they argue that the changes are associated with variation in mechanical loading, rather than directly with nutrition. In the former study, the shift to agriculture is thought to have lowered habitual levels of physical activity, thus promoting a lower bone mass. In the latter study, they note that the greater carbohydrate content of a maize diet—said to achieve its highest levels in the diet of post-contact "missionized" natives—might translate into a higher ponderal (weight for height) index for these individuals. This relatively greater body weight would confer larger levels of functional strain leading to a greater bone mass.

Bridges (1989) reports finding no subsistence-based difference in bone mass between two samples from eastern North America, an early group of hunter-gatherers (6000-1000 B.C.) and a later group of hunters-cum-maize farmers (A.D. 1200-1500). Comparing cortical cross-sectional area data from computed tomographic scans of adult male and female humeri and femora, she found that the more recent sample had generally larger femoral cortices. No significant differences were noted for the humerus in either sex. She concluded that the mixed hunting-farming diet was sufficient not only to maintain bone mass, but to produce larger bones if necessary.

Over two decades ago, Frost (1966) argued that bone histology could be useful to study human paleopathology, and a number of studies have followed that approach. Weinstein *et al.* (1981) examined transilial bone histomorphometry in a middle-aged male pre-Columbian mummy from Peru (A.D. 400-1600). The tissue showed excellent preservation, appearing normal in microradiographic appearance with regard to architecture and mineral density. Trabecular bone volume and mean trabecular diameter were abnormally low. The authors suggest an etiology of secondary hyperparathyroidism due to the low Ca:P ratio of a maize-based diet. However, much of the phosphorus in maize exists as insoluble phytin (Pfeiffer and King, 1983), making it unlikely that a maize diet alone could establish a Ca:P ratio conducive to secondary hyperparathyroidism.

Stout (1978) examined rib histomorphometry of two transitional populations from Illinois and showed the more agricultural samples to have elevated rates of bone turnover relative to the pre-agricultural groups. Martin and Armelagos (1979) found an increase in the frequency of resorption spaces and forming osteons in microradiographs of young Nubian female skeletons dating from A .D. 350-550, along with smaller femoral cortical thickness and percent cortical area. They surmise that this group was experiencing physiological stress from pregnancy and lactation in addition to the more general nutritional deficiency which is characteristic of this population (Armelagos *et al.*, 1972).

The study of juvenile skeletal remains may elucidate prehistoric behavior and nutritional status, since differential access to dietary resources by age (and sex) is a common feature of human societies, and calcium absorption efficiency in rapidly growing children is not much greater than in adults (Peacock, 1991). Nevertheless, the studies available to date are few in number; juveniles are usually under-represented in archaeological samples, through various combinations of low mortality, poor preservation or selective burial practices (Saunders and Melbye, 1990).

A consistent finding in studies of juveniles has been the maintenance of growth in bone length and total cross-sectional area at the expense of percent cortical area (Figure 3). Bones might appear outwardly normal yet have comparatively low bone mass due to excessive endosteal resorption (Cook, 1979; Hummert, 1983; Saunders and Melbye, 1990). This pattern is consistent with that reported for modern populations suffering chronic protein-calorie malnutrition (Garn *et al.*, 1964; Himes, 1978). The pronounced deficit at ca. two years of age (Figure 3) suggests that chronic malnutrition is compounded by weaning stress.

### 4.2. Effects of Animal Protein in Prehistoric Populations

Mazess (1966) and Mazess and Jones (1972) measured "anatomical" ash density (fat-free weight ÷ displacement volume) of three cortical bone sections from several long bones of Sadlermiut Inuit (A.D. 1500-1900), and found that older bones had lower density as a function of reduced weight (volume was unchanged). Compared to data for American whites and blacks, middle aged and older Sadlermiut adults had lower bone mass and an earlier onset of loss. These results correlated with Merbs' report (1983) of a high incidence of vertebral compression fractures, with both sexes and all ages affected. However, Merbs

(1983) interprets these fractures as the traumatic outcome of riding in sleds over very rough terrain; the contribution of low vertebral bone mass to fracture incidence is unclear, and these data should not be used uncritically as an indicator of low bone mass in this skeletal population.

Most research concerning Arctic population bone mass has relied on "bone core analysis" (Laughlin *et al.,* 1979), in which photon absorptiometric, cortical thickness, and histomorphometric data are collected from a core or wedge extracted from the anterior femoral shaft. This sampling method minimizes destruction of the specimen. These studies (Thompson and Guiness-Hey, 1981; Thompson *et al.*, 1981, 1983; Thompson and Cowen, 1984) find that Eskimo and Inuit bones, dating from 700 B. C. to contact, have thinner cortices (Figure 4), a lower bone mineral index, and elevated levels of Haversian remodeling (i.e., greater osteon density). A latitudinal gradient appears as well, with the more northerly Inupiaq populations having thinner cortices than the sub-Arctic Yupik Eskimos (Thompson and Guiness-Hey, 1981). Thompson and Cowen (1984) report reductions in iliac trabecular bone volume of 40 and 27% in two Barrow, Alaska female mummies aged 24 and 42 years, respectively, compared to age-matched Caucasians.

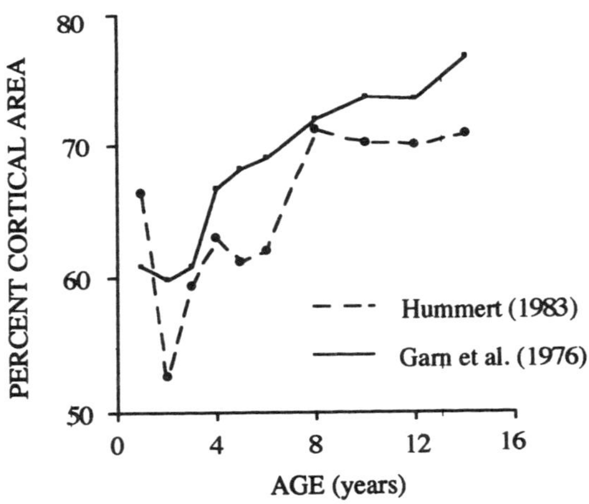


Figure 3. Percent cortical area for nutritionally stressed Nubian (--, tibial data) and normal modern (—, 2nd metacarpal data) subadults. The stressed individuals show weaning stress at ca. 2 years of age, and a failure to gain bone mass between ages 8 and 16 years.

In all these studies, the data show considerable variation. While mean values consistently demonstrate low bone mass in Eskimos, individual values often overlap the normal Caucasian range (Figure 4). Stout and Teitlebaum

(1976), for example, examined rib cortex of a 50+ years old Eskimo female dated to ca. A.D. 350 and found a mean annual osteon creation frequency and annual Haversian bone formation rate comparable to that of age-matched normal Caucasians.

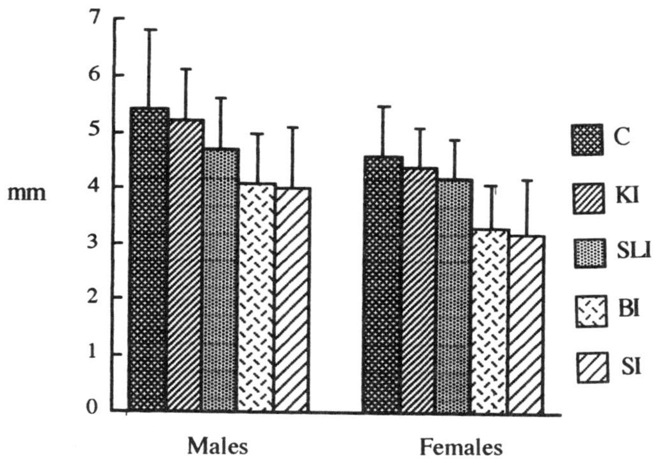


Figure 4. Lower femoral cortical thickness (Mean ± 1 SD) in Eskimo and Inuit archaeological populations than in modern, similarly aged Caucasians (C). Yupik-speaking "southern" groups from Kodiak (KI) and St. Lawrence (SLI) Islands; Inupiak-speaking "northern" groups from Baffin (BI) and Southampton (SI) Islands. Data from Thompson and Guiness-Hey (1981).

## 5. The Etiology of Low Bone Mass in Aboriginal Populations

Studies of bone mass within anthropological populations lack the possibility of follow-up available to modern clinical researchers, and they are often methodologically inconsistent. All the same, it is clear that humans have experienced conditions of lower bone mass for several millennia, when compared to either prior hunting and gathering or subsequent industrial populations. It is noteworthy that osteopenia occurs in aboriginal populations from a variety of geographic locales, and not solely among the ancestors of ethnic groups currently identified as being at risk.

The most common explanation offered for the occurrence of osteopenia in aboriginal populations is a nutritional one: relevant diets are heavily biased toward either carbohydrates (maize farmers) or animal protein (Arctic hunters). These diets have been presumed to restrict, directly or indirectly, calcium availability. However, such scenarios are at variance with epidemiological data which demonstrate a direct relationship between calcium intake and osteoporotic fracture incidence (Hegsted, 1986). They also fail to incorporate physiological observations that low calcium diets can be accommodated through increased absorption efficiency, though this response is diminished with age (Avioli, 1988). Our paleopathological explanations must be consistent with contemporary causal relationships; thus, it appears that we may need to reconsider the utility of nutrition as an explanatory model.

At the same time, it is also clear that osteopenia as seen in the archaeological record and in historic Inuit and Eskimo populations is a population-level phenomenon. Alternatives to a nutritional model thus would need to act at that level, rather than idiosyncratically. We cannot, for example, cite a causative factor such as parasitic load impairing or exacerbating calcium absorption or excretion. While parasites would have been a common pathogen for prehistoric groups as well as historic Arctic populations (Oswalt, 1967), it is unlikely that they would have affected an aboriginal population as pervasively as is suggested by observed patterns of bone loss. We would expect a good deal more variance in age- and sex-matched measures of bone mass than studies actually report. Moreover, parasitic load would likely have been as high or higher prior to the onset of agriculture when animal protein would have constituted a larger part of the diet. In any case, human-parasite coexistence is long-standing, and indigenous populations are usually well-adapted to local varieties (Reinhard, 1992). A similar logic would apply to infectious disease as an arbiter of bone mass.

An argument that the differences are genetic in origin is similarly unsatisfying, given the relative genetic homogeneity of many of the populations showing temporal change in bone mass. The rapidity of change, often on the order of 1,000 or 2,000 years (50 to 100 generations), also suggests a nongenetic basis for the observed differences between populations.

Accepting that genetic differences across populations may influence variation in bone mass, three possible explanations remain. First, the differences are truly nutritional in origin, though perhaps not acting through the mechanisms suggested to date. The focus on factors dealing with calcium homeostasis may need to be shifted, and evolutionary models that postulate a positive causal relationship between a calcium-rich diet and skeletal robusticity (Eaton and Nelson, 1991) may be missing the mark. At this time, we are not prepared to suggest what other nutritional factors may be responsible for variation in bone mass in different aboriginal populations through time and space.

The second alternative is to posit that lower bone mass does not mean inadequate bone mass. This line of reasoning follows that used by Ruff *et al.* (1984) and Ruff and Larsen (1990). Under this model, changes in bone mass are more indicative of population variation in activity level than diet. Farmers were less active (Ruff *et al.*, 1984) or at least *differently* active (Bridges, 1989) than hunters and gatherers. The general absence of symptomatic fracture in the archaeological (and historic Arctic) record may suggest that populations were simply "living within the means of their skeletons."

The last possible explanation relies on dietary and activity differences acting synergistically across populations through time and space. It has been recently argued that biological anthropologists must move to a much more stringent level of hypothesis testing, especially when those hypotheses relate to the effects of domestication and sedentary life on human populations (Wood *et al.*, 1992). The challenge of explaining low bone mass in anthropological populations is a fine example of an intriguing but convoluted problem, requiring focused investigation.

## References


Armelagos, G. J., Mielke, J. H., Owen, K. H., Van Gerven, D. P., Dewey, J. R., and Mahler, P. E., 1972, Bone growth and development in prehistoric populations from Sudanese Nubia, *J. Hum. Evol.* **1**:89.

Avioli, L. V., 1988, "Calcium and phosphorus," in: *Modern Nutrition in Health and Disease.* (M. E. Shils and V. R. Young, eds.), pp. 142-158, Lea and Febiger, Philadelphia.

Barnett, E., and Nordin, B. E. C., 1960, The radiological diagnosis of osteoporosis: a new approach, *Clin. Radiol.* **11**:166.

Bennike, P., and Bohr, H., 1990, "Bone mineral content in the past and present," in: *Osteoporosis 1990: Proceedings of the 3rd International Symposium on Osteoporosis, Copenhagen, Denmark.* (C. Christiansen and K. Overgaard, eds.), pp. 89-91, Osteopress Aps, Copenhagen.

Bridges, P. S., 1989, Bone cortical area in the evaluation of nutrition and activity levels, *Am. J. Hum. Biol.* **1**:785.

Burr, D. B., 1980, The relationships among physical, geometrical and mechanical properties of bone, with a note on the properties of nonhuman primate bone, *Yearbk. Phys. Anthropol.* **23**: 109.

Carlson, D. S., Armelagos, J. G., and Van Gerven, D. P., 1976, Patterns of age-related cortical bone loss (osteoporosis) within the femoral diaphysis, *Hum. Biol.* **48**:295.

Cassidy, C. M., 1984, "Skeletal evidence for prehistoric subsistence adaptation in the central Ohio River valley," in: *Paleopathology at the Origins of Agriculture* (M. N. Cohen and G. J. Armelagos, eds.), pp. 307-345, Academic Press, New York.

Charles, P., 1992, Calcium absorption and calcium bioavailability, *J. Int. Med.* **231**:161.

Cohen, M. N., 1989, *Health and the Rise of Civilization,* Yale University Press, New Haven.

Cook, D. C., 1979, Subsistence base and health in prehistoric Illinois valley: evidence from the human skeleton, *Med. Anthropol.* **4**:109.

Dewey, J. R., Armelagos, G. J. and Bartley, M. H., 1969a, Femoral cortical involution in three Nubian archaeological populations, *Hum. Biol.* **41**:13.

Dewey, J. R., Bartley, M. H., and Armelagos, G. J., 1969b, Rates of femoral cortical bone loss in two Nubian populations, *Clin. Orthopaed.* **65**:61.

Draper, H. H., 1986, The nutritional health of Eskimos, *Coll. Anthropol.* **10**:221.

Eaton, S. B. and Nelson, D. A., 1991, Calcium in evolutionary perspective, *Am. J. Clin. Nutr.* **54**:281S.

Eaton, S. B., Konner, M. J., and Shostak, M., 1988, Stone-agers in the fast lane: chronic degenerative diseases in evolutionary perspective, *Am. J. Med.* **84**:739.

Ericksen, M. F., 1976, Cortical bone loss with age in three native American populations, *Am. J. Phys. Anthropol.* **45**:443.

Ericksen, M. F., 1980, "Patterns of microscopic bone remodeling in three aboriginal American populations," in: *Early Native Americans: Prehistoric Demography, Economy and Technology* (D. L. Browman, ed.), pp. 239-270, Mouton, The Hague.

Frost, H. M., 1966, "Morphometry of bone in paleopathology," in: *Human Paleopathology* (S. Jarcho, ed.), pp. 131-150, Yale University Press, New Haven.

Garn, S. M., 1970, *The Earlier Gain and Later Loss of Cortical Bone in Nutritional Perspective,* Charles C. Thomas, Springfield.

Garn, S. M., Rohmann, C. G., Behar, M., Viteri, F., and Guzman, M. A., 1964, Compact bone deficiency in protein-calorie malnutrition, *Science* **145**:1444.

Garn, S. M., Poznanski, A. K., and Larson, K., 1976, "Metacarpal lengths, cortical diameters and areas from the 10-state nutritional survey," in: *Proceedings of the First Workshop on Bone Morphometry* (Z. F. G. Jaworski, ed.), pp. 367-391, University of Ottawa Press, Ottawa.

Hancock, R. G. V., Grynpas, M. D., and Pritzker, K. P. H., 1989, The abuse of bone analyses for archaeological dietary studies, *Archaeometry* **31**:169.

Hanson, D. B., and Buikstra, J. E., 1987, Histomorphological alteration in buried human bone from the lower Illinois Valley: implications for paleodietary research, *J. Arch. Sci.* **14**:549.

Harper, A. B., Laughlin, W. S. and Mazess, R. B., 1984, Bone mineral content in St. Lawrence Island Eskimos, *Hum. Biol.* **56**:63.

Hegsted, D. M., 1986, Calcium and osteoporosis, *J. Nutr.* **116**:2316.

Hildes, J. A. and Schaefer, O., 1972, Health of Igloolik Eskimos and changes with urbanization, *J. Hum. Evol.* **2**:241.

Himes, J. H., 1978, Bone growth and development in protein-calorie malnutrition, *World Rev. Nutr. Diet.* **28**:143.

Hummert, J. R., 1983, Cortical bone growth and dietary stress among subadults from Nubia's Batn el Hajar, *Am. J. Phys. Anthropol.* **62**:167.

Huss-Ashmore, R., Goodman, A. H., and Armelagos, G. J., 1982, "Nutritional inference from paleopathology," in: *Advances in Archaeological Method and Theory* (M. B. Schiffer, ed.), pp. 395-474, Academic Press, New York.

Keith, M. S., 1981, "Cortical bone loss in juveniles of Dickson Mounds," in: *Biocultural Adaptation: Comprehensive Approaches to Skeletal Analysis* (D. L. Martin and M. P. Bumstead, eds.), pp. 64-79, Research Reports No. 20, Dept. of Anthropology, University of Massachusetts, Amherst.

Laughlin, W. Q., Harper, A. B., and Thompson, D. D., 1979, New approaches to the pre- and post-contact history of Arctic peoples, *Am. J. Phys. Anthropol.* **51**:579.

Leonard, W. R., and Robertson, M. L., 1992, Nutritional requirements and human evolution: a bioenergetics model, *Am. J. Hum. Biol.* **4**:179.

Martin, D. L., 1983, Paleophysiological aspects of bone remodeling in the Meroitic, X-group and Christian populations from Sudanese Nubia, *Am. J. Phys. Anthropol.* 60: 83.

Martin, D. L., and Armelagos, G. J., 1979, Morphometrics of compact bone: an example from Sudanese Nubia, *Am. J. Phys. Anthropol.* **51**:571.

Martin, D. L., Armelagos, G. J., and Van Gerven, D. P., 1984, "The effects of socioeconomic change in prehistoric Africa: Sudanese Nubia as a case study," in: *Paleopathology at the Origins of Agriculture* (M. N. Cohen and G. J. Armelagos, eds.), pp. 193-214, Academic Press, New York.

Martin, D. L., Goodman, A. H., and Armelagos, G. J., 1985, "Skeletal pathologies as indicators of quality and quantity of diet," in: *The Analysis of Prehistoric Diets* (R. I. Gilbert, Jr. and J. H. Mielke, eds.), pp. 227-279, Academic Press, New York.

Mazess, R. B., 1966, Bone density in Sadlermiut Eskimo, *Hum. Biol.* **38**:42.

Mazess, R. B., and Cameron, J. R., 1974, "Bone mineral content in normal U.S. whites," in: Proceedings of the International Conference on Bone Mineral Measurement. (R. B. Mazess, ed.). pp. 228-237, U.S. Government Printing Office, Washington, D.C.

Mazess, R. B., and Jones, R., 1972, Weight and density of Sadlermiut Eskimo long bones, *Hum. Biol.* **44**:537.

Mazess, R. B., and Mather, W., 1974, Bone mineral content of North Alaskan Eskimos, *Am. J. Clin. Nutr.* **27**:916.

Mazess, R. B., and Mather, W., 1975, Bone mineral content in Canadian Eskimos, *Hum. Biol.* **47**:45.

McHenry, H. M., 1992, Body size and proportions in early hominids, *Am. J. Phys. Anthropol.* **87**:407.

Melton, III, L. J., and Wahner, H. W., 1989, Defining osteoporosis, *Calcif. Tiss. Int.* **45**:263.

Merbs, C., 1983, *Patterns of Activity-Induced Pathology in a Canadian Inuit Population,* National Museums of Canada, Mercury Series Publ. No. 119, Ottawa.

Mielke, J. H., Armelagos, G. J., and Van Gerven, D. P., 1972, Trabecular involution in femoral heads of a prehistoric (X-group) population from Sudanese Nubia, *Am. J. Phys. Anthropol.* **36**:39.

Mosekilde, Lis, 1990, Consequences of the remodelling process for vertebral trabecular bone structure: a scanning electron microscope study (uncoupling of unloaded structures), *Bone and Mineral* **10**:13.

Nelson, D. A., 1984, Bone density in three archaeological populations, *Am. J. Phys. Anthropol.* **63**:198 (abstr.).

Oswalt, W. H., 1967, *Alaskan Eskimos*, Chandler Publishing Co., San Francisco.

Pawson, I. G., 1974, Radiographic determination of excessive bone loss in Alaskan Eskimos, *Hum. Biol.* **46**:369.

Peacock, M., 1991, Calcium absorption efficiency and calcium requirements in children and adolescents, *Am. J. Clin. Nutr.* **54**:261S.

Perzigian, A. J., 1973a, Osteoporotic bone loss in two prehistoric Indian populations, *Am. J. Phys. Anthropol.* **39**:87.

Perzigian, A. J., 1973b, The antiquity of age-associated bone demineralization in man, *J. Am. Geriat. Soc.* **21**:100.

Pfeiffer, S., 1990, The evolution of human longevity: distinctive mechanisms? *Can. J. Aging* **9**:95.

Pfeiffer, S., and Fairgrieve, S. I., 1994, "Evidence from ossuaries: the effect of contact on the health of Iroquoians," in: *In the Wake of Contact: Biological Responses to Conquest* (C. S. Larsen and G. Milner, eds.), pp. 47-61, Wiley-Liss, New York.

Pfeiffer, S., and King, P., 1983, Cortical bone formation and diet among protohistoric Iroquoians, *Am. J. Phys. Anthropol.* **60**:23.

Reinhard, K. J., 1992, Parasitology as an interpretive tool in archaeology, *Am. Antiq.* **57**:231.

Richman, E. A., Ortner, D. J., and Schulter-Ellis, F. P., 1979, Differences in intracortical bone remodeling in three aboriginal American populations, *Calcif. Tissue Int.* **28**:209.

Roberts, C., and Wakely, J., 1992, Microscopical findings associated with the diagnosis of osteoporosis in paleopathology, *Int. J. Osteoarch.* **2**:23.

Roosevelt, A. N., 1984, "Population, health, and the evolution of subsistence: conclusions from the conference," in: *Paleopathology at the Origins of Agriculture* (M. N. Cohen and G. J. Armelagos, eds.), pp. 559-583, Academic Press, New York.
Ruff, C. B., 1988, Hindlimb articular surface allometry in Hominoidea and *Macaca*, with comparisons to diaphyseal scaling, *J. Hum. Evol.* **17**:687.
Ruff, C. B., 1992, "Biomechanical analyses of archaeological human skeletal samples," in: *Skeletal Biology of Past Peoples: Research Methods* (S. R. Saunders and M. A. Katzenberg, eds.), pp. 37-52, Wiley-Liss, New York.
Ruff, C. B., and Hayes, W. C., 1984, Bone-mineral content of the lower limb, *J. Bone Jt. Surg.* **66A**:1024
Ruff, C. B. and Larsen, C. S., 1990, "Postcranial biomechanical adaptations to subsistence strategy changes on the Georgia coast," in: *The Archaeology of Mission Santa Catalina de Guale: 2. Biocultural Interpretations of a Population in Transition* (C. S. Larsen, ed.), pp. 94-120, Anthropological Papers of the American Museum of Natural History, No. 68, New York.
Ruff, C. B., Larsen C. S., and Hayes, W. C., 1984, Structural changes in the femur with the transition to agriculture on the Georgia coast, *Am. J. Phys. Anthropol.* **64**:125.
Sambrook, P. N., Browne, C. D., Eisman, J. A., and Bourke, S. J., 1986, A case of crush-fracture osteoporosis from late Roman Pella in Jordan. *OSSA* **13**:167.
Saunders, S. R., and Melbye, F. J., 1990, Subadult mortality and skeletal indicators of health in Late Woodland Ontario Iroquois, *Can. J. Arch.* **14**:61.
Schaefer, O., Timmermans, J. F. W., Eaton, R. D. P., and Matthews, A. R., 1980, General and nutritional health in two Eskimo populations at different stages of acculturation, *Can. J. Publ. Health* **71**:397.
Smith, P., Bloom, R. A., and Berkowitz, J., 1983, Diachronic trends in humeral cortical thickness of Near Eastern populations, *J. Hum. Evol.* **13**:603.
Southern, R. A., 1990, *Cortical Bone Quality Among Pre-Iroquoian and Iroquoian Populations of the Lower Great Lakes Region,* Masters thesis, Department of Anthropology, McMaster University, Hamilton, Ontario.
Stout, S. D., 1978, Histological structure and its preservation in ancient bone, *Curr. Anthropol.* **19**:601.
Stout, S. D., and Teitlebaum, S. L., 1976, Histomorphometric determination of formation rates of archaeological bone, *Calcif. Tissue Res.* **21**:163.
Thompson, D. D., and Cowen, K. S., 1984, Age at death and bone biology of the Barrow mummies, *Arctic Anthropol.* **21**:83.
Thompson, D. D., and Guiness-Hey, M., 1981, Bone mineral-osteon analysis of Yupik-Inupiak skeletons, *Am. J. Phys. Anthropol.* **55**:1.
Thompson, D. D., Slater, E. M., and Laughlin, W. S., 1981, Bone core analysis of Baffin Island skeletons, *Arctic Anthropol.* **18**:87.
Thompson, D. D., Posner, A. S., Laughlin, W. S., and Blumenthal, N. C., 1983, Comparison of bone apatite in osteoporotic and normal Eskimos, *Calcif. Tissue. Int.* **35**:392.
Van Gerven, D. P., 1973, Thickness and area measurements as parameters of skeletal involution of the humerus, femur and tibia, *J. Gerontology* **28**:40.
Van Gerven, D. P., and Armelagos, J. G., 1970, Cortical involution in prehistoric Mississippian femora, *J. Gerontology* **25**:20.
Van Gerven, D. P., Hummert, J. R., and Burr, D. B., 1985, Cortical bone maintenance and geometry of the tibia in prehistoric children from Nubia's Batn el Hajar, *Am. J. Phys. Anthropol.* **66**:275.

Van Gerven, D. P., Hummert, J. R., Prendergast Moore, K., and Sandford, M. K., 1990, "Nutrition, disease, and the human life cycle: a bioethnography of a medieval Nubian community," in: *Primate Life History and Evolution* (J. DeRousseau, ed.), pp. 297-323, Wiley-Liss, New York.

Weinstein, R. S., Simmons, D. J., and Lovejoy, C. O., 1981, Ancient bone disease in a Peruvian mummy revealed by quantitative skeletal histomorphometry. *Am. J. Phys. Anthropol.* **54**:321.

Wood, J.W., Milner, G.R., Harpending, H.C., and Weiss, K.M. 1992, The osteological paradox: problems of inferring prehistoric health from skeletal samples, *Current Anthropol.* **33**:343.

Yuen, D. E., Draper, H. H., and Trilok, G., 1984, Effect of dietary protein on calcium metabolism in man, *Nutr. Abstr. and Rev. Clin. Nutr.* **54**:447.

Chapter 3

# Bone Loss in Animals

H. H. Draper

## 1. Introduction

When Garn *et al.* (1967) concluded that aging bone loss is universal, they were referring to the human species. Subsequent research has shown that this conclusion extends to all other species of vertebrates so far examined, from laboratory rodents to non-human species living in the wild. Although the pattern of bone loss differs among species and no animal model has been identified in which the pattern of loss exactly simulates that in humans, studies on aging bone loss in animals have yielded valuable information on the causes of bone loss in human subjects, particularly about the effects of physical activity and nutrition. Much of this information is summarized in a publication by the U.S. National Academy of Sciences on mammalian models for research on aging (1981).

## 2. Rodents

### 2.1. Aging Bone Loss

The finding that mature rodents do not actively remodel bone (Jowsey, 1966) led to a prevalent impression that they are not subject to aging bone loss. This impression stems from several sources: comparisons of young adults with middle aged rather than old adults (e.g., 6-months-old rats with rats 12 months

H. H. Draper • Department of Nutritional Sciences, University of Guelph, Guelph, Ontario, Canada, N1G 2W1

*Advances in Nutritional Research,* Vol. 9
Edited by Harold H. Draper
Plenum Press, New York, 1994

rather than 24 months old); slow accretion of bone mineral well beyond reproductive age in both females and males; and failure to distinguish between trabecular and compact bone loss. On the basis of histological observations on the bones of rats one year of age (equivalent to about 35 years in man), Jowsey (1966) concluded that the epiphyseal plate in this species does not close. However, examination of the long bones of rats 18 or 24 months of age shows that the epiphysis and diaphysis are firmly fused by a bridge of mineralized cartilage and can be separated only with the use of bone cutting tools. However, in contrast to humans, who cease to accumulate bone at about 35 years of age, aging rats continue to deposit bone periosteally while resorbing it endosteally, with the result that there is a marked decrease in cortical thickness. Cortical thinning of the femur and lumbar vertebrae of mice is illustrated in the data shown in Figure 1, and an increase in bone width with little or no change in total bone mineral in the findings shown in Table 1 (Rao and Draper, 1969). It is noteworthy that no increase in the length of the femur was found after 6 months of age, indicating that any further "growth" was limited to consolidation of existing bone.

Table 1. Characteristics of Mouse Femurs at Four Adult Ages[a]

| Age (mo) | 6 | 18 | 30 | 40 |
|---|---|---|---|---|
| Length (mm) | 15.05 ± .55[b] | 15.00 ± .39 | 15.41 ± .24 | 14.93 ± .40 |
| Width (mm) | 1.31 ± .10 | 1.39 ± .15 | 1.45 ± .11 | 1.55 ± .27 |
| Cortical Width (mm) | .29 ± .13 | .15 ± .03 | .11 ± .03 | .11 ± .04 |

[a] Stock diet [b] Mean ± SD
Reproduced from Rao and Draper (1969) with permission

Bar-Shira-Maymon *et al.* (1989) carried out a histomorphometric study of age-related bone loss from the lumbar vertebrae of CW-1 female mice. Peak bone mass, both cortical and trabecular, was observed at 13.5 months of age. Thereafter, there was a progressive decline of as much as 60% in the mass of both types of bone that was significant in males as well as females. This study demonstrated that there is a loss of trabecular as well as compact bone from the vertebral bodies in the course of aging osteopenia in mice.

Extensive studies on age-related bone changes in the rat have been carried out by Kiebzak and coworkers (1988a, 1988b, 1991). Diaphyseal Ca, P and osteocalcin did not change between 6 and 12 months of age, but were significantly lower at 24 months (Kiebzak *et al.*, 1988a). As found in the aged mouse (Rao and Draper, 1969), cortical porosities were observed in the femur, as well

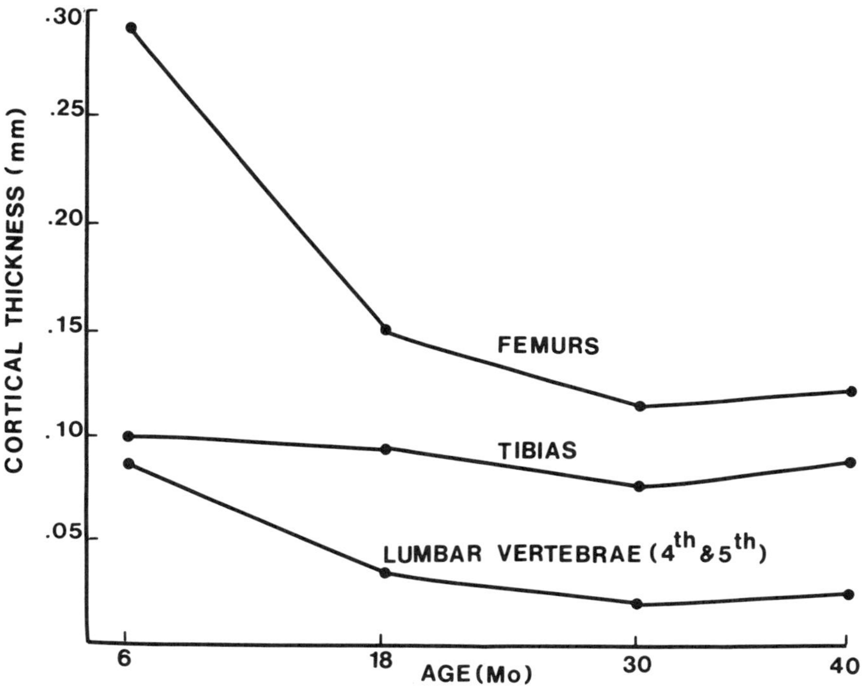


Figure 1. Age-related changes in the bones of adult mice. Reproduced from Rao and Draper (1969) with permission.

as a scalloped appearance at the endosteal surface. There was a nearly two-fold increase in serum parathyroid hormone and, probably as a consequence, a rise in serum osteocalcin and a decrease in serum phosphate. Although measurements of "ultimate stress," a parameter that normalizes for differences in bone geometry and size, decreased by 35% in the femur between 12 and 24 months of age, breaking strength was not affected. It was postulated that the strength of the femur was maintained by an architectural compensation that involved a two-fold increase in the areas of the cortex and medulla. Single photon absorptiometry of the senescent rat femur revealed that the bone mineral content of the distal metaphysis, representing trabecular bone, decreased markedly between 1 and 2 years of age (Kiebzak *et al.*, 1988b).

The femoral and lumbar cortices of female mice of the B6D2F1 Bar Harbor strain undergo a decrease in thickness of about 50% between the ages of 6 and 18 months (Figure 1), most of it during the latter half of this period (Rao and Draper, 1969). Females become sterile at about 12 months of age, indicating that, as in the human female, bone loss occurs most rapidly during the period immediately following cessation of ovarian function. There is little further decrease in cortical thickness from 18 to 30 months of age and none between the ages of 30 and 40 months. The bone homeostasis during the terminal period of life is analogous to that seen in so-called burnt out human osteoporotics. Bone loss is also greater in the female mouse than in the male (Silberberg and Silberberg, 1962). Hence there is an analogy between the physiological age-adjusted patterns of bone loss (though not the sites of loss) in mice and humans. As for other animals, there is no evidence of spontaneous osteoporotic bone fractures in rodents comparable to those that occur in humans; however, the susceptibility of the femoral cortex of aged mice to fracture during surgical removal suggests that trauma fractures in this and other species of animals may occur under natural conditions.

Ovariectomy of young adult mice results in increased bone resorption and a reduction in bone mass in advanced age that can be prevented by estrogen administration (Shin *et al.*, 1976). Osteopenia in ovariectomized rats is associated with increased bone turnover; estrogen replacement results in suppression of bone turnover and prevention of bone loss (Wronski *et al.*, 1988a). Both bone loss and bone turnover subside with increasing adult age, indicating that there is a temporal relationship between these events (Wronski *et al.*, 1988b). These observations are consistent with those of Cruess and Hong (1979), which indicated that oophorectomy in rats causes an increase in both bone resorption and bone formation resulting in a net decrease in mineral content. Administration of estrogen prevented these changes over a 12-month observation period. Similar findings have been reported by Kalu *et al.* (1989), who also observed a 30% reduction in intestinal calcium absorption in ovariectomized adult rats.

## 2.2. Calcium and Phosphorus Intake

The calcium requirement of laboratory and domestic animals traditionally has been expressed in terms of a ratio to phosphorus in the diet. This practice is based on well documented evidence that the dietary concentration of calcium required to maximize bone accretion during growth and maintain bone homeostasis in adulthood is influenced by the level of phosphorus in the diet. During active growth, the optimum ratio of Ca to P in a diet that contains 0.6% Ca (slightly in excess of the level required for maximal bone mineralization) approaches that in the bone mineral being deposited, i.e., 2:1. As the animal matures, the optimum ratio shifts in favor of phosphorus, reflecting the lower amount of Ca required for the synthesis of bone (where 99% of body Ca resides) and the increased amount of P required to support its diverse functions in the soft tissues (where 20% of body P resides). The recommended Ca content of the diet of the laboratory rat is 0.50% for both growth and maintenance (National Academy of Sciences, 1978). Because some strains of rats are susceptible to

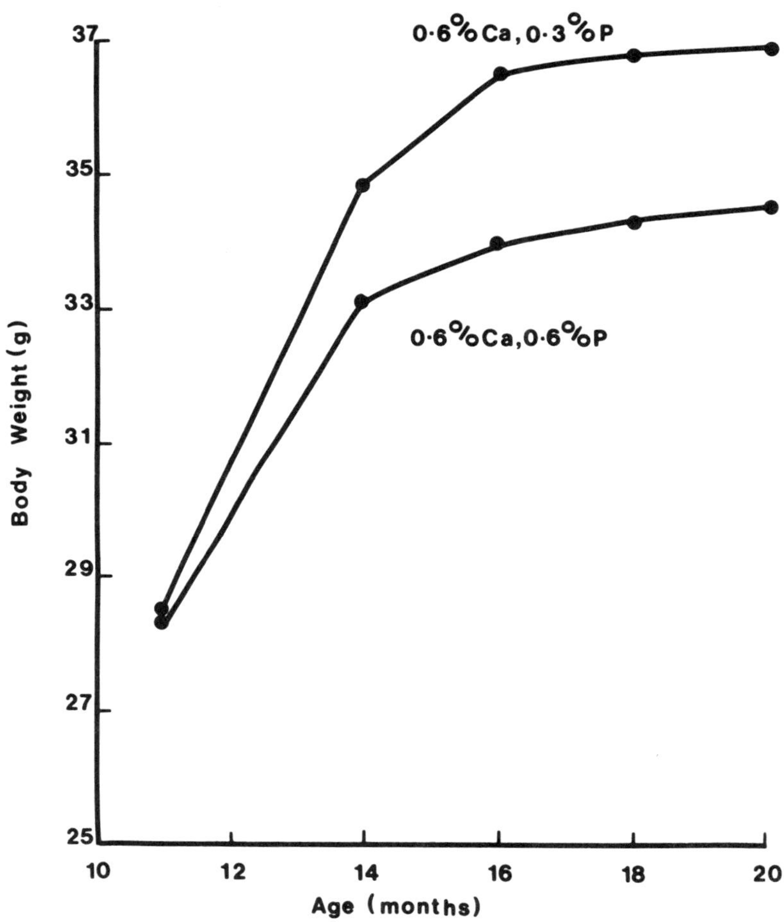


Figure 2. Influence of dietary phosphorus on weight gain in adult mice. Reproduced from Shah *et al* (1967) with permission.

nephrocalcinosis caused by excess dietary phosphorus, it has been recommended that the intake of this element be limited to 0.30% of the diet.

Growing rats offered two diets simultaneously, both containing adequate Ca (0.64%) but markedly different concentrations of P (0.1% vs. 1.8%; 0.1% vs. 1.2%; 0.1% vs 0.6%) selected mixtures of all three pairs of diets that provided almost identical concentrations of phosphorus (0.36 ± 0.01%) (Mean ± SEM) for a Ca:P ratio of 1.8:1 (Siu *et al.*, 1981). Adults offered the same diets chose mixtures with a P concentration of 0.77 ± 0.02% (Ca:P ratio 1:1.2) (Siu *et al.*, 1984). The propensity of rats to self-select dietary P in accordance with their physiological needs appears to be related to the maintenance of serum Ca homeostasis. Vitamin D deficient growing rats, which are susceptible to phosphate-induced hypocalcemia, down-regulated their P intake from 0.24% to 0.19% ($P<0.01$) when the diet contained 0.55% calcium. Parathyroidectomized rats, which are highly vulnerable to hypocalcemic tetany caused by excess phosphate, rejected diets containing more than 0.1% phosphorus. The observation that there is a 24-hour period of inappetance following intracerebroventricular administration of the hypocalcemic hormone calcitonin (Free *et al.*, 1980) suggests that the hypocalcemic action of excess phosphate is in some way responsible for the associated depression of appetite.

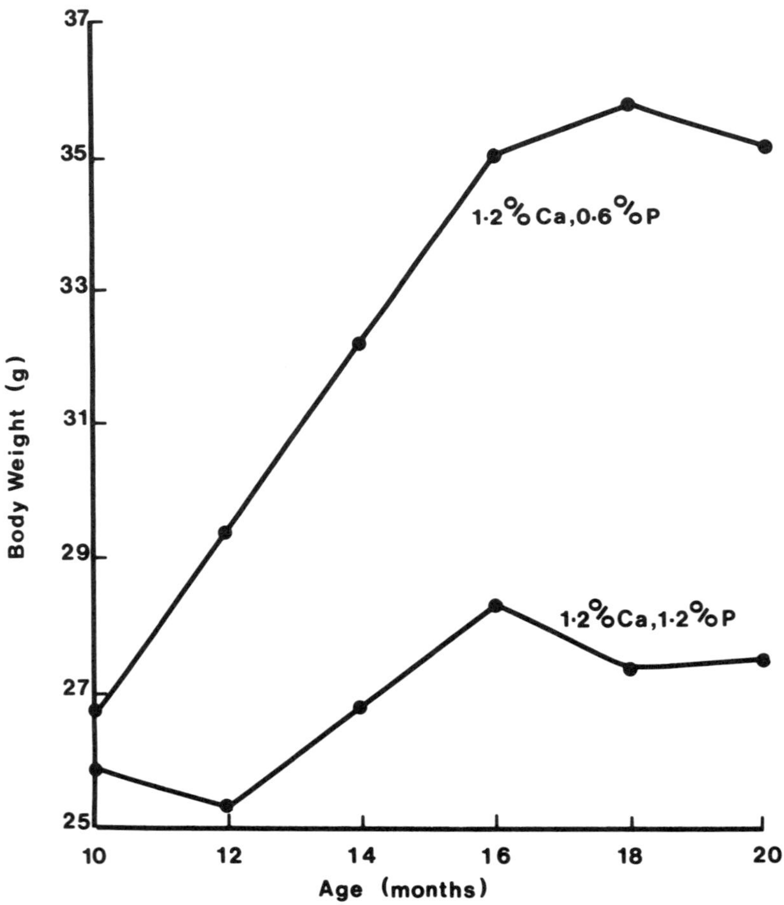


Figure 3. Suppression of adult weight gain by excess dietary phosphate in mice fed a high calcium diet. Reproduced from Shah *et al.* (1967) with permission.

Aging female mice fed a diet containing a moderate excess of phosphorus (0.6% as opposed to the estimated requirement of 0.3%) reduce their food intake and gain less weight, even when their diet contains a liberal amount of calcium (0.6%) (Figure 2) (Shah *et al.*, 1967). When these concentrations of Ca and P are doubled, i.e., to 1.2% Ca and 1.2% P, the appetite of the high P animals is markedly depressed and there is little weight gain during the period of normal adipose tissue accumulation following loss of reproductive capacity at about one year of age (Figure 3). The failure of doubling Ca intake to offset the effect of doubling phosphate intake is attributable to the fact that the fractional absorption of Ca decreases markedly when dietary Ca is increased from 0.6% to 1.2%, whereas the fractional absorption of phosphate remains essentially unchanged at about 75% of intake. Hence, the ratio of absorbed P to absorbed Ca increases as the intake of both elements is raised proportionately and the consequent increase in serum phosphate results in a depression of serum calcium (Draper, 1985).

These animal experiments demonstrate that, at least in the rat, moderate excesses of dietary phosphate can have pervasive metabolic effects, including depression of appetite and body weight loss. The relevance of these findings to humans is unknown, but they provide a base of speculation about the response to dietary phosphate, for example, in the case of high protein diets prescribed for weight control. Is the effect of these diets due to their high protein content *per se* or to the phosphorus that is intimately associated with protein in the diet? In the present context, however, the important question relates to the effect of phosphate intake on bone homeostasis.

Radiocalcium techniques have clearly shown that high P diets increase bone resorption in rodents (Figure 4) (Sie *et al.*, 1974). This increase occurs as a result of enhanced parathyroid hormone synthesis induced by a depression of serum Ca in the presence of excess serum phosphate (so-called secondary hyperparathyroidism). Parathyroid stimulation in rats fed a high P diet has been observed using several indirect indices and is prevented by parathyroidectomy (Anderson and Draper, 1972). The decrease in serum Ca (Figure 4) is caused by increased secretion of Ca into the intestine, presumably as a complex with phosphate formed in the presence of excess serum phosphorus. Urinary Ca is decreased as a result of enhanced parathyroid-dependent reabsorption of Ca from the renal tubules. As the urine is a minor pathway of Ca loss in the rat ($\leq$ 5% of total endogenous Ca excretion), this decrease does not compensate for the increased fecal loss of endogenous Ca on diets containing excess phosphate. Consequently, there is reduced bone accretion in growing rats and increased bone loss in adults fed high P diets (Figure 5) (Shah *et al.*, 1967). The effect of inadequate dietary Ca or excess P on bone homeostasis in adults decreases with aging until, in animals beyond 16 months of age, the intake of these elements has little influence. There is also no effect of ovariectomy or estrogen administration.

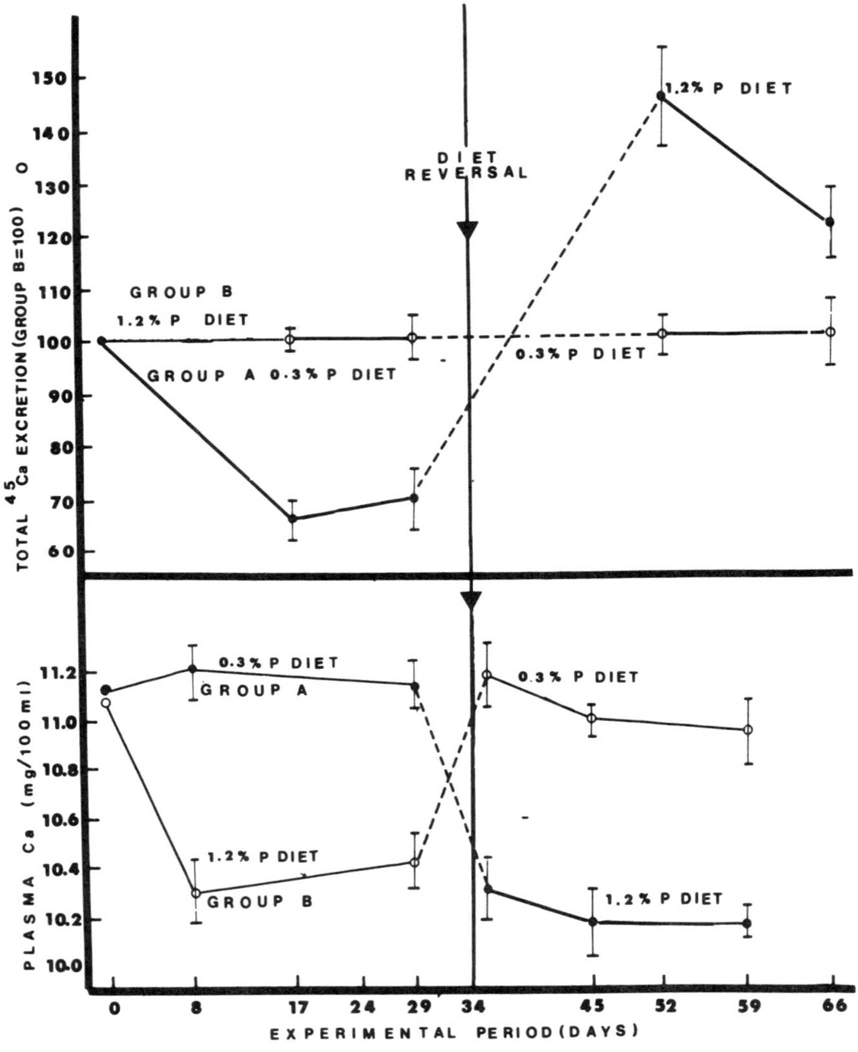


Figure 4. Decrease in plasma calcium and increase in parathyroid hormone-mediated bone resorption in adult rats fed a high phosphorus diet. Reproduced from Sie *et al.* (1974) with permission.

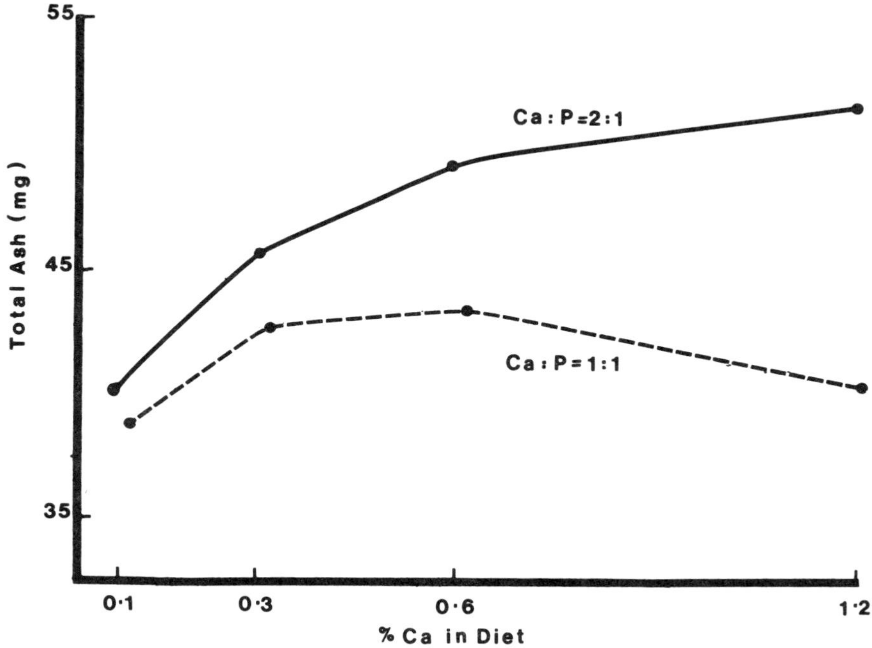


Figure 5. Effect of the ratio of calcium to phosphorus in the diet of mice between 11 and 26 months of age on the mineral content of the femur. Reproduced from Shah *et al.* (1967) with permission.

## 2.3. The Dietary Ca:P Ratio

The effect of feeding adult rats from 10 to 26 months of age diets containing levels of Ca from 0.1% to 1.2% at two ratios to phosphorus (2:1 vs. 1:1) on their femur ash content is illustrated in Figure 5 (Shah *et al.*, 1967). Similar effects were seen on fat-free dry matter, indicating that there was no effect of diet on bone composition. The results show that bone mineral mass was depressed by either a low Ca intake or a high P intake. Maximum bone mineral required at least 0.6% Ca in the diet, even at the more favorable Ca to P ratio of 2:1. At this level of Ca intake, depression of bone mineral was evident at the 0.6% P level and was aggravated at the 1.2% P level.

It is important to note that the effect of the Ca:P ratio on bone mineral depends on the intake of both elements (Figure 5). When the diet contained 0.3% Ca the difference in bone ash at the two ratios was nonsignificant. The differences seen at the higher intakes of Ca and P are attributable to the increasing disparity in their fractional absorption that occurs as the intake of both elements is increased proportionally. This disparity results in a marked shift in the ratio of absorbed Ca to absorbed P in favor of the latter. Neither raising the Ca content of the 1.2% P diet to 2.4% to restore a 2:1 ratio nor forced exercise fully offset the depression of bone mass caused by feeding this diet (Table 2) (Bell *et al.*, 1980). The optimum ratio of Ca:P changes with intake, and therefore the concept of an ideal ratio is practical only when the intake of both elements is controlled, as in the feeding of domestic, laboratory and companion animals. The question of a possible adverse effect of the increased bone resorption demonstrated in human adults consuming a high P diet on bone homeostasis (Calvo *et al.*, 1988) is better dealt with in terms of the absolute intakes of both elements rather than the ratio of Ca to P in the diet. It is noteworthy that, as in humans, bone loss in rodents decreases in old age.

Table 2. Influence of Dietary Ca, P and Exercise on Aging Bone Loss in Mice

| Diet | | Body Wt (g)[a] | | FFDM (mg)[b] | |
|---|---|---|---|---|---|
| % Ca | %P | + Ex | -Ex | + Ex | - Ex |
| 0.6 | 0.3 | 45.0 ± 5.8 | 52.4 ± 4.3* | 46.0 ± 2.8 | 44.6 ± 1.9 |
| 0.6 | 1.2 | 36.5 ± 3.9 | 33.8 ± 4.5 | 34.2 ± 1.9 | 33.2 ± 2.3 |
| 1.2 | 1.2 | 41.9 ± 7.1 | 48.7 ± 5.1* | 35.5 ± 2.8 | 34.7 ± 1.0 |
| 2.4 | 1.2 | 41.3 ± 6.9 | 49.2 ± 4.6* | 41.9 ± 3.8 | 41.7 ± 2.4 |

[a]Mean ± SD [b]Fat-free dry matter *$P < .01$
Bell, Tzeng and Draper (1980)

## 2.4. Phosphate-Induced Nephrocalcinosis

Some strains of rats and mice are susceptible to phosphate-induced nephrocalcinosis and, for this reason, some investigators have recommended that the concentration of P in the diet of the rat be limited to 0.3% in the presence of the nominal requirement of 0.5% calcium. Renal calcification has been observed in female rats (but not in males) fed a diet containing as little as 0.4% P, even in the presence of 0.8% calcium (Shah and Belonje, 1991). A level of 1.2% Ca was required for its prevention. Similarly, Hoek *et al.* (1988) found that 0.75% Ca was required for the prevention of nephrocalcinosis in rats fed 0.4% P in the diet. Calcinosis is also prevented by feeding a diet with a high protein content (Sterck *et al.*, 1992) or by the addition of acid salts (Kootstra *et al.*, 1991), both of which increase renal acid excretion, as well as by the addition of fluoride (Shah *et al.*, 1980; Grooten *et al.*, 1991). A close association between the phosphorus and protein content of foodstuffs may provide protection against any risk of nephrocalcinosis caused by excess phosphate intake in humans, as it does against the hypercalciuria caused by excess protein intake.

## 2.5 Protein Intake

Human adults fed purified diets containing a high level of isolated proteins exhibit a marked increase in urinary Ca and a negative Ca balance (Linkswiler *et al.,* 1974). This increase is due largely to the excretion of an endogenous acid load generated by the oxidation of excess sulfur amino acids in the diet (Yuen *et al.,* 1984). Parathyroid hormone-dependent reabsorption of Ca from the renal tubules is reduced by urine acidification, resulting in increased Ca loss in the urine. On a high protein diet consisting of normal foodstuffs, this loss is modulated by the parathyroid hormone-stimulating effect of excess phosphorus naturally associated with protein in the diet. Whether the hypercalciuric effect of excess protein is fully offset by the hypocalciuric effect of excess phosphorus without any compromise of bone homeostasis is a question of current interest with respect to the high incidence of osteoporotic bone fractures among consumers of the high protein diet of industrialized countries.

Table 3. Contrasting Effects of Excess Diet Protein and P on the Femur of Adult Mice[a]

| Diet (0.6 % Ca) | | Dry Fat-Free wt (mg) | Ash (mg) |
|---|---|---|---|
| % Pro | % P | | |
| 15 | 0.3 | 48.4 ± 0.7[b] | 30.5 ± 0.5[b] |
| 15 | 1.2 | 38.6 ± 2.5[c] | 25.1 ± 0.3[c] |
| 30 | 0.3 | 47.5 ± 1.2[b] | 29.7 ± 1.0[b] |
| 30 | 1.2 | 40.5 ± 1.4[c] | 25.1 ± 1.0[c] |

[a] Diets fed from 4 to 16 months of age. N = 10

[b,c] Mean ± SEM. Means with different superscripts are significantly different. ($P < 0.01$

Yuen and Draper (1983)

When this question was addressed using adult mice (Yuen and Draper, 1983), the answer was unequivocal; excess dietary protein had no effect on femoral bone mass, whereas excess phosphorus had a depressing effect that was independent of protein intake (Table 3). Further, at the same level of P intake, high protein feeding had no significant effect on urinary Ca loss. The difference between humans and mice in their response to excess dietary protein is attributable to the much smaller fraction of endogenous Ca excreted in the urine of mice and to their greater capacity to buffer endogenous acid (Upton and

L'Estrange, 1977). A similar study on adult rats "deep-labelled" with radiocalcium (Whiting and Draper, 1981) showed that feeding a high protein diet for 10 months had no effect on the amount of residual radioactivity in the femur, tibia or mandible. There was also no difference in bone mass. With respect to the effect of protein intake on Ca metabolism and bone homeostasis in human adults, the low fractional excretion of Ca in the urine of rodents makes them poor animal models.

## 2.6. Fluoride Intake

Adult mice fed a stock diet containing 19 ppm fluoride and apparently optimal concentrations of all osteogenic nutrients including Ca, P and protein, exhibit progressive aging bone loss (Figure 1). As fluoride is capable of increasing bone mass in osteoporotic women when administered in pharmacological amounts, it has been speculated to have a possible protective effect against osteoporosis, as it does against dental caries, at concentrations naturally present in some food and water supplies. Administration of fluoride in the drinking water of mice at a level of 10 ppm (10 times the concentration recommended for the prevention of dental caries in children) had no effect on the mass or composition of the femurs or of the lumbar vertebrae at 25 months of age, judged by comparison with controls provided with a fluoride-free diet and water supply (Rao *et al.*, 1972). Reports that fluoride administration to osteoporotic patients fails to reduce (and may increase) the risk of fractures, despite an increase in bone mass, because of the greater brittleness of fluorotic bone, prompted Einhorn *et al.* (1992) to investigate the effect of its administration on the appendicular skeleton of growing rats. Although fluoride was actively incorporated into the tibia and femur, no changes were observed in histomorphometric indices of bone formation and turnover in these bones nor in their capacity to withstand mechanical loads. It is worthy of note that rigorous experiments have failed to demonstrate that fluoride is required in the diet of rodents for life or health.

## 2.7. Exercise

The effect of forced exercise (1 hour on a treadmill per day for 50 weeks) on the bones of mature mice fed diets adequate in Ca (0.6%) and 0.3% vs. 1.2% P was investigated by Bell and coworkers (1980). The femur and tibia of the unexercised mice fed the high P diet had a significantly lower weight, mineral content and cortical width. Increasing dietary Ca to 1.2% or 2.4% partially offset these decreases, but did not achieve the values observed in the animals fed the 0.3% P diet. Exercise tended to increase bone weight and mineral content in all groups, but the increases were generally small, being significant only for tibia weight and for cortical thickness, cortical area and percent cortical area of the femur. Similar results were obtained by Bauer and Griminger (1983) using weanling female rats subjected to experimental regimens lasting 22 to 43 weeks, i.e., a 1.2% P diet containing 0.5% Ca depressed bone mass (as well as causing renal calcinosis), whereas forced exercise and access to a running wheel had a modest beneficial effect on the mass of the long bones, though not of the vertebral atlas bone. Treadmill exercise has been found to increase the mass of the maturing skeleton of rats as determined by neutron activation analysis for total body calcium (Yeh *et al.*, 1989), and swimming to enhance bone growth and development in growing rats (Swissa-Sivan et al., 1989). Immobilization has the opposite effect (Okumura *et al.*, 1986; Yeh *et al.*, 1989). Yeh and Aloia (1990) concluded that exercise must be continuous to sustain any increase it produces in bone mineral in the rat; deconditioning for as little as 2 weeks resulted in an increase in bone resorption and a decrease in bone formation.

## 3. Dogs

Much information on the composition, morphology and metabolism of the canine skeleton was gathered during the 1950s and 1960s when this species was used to assess the risk to humans of bone-seeking radionuclides released during atmospheric atomic bomb testing. In many respects, dog bone is an excellent model for studying age-related changes in human bone. The adult skeleton of both species contains a similar fraction of compact bone (~80%), is similar in composition and microscopic characteristics, exhibits more rapid bone loss in the female, and shows similar responses to factors that influence bone metabolism such as hormones, drugs and immobilization (Jee *et al.*, 1978). The turnover rate of adult cortical and trabecular bone in beagles has been estimated at 10% and 50% per year, respectively. In accordance with their shorter life span, the rate of bone loss in dogs, expressed on an annual basis, is four times faster than in man. As for other animals, there is no record of osteoporotic bone fractures in dogs, though it seems probable that such fractures occur in aged animals that suffer physical trauma. A comparison of kinetic and morphological data on adult canine and human bone is shown in Table 4 (National Academy of Sciences, 1981).

Reversible bone loss in adult dogs can be induced experimentally by feeding a diet either deficient in calcium or high in phosphorus (Krook *et al.*, 1971; Jowsey *et al.*, 1974). The effect of excess phosphate was shown to be due to secondary hyperparathyroidism. The reversibility of the bone loss produced in dogs by feeding a high P diet may be attributable to their rapid rate of bone turnover. The decreased bone mass caused by feeding excess P to adult rats, which do not actively remodel bone, can be arrested by lowering dietary phosphate, but the bone previously lost is not recovered (Figure 6) (Krishnarao and Draper, 1972). However, it is noteworthy that feeding high P diets has been shown to cause bone loss in animals with remodelling rates either faster or slower than those of humans.

Table 4. Typical Kinetic and Morphological Data on Adult Canine and Human Bone[a]

| | Canine | Human |
|---|---|---|
| Bone Mass at Maturity | | |
| Vertebral trabecular bone (%) | 30 | 20 |
| Turnover Rate (%/yr) | | |
| Cortical bone | 10 | 3 |
| Trabecular bone | 50 | 10 |
| Bone Loss (%/decade) | | |
| Cortical bone | | |
| Males | 5 | 3 |
| Females (all ages) | 10 | 5 |
| Postmenopausal and elderly | --- | 3 |
| Postmenopausal (0-10 yr) | --- | 10 |
| Trabecular bone | | |
| Males | 10 | 5 |
| Females (all ages) | 20 | 8 |
| Postmenopausal and elderly | --- | 6 |
| Postmenopausal (0-10 yr) | --- | 10 |
| Effect of low Ca diet | + | ? |
| Effect of high P diet | + | ? |
| Effect of immobilization, weightlessness | + | + |
| Bone Fractures | ? | + |

[a]National Academy Press (1981)

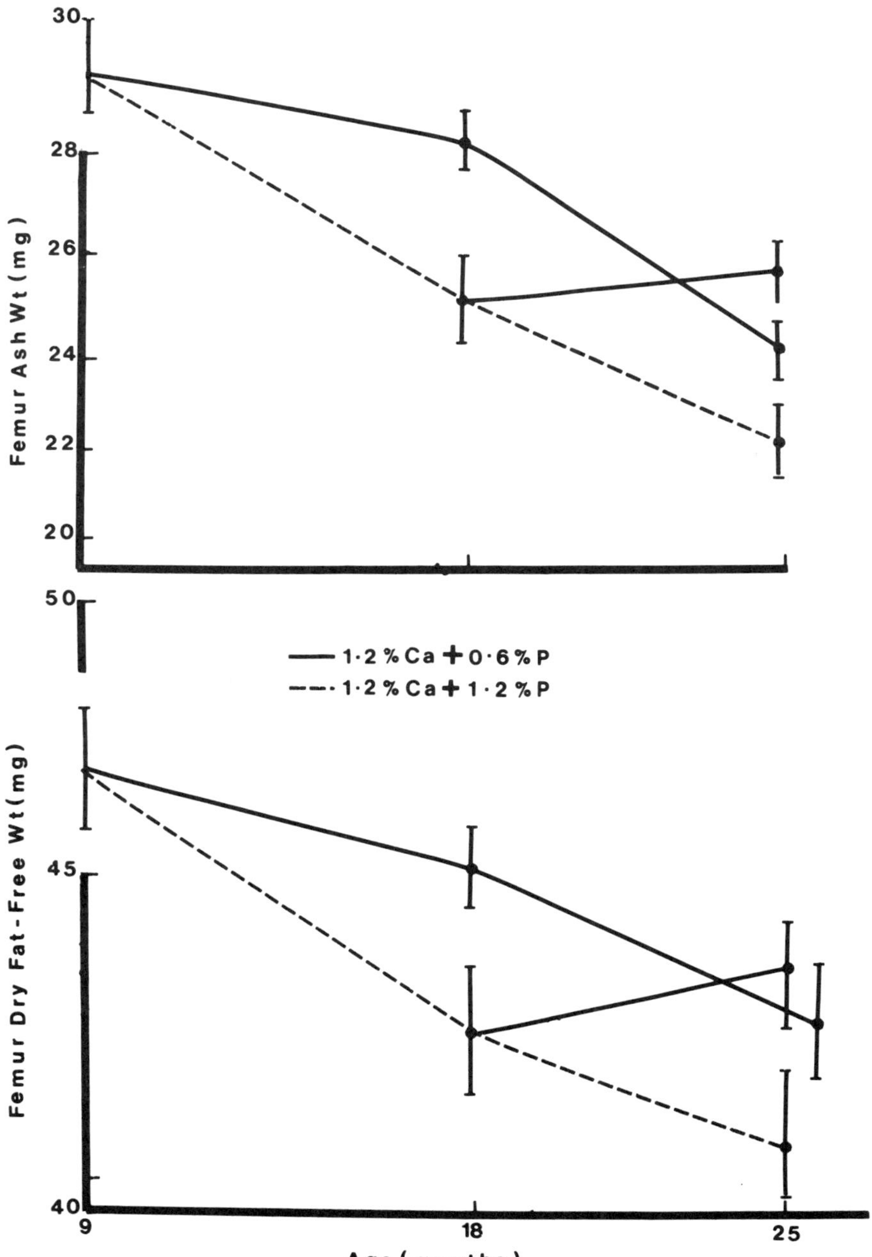


Figure 6. Effect of the dietary Ca:P ratio on the ash and dry matter content of the femur of adult mice fed a high Ca diet. Reproduced from Krishnarao and Draper (1972) with permission.

## 4. Cats

Aging bone loss in cats has been poorly characterized and, as obligate carnivores, experimental manipulation of their diet is seriously restricted. Nevertheless, it has been shown that a deficiency of dietary calcium (Jowsey and Gershon-Cohen, 1964) or an excess of phosphorus (Jowsey et al., 1974) results in osteopenia. Bone loss also has been shown to be accelerated by immobilization and by parathyroid hormone administration (Burkhart and Jowsey, 1967).

## 5. Non-human Primates

There is little information on aging bone loss in non-human primates, but a cross-sectional study on pig-tailed macaques ranging in age from 4 to 20 years revealed a pattern of osteopenia resembling that in humans (Bowden *et al.*, 1979). This study was focussed on cortical thinning of the second metacarpal and the tibia. The rate of cortical bone loss appeared to be greater than in humans; no information was obtained on the comparative rates of trabecular bone loss. Pope *et al.* (1989) obtained normative data on cortical and trabecular bone for large samples of male and female rhesus monkeys of various ages. Bone density in all bones increased to a plateau at 3 to 4 years. The density of the humerus was lower in females over 30 years of age, but not in males, and there was evidence of decreased vertebral bone density at advanced age in both sexes. It was concluded that the rhesus monkey undergoes a natural pattern of change in bone mineral content paralleling that seen in humans. With respect to the effect of excess dietary phosphate on bone loss, a study on Cebus monkeys indicated that, like humans, they appear to be more tolerant of excess dietary phosphate than is the laboratory rat (Anderson *et al.*, 1977).

## 6. Hamsters

Total bone loss during aging has been reported to be greater in Syrian hamsters than in rats (Lovelace *et al.*, 1958). To what extent it involves cortical thinning, which is extensive in rats, is unknown, nor is it known to what extent this species difference is attributable to the slow accretion of bone that extends into late adulthood in rats. It appears that, as for rodents, bone loss in hamsters can be modified, but not prevented, by diet.

## References

Anderson, G.H., and Draper, H.H., 1972, Effect of dietary phosphorus on calcium metabolism in intact and parathyroidectomized adult rats, *J. Nutr.* **102**:1123.

Bar-Shira-Maymon, B., Coleman, R., Cohen, A., Steinhagen-Thiessen, E., and Silbermann, M., 1989, Age-related bone loss in lumbar vertebrae of CW-1 female mice: a histomorphometric study, *Calcif. Tissue Int.* **44**:36.

Bauer, K.D., and Griminger, P., 1983, Long-term effects of activity and of calcium and phosphorus intake on bones and kidneys of female rats, *J. Nutr.* **113**:2011.

Bell, R. Raines, Tzeng, D.Y., and Draper, H.H., 1980, Long-term effects of calcium, phosphorus and forced exercise on the bones of mature mice, *J. Nutr.* **110**:1161.

Bowden, D.M., Teets, C., Witkin, J., and Young, D.M., 1979, "Long bone calcification and morphology," in: *Aging in Nonhuman Primates* (D.M. Bowden, ed.), pp. 335-347, Van Nostrand Reinhold, New York.

Burkhart, J.M., and Jowsey, J., 1967, Parathyroid and thyroid hormones in the development of immobilization osteoporosis, *Endocrinol.* **81**:1053.

Calvo, M.S., Kumar, R., and Heath, H., 1988, Elevated secretion and action of serum parathyroid hormone in young adults consuming high phosphorus, low calcium diets assembled from common foods, *J. Clin. Endocrinol. Metab.* **66**:823.

Cruess, R.L., and Hong, K.C., 1979, The effect of long-term estrogen administration on bone metabolism in the female rat, *Endocrinol.* **104**:1188.

Draper, H.H., 1985, "Similarities and differences in the response of animals and man to factors affecting calcium needs," in: *Calcium in Biological Systems* (R.P. Rubin, G.B. Weiss and J.W. Putney, eds.), pp. 575-581, Plenum, New York.

Einhorn, T.A., Wakley, G.K., Linkhart, S., Rush, E.B., Maloney, S., Faierman, E., and Baylink, D.J., 1992, Incorporation of sodium fluoride into cortical bone does not impair the mechanical properties of the appendicular skeleton in rats, *Calcif. Tissue Int.* **51**:127.

Freed, W.J., Perlow, M.J., and Watt, R.J., 1980, Calcitonin: inhibitory effect on eating in rats, *Science* **206**:850.

Garn, S.M., Rohmann, C.G., and Wagner, B., 1967, Bone loss as a general phenomenon in man, *Fed. Proc.* **26**:1729.

Grooten, H.N.A., Ritskes-Hoitinga, J., Mathot, J.N.J.J., Lemmens, A.G., and Beyen, A.C., 1991, Dietary fluoride prevents phosphorus-induced nephrocalcinosis in female rats, *Biol. Trace Elem. Res.* **9**:147.

Hoek, A.C., Lemmens, A.G., Mullink, J.W.M.A., and Beyen, A.C., 1988, Influence of dietary calcium:phosphorus ratio on mineral accretion and nephrocalcinosis in female rats, *J. Nutr.* **118**:1210.

Jee, W.S.S., Smith, J.M., Kimmel, D.B., Miller, S.C., VanDura, C., Smith, C., and Dell, R., 1978, "Preliminary surface/volume ratios and bone turnover rates of trabecular bone in young adult Beagles," in: *Research in Radiobiology*, University of Utah Radiobiology Laboratory Report C00-119-253, pp. 224-230, University of Utah, Salt Lake City, Utah.

Jowsey, J., Studies of haversian systems in man and some animals, 1966, *J. Anat.* **100**:857.

Jowsey, J., and Gershon-Cohen, J., 1964, Effect of dietary calcium levels on production and reversal of experimental osteoporosis in cats, *Proc. Soc. Exp. Med.* **116**:437.

Jowsey, J., Reiss, E., and Canterbury, J.M., 1974, Long-term effect of high phosphate intake on parathyroid hormone levels and bone metabolism, *Acta Orthop. Scand.* **45**:801.

Kalu, D.N., Liu, C., Hardin, R.R., and Hollis, B.W., 1989, The aged rat model of ovarian hormone deficiency bone loss, *Endocrinol.* **124**:7.

Kiebzak, G.M., 1991, Age-related bone changes, *Exp. Gerontol.* **26**:171.

Kiebzak, G.M., Smith, R., Gundber, C.C., Howe, J.C., Sacktor, B., 1988a, Bone status of senescent male rats: chemical, morphometric, and mechanical analysis, *J. Bone Min. Res.* **3**:37.

Kiebzak, G.M., Smith, M., Howe, J.C., and Sacktor, B., 1988b, Bone mineral content in the senescent rat femur: an assessment using single photon absorptiometry, *J. Bone Min. Res.* **3**:311.

Koostra, Y., Ritskes-Hoitinga, J., Lemmens, A.G., and Beyen, A.C., 1991, Diet-induced calciuria and nephrocalcinosis in rats, *Int. J. Vit. Nutr. Res.* **61**:100.

Krishnarao, G.V.G., and Draper, H.H., 1972, Influence of dietary phosphate on bone resorption in senescent mice, *J. Nutr.* **102**:1143.

Krook, L., Lutwak, L., Henrikson, P.A., Kallfelz, F., Hirsch, C., Romanus, B., Belanger, L.F., Marier, J.R., and Sheffy, B.E., 1971, Reversibility of nutritional osteoporosis: Physicochemical data on bones from an experimental study in dogs, *J. Nutr.* **101**:233.

Linkswiler, H.M., Joyce, C.L., and Anand, C.R., 1974, Calcium retention in adult males as affected by level of protein and of calcium intake, *Trans. New York Acad. Sci., Ser.* 2, **36**:333.

Lovelace, F., Will, L., Sperling, G., and McCay, C.M., 1958, Teeth, bones and aging of Syrian hamsters, *J. Gerontol.* **13**:27.

National Academy of Sciences, National Research Council, 1978, *Nutrient Requirements of Laboratory Animals*, National Academy Press, Washington, D.C., 96 pp.

National Academy of Sciences, National Research Council, 1981, *Mammalian Models for Research on Aging*, National Academy Press, Washington, D.C., 587 pp.

Okumura, H., Yamamuro, T., Kasai, R., Ichisaka, A., Hayashi, T., and Matsushita, M., 1986, The effect of immobilization on osteoporosis in rats, *J. Bone Min. Met*ab. **4**:75.

Pope, N.S., Gould, K.G., Anderson, D.C., and Mann, D.R., 1989, Effect of age and sex on bone density in the rhesus monkey, *Bone* **10**:109.

Rao, G.V.G. Krishna, and Draper, H.H., 1969, Age-related changes in the bones of adult mice, *J. Gerontol.* **24**:149.

Rao, G.V.G. Krishna, Ts'ao, K., and Draper, H.H., 1972, The effect of fluoride on some physical and chemical characteristics of the bones of aging mice, *J. Gerontol.* **27**:183.

Shah, B.G., and Belonje, B., 1991, Different dietary calcium levels required to prevent nephrocalcinosis in male and female rats, *Nutr. Res.* **11**:385.

Shah, B.G., Krishnarao, G.V.G., and Draper, H.H., 1967, The relationship of Ca and P nutrition during adult life and osteoporosis in aged mice, *J. Nutr.* **92**:30.

Shah, B.G., Belonje, B., and Nera, E.A., 1980, Reduction of nephrocalcinosis in female rats by additional magnesium and by fluoride, *Nutr. Rep. Int.* **22**:957.

Shin, K.S., Bell, R.R., and Draper, H.H., 1976, Effect of estrogen on bone resorption induced by excess dietary P in mature and aged rats, *Fed. Proc.* **35**:499 (Abst.).

Sie, Ten-Lin, Draper, H.H., and Bell, R.R., 1974, Hypocalcemia, hyperparathyroidism and bone resorption induced by dietary phosphate, *J. Nutr.* **104**:1195.

Silberberg, M., and Silberberg, R., 1962, Osteoarthrosis and osteoporosis in senile mice, *Gerontologia* **6**:91.

Siu, G.M., Hadley, M., and Draper, H.H., 1981, Self-regulation of phosphate intake by growing rats, *J. Nutr.* **111**:1681.

Siu, G.M., Hadley, M., Agwu, D.E., and Draper, H.H., 1984, Self-regulation of phosphate intake in the rat: The influence of age, vitamin D and parathyroid hormone, *J. Nutr.* **114**:1097.

Sterck, J.G.H., Ritskis-Hoitinga, J., and Beyen, A.C., 1992, Inhibitory effect of high protein intake on nephrocalcinosis in female rats, *Br. J. Nutr.* **67**:223.

Swissa-Sivan, A., Simkin, A., Leichter, I., Nyska, A., Nyska, M., Statter, M., Bivas, A., Menczel, J., and Samueloff, S., 1989, Effect of swimming on bone growth and development in young rats, *Bone and Mineral* **7**:91.

Upton, P.K., and L'Estrange, J.L., 1977, Effects of chronic hydrochloric and lactic acid administration on food intake, blood acid-base balance and bone composition of the rat, *Quart. J. Exp. Physiol.* **62**:223.

Whiting, S.J., and Draper, H.H., 1981, Effect of chronic high protein feeding on bone composition in the adult rat, *J. Nutr.* **111**:178.

Wronski, T.J., Citron, M., Doherty, A.L., and Dann, L.M., 1988a, Estrogen treatment prevents osteopenia and depresses bone turnover in ovariectomized rats, *Endocrinol.* **123**:681.

Wronski, T.J., Citron, M., and Dann L.M., 1988b, Temporal relationship between bone loss and increased turnover in ovariectomized rats, *Calcif. Tissue Int.* **43**:179.

Yeh, J.K., and Aloia, J.F., 1990, Deconditioning increases bone resorption and decreases bone formation in the rat, *Metabolism* **39**:659.

Yeh, J.K., Aloia, J.F., and Yasumura, S., 1989, Effect of physical activity on calcium and phosphorus metabolism in the rat, *Am. J. Physiol.* **256** (*Endocrinol. Metab.* **19**):E1-E6.

Yuen, D.E., and Draper, H.H., 1983, Long-term effects of excess protein and phosphorus on bone homeostasis in adult mice, *J. Nutr.* **113**:1374.

Yuen, D.E., Draper, H.H., and Trilok, G., 1984, Effect of dietary protein on calcium metabolism in man, *Rev. Clin. Nutr., Nutr. Abst. and Rev.* **54**:447.

Chapter 4

# The Significance of Habitual Calcium Intake in the Pathogenesis of Peri- and Early Postmenopausal Bone Loss

E. C. H. van Beresteijn

## 1. Introduction

The first hypothesis for the pathogenesis of postmenopausal osteoporosis was postulated by Albright and colleagues (1941). They suggested a cause and effect relationship between estrogen withdrawal and the occurrence of crush fractures of the vertebrae resulting from loss of bone. Since then ovarian failure has been considered to be the most important determinant of the accelerated bone loss in women during the early postmenopause, and the evidence that the adverse skeletal changes can be prevented with estrogen replacement therapy seems reasonably firm (Weiss *et al.*, 1980; Kreiger *et al.*, 1982; Ettinger *et al.*, 1985).

The role of calcium deficiency as a causal factor in osteoporosis was brought to attention by Nordin (1960), who reviewed studies linking calcium deficiency with osteoporosis in animals. At the end of the 1970s the population study of Matkovic *et al.* (1979) and the balance studies of Heaney *et al.* (1978) on pre- and postmenopausal women stressed the importance of a high calcium

E. C. H. van Beresteijn • Department of Nutrition, Netherlands Institute for Dairy Research (NIZO), P.O. Box 20, 6710 BA Ede, the Netherlands

*Advances in Nutritional Research,* Vol. 9
Edited by Harold H. Draper
Plenum Press, New York, 1994

intake and, consequently, an increase in the Recommended Daily Allowance (RDA) from 800 to 1500 mg per day for postmenopausal women was advocated by the consensus conference organized by the U.S. National Institutes of Health (1984). Since then calcium has been widely recommended to postmenopausal women as a means of preventing osteoporosis and the sale of calcium supplements has increased tremendously in recent years. However, the evidence linking calcium deficiency to osteoporosis is less strong than that for ovarian failure and, as a consequence, the role of dietary calcium and the value of calcium supplementation are still controversial (Kanis and Passmore, 1989a; Kanis and Passmore, 1989b; Nordin and Heaney, 1990; Cumming, 1990).

Most studies relating dietary calcium intake or calcium supplementation to bone loss have focused on measurements of bone mass and dietary intake. These mostly cross-sectional and short-term intervention studies have been extensively reviewed in recent years (Kanis and Passmore, 1989a; Kanis and Passmore, 1989b; Nordin and Heaney, 1990; Cumming, 1990; Dawson-Hughes, 1991; Heaney, 1992). However, it is well known that these study designs have a number of limitations; the rate of bone loss cannot be accurately calculated and short-term effects of calcium supplementation may not predict its effect in the long run. Therefore, in order to further elucidate the mechanisms involved in the pathogenesis of postmenopausal bone loss, the results of long-term prospective studies are sorely needed. This chapter will focus on the significance of habitual calcium intake in the pathogenesis of peri- and postmenopausal bone loss on the basis of the results of a 10-year longitudinal study.

## 2. Bone Mass and Menopause

Although it has been recognized for several decades that menopause is characterized by increased bone turnover and bone resorption, the pattern of bone loss at different skeletal sites has not been well established. Bone loss from the predominantly cortical bone of the forearm follows an exponential function; bone loss starts several years before menopause, increases rapidly for about five years after menopause and slows down to a more gradual loss thereafter (Mazess, 1982; Geusens *et al.*, 1986). Most studies (Geusens *et al.*, 1986; Elders *et al.*, 1988; Schaadt and Bohr, 1988), but not all (Riggs *et al.*, 1981; Hansson and Roos, 1986), indicate the same pattern of bone loss from the spine, but the accelerated bone loss following the menopause appears to be more rapid from the spine than from the forearm. The pattern of postmenopausal bone loss from the hip is even more uncertain. Although age-related bone loss from the hip has been demonstrated in cross-sectional studies (Schaadt and Bohr, 1988; Stevenson *et al.*, 1989), an accelerated loss after menopause could not be demonstrated when measured longitudinally (Harris and Dawson-Hughes, 1992).

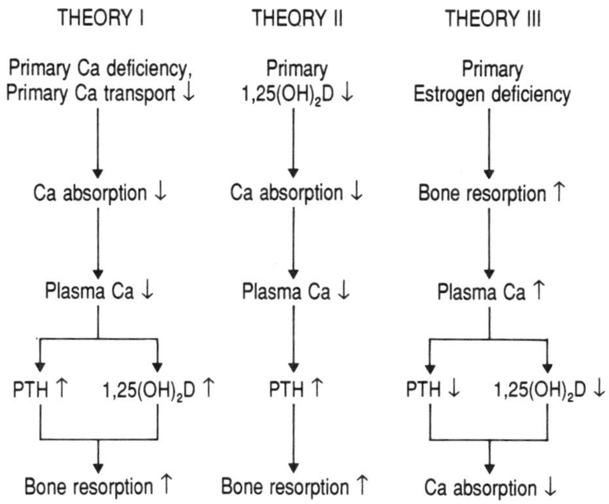


Figure 1. Outline of the hypothetical sequences for the pathogenesis of postmenopaual osteoporosis.

This biphasic pattern has led to the suggestion that there are two major components of postmenopausal bone loss: a first (transient) rapid component induced by reduced estrogen production, superimposed on a slower protracted age-related component (Riggs and Melton, 1986). The latter, occurring in both sexes, is probably due to factors such as decreased physical activity, decreased muscle mass, decline in osteoblastic function, impaired calcium absorption and possibly nutritional deficiency. The significance of habitual calcium intake in the pathogenesis of peri- and postmenopausal bone loss may, therefore, depend on the relative contributions of menopause and age to peri- and postmenopausal bone loss. These may change with time after menopause and, as a consequence, the significance of habitual calcium intake may change as well (Riggs and Melton, 1990).

## 3. Pathogenesis of Bone Loss

Several hypotheses have been proposed for the pathogenetic mechanisms responsible for the increase in bone resorption following the menopause (Figure 1). One concept (Theory I) suggests that malabsorption of calcium is the primary defect, necessitating mobilization of bone calcium in order to maintain plasma calcium levels (Nordin, 1960). This could be due either to a dietary deficiency or to an impairment of intestinal calcium transport or to both. The plasma calcium level would tend to fall, with homeostatic rises in parathyroid hormone (PTH) and 1,25-dihydroxyvitamin D, effecting mobilization of calcium from bone. Theory II postulates renal endocrine failure leading to secondary hyperparathyroidism and hence to bone loss, the so-called type 2 osteoporosis (Riggs and Melton, 1986). Theory III is a modification of the original proposal of Albright *et al.* (1941): loss of ovarian function itself leads to bone loss by either direct or indirect means. Plasma calcium tends to rise, causing secondary suppression of both PTH and 1,25-dihydroxyvitamin D secretion. Theories I and II imply a systemic calcium deficiency and one would anticipate improvement by increasing the dietary calcium intake. Conversely, if Theory III is correct, increasing dietary calcium intake would exacerbate the putative systemic calcium excess, leading to a homeostatic decrease in calcium absorption. As a consequence, one would not expect that manipulating calcium intake would have much influence on bone loss. Thus, exploring the biochemical and hormonal changes occurring around and after menopause may give more insight into the significance of dietary calcium intake in peri- and postmenopausal bone loss.

## 4. Methodological Considerations

Studies designed to assess the relation of dietary calcium intake to peri- and postmenopausal bone loss meet a number of methodological problems. One is the problem of measuring both variables accurately. During the last 20 years technologic advances in bone mineral densitometry have resulted in increasingly precise and accurate methods of estimating bone mineral density at different sites. However, bone loss—if related at all to calcium intake—is slow compared to the bone loss that follows ovarian failure, and with the techniques presently available such a loss cannot be reliably detected in an individual woman in less than a 4– to 5–year period (Heaney, 1986; Davis *et al.,* 1991). In contrast to bone mass measurements, accurate assessment of dietary calcium intake in free-living individuals is extremely difficult. Moreover, the important determinant of bone loss is habitual dietary calcium intake, averaged over many years, rather than a single estimate of current calcium intake. As current intake is not highly correlated with prior or future intake (Heaney, 1991), it is important to assess intake periodically over time.

Other problems are the phenomena of long-term adaptation to changes in calcium intake at the level of absorption and excretion (Kanis, 1991) as well as at the level of bone remodeling transients (Kanis, 1984; Heaney, 1986; Kanis, 1991). Since bone turnover is slow, the remodeling transient can persist for up to 3 years. These adaptation problems can be avoided when the habitual calcium intake of individual subjects is studied but play an important role when calcium intake is manipulated as in intervention trials.

Variability in estimated rates of bone loss can be introduced by several physiologic conditions. Unrelated radiographic abnormalities in the spine scan field (e.g., calcification of the aorta, calcified lymph nodes) have been shown to give low estimates of bone loss in the spine when measured by dual photon absorptiometry (Dawson-Hughes and Dallal, 1991), and seasonal variations of up to 3% in bone mineral density have been demonstrated (Krølner, 1982).

Calcium deficiency may be due not just to a low calcium intake, but also to poor absorption or to excessive excretory loss, and neither of these factors can be easily measured in an epidemiologic set-up. Finally, individual changes in biochemical and hormonal parameters are small and difficult to detect. These parameters are particularly prone to biological variations and the sensitivity of hormone assays may be too low to detect them (e.g., serum iPTH assay, mid-region fragment).

It is clear that the question whether dietary calcium is related to peri- and postmenopausal bone loss can be answered only by longitudinal studies. Cross-sectional studies are not suitable for accurate evaluation of individual changes over time because of possible cohort effects. Moreover, calcium intake is usually restricted to an estimate of current intake, which in comparison to longitudinal estimates, is more prone to bias arising from confounding factors. Calcium intervention trials should cover a period of at least 2 to 3 years; extending the length of follow-up will give better-substantiated results. On the other hand, longitudinal studies require high data quality because the validity of repeated measurements of dietary intake, bone mass and biochemical and hormonal parameters can be jeopardized by many factors (van Beresteijn *et al.,* 1986).

## 5. A 10-year Prospective Study of Habitual Calcium Intake in Relation to Bone Loss

In view of the importance of prevention and treatment of osteoporosis and because results of long-term studies were not available, a longitudinal prospective study lasting 10 years on a group of healthy peri- and early postmenopausal women born between 1922 and 1931 was started in 1979. The women were not on any medication known to influence calcium or bone metabolism and they were still menstruating regularly or had been postmenopausal for not more than two years. The group was subjected to annual measurements of dietary calcium intake (cross-check dietary history method), cortical bone mineral density of the radius (BMD radius, single photon absorptiometry), body weight and body height, and biochemical and hormonal parameters were determined in annual samples of serum and 24–hour urine collections. To exclude seasonal variations, the annual measurements for individual subjects were taken in the same month of the year (between February and June). In 1987, measurements of bone mass of the lumbar spine and femoral neck (dual photon absorptiometry) were performed once in a subgroup of women selected from the study population. During the follow-up period, food habits did not change very much in the study population; habitual calcium intake, calculated as the mean of the annual consecutive dietary intake estimations, varied from 560 to 2580 mg per day.

The results showed that at the end of the 10–year follow-up the rate of bone loss from the radius was slightly lower, and the bone mineral density of the

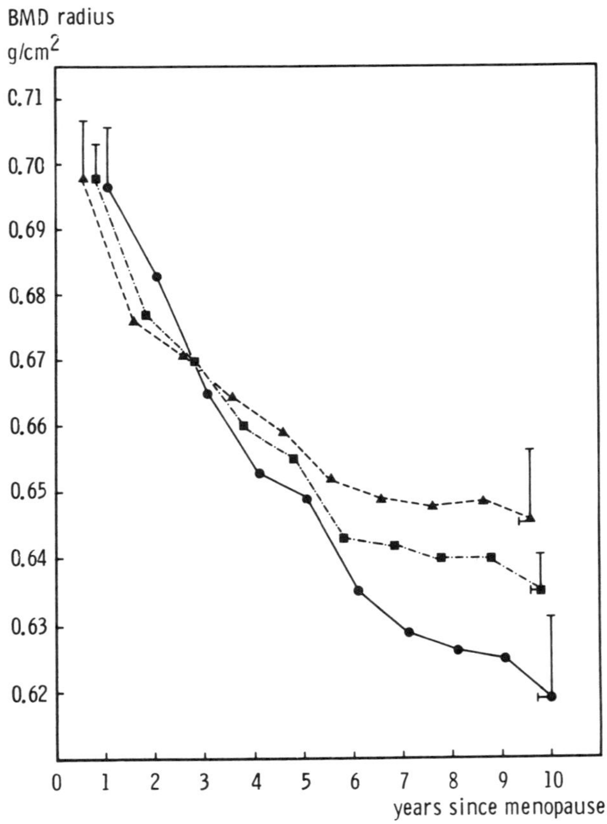


Figure 2. Changes in BMD radius as a function of the number of years since menopause in 154 perimenopausal women, subdivided according to habitual dietary calcium intake. ▲<800 mg/day (n = 28), ■ 800-1350 mg/day (n = 95), ● > 1350 mg/day (n = 31). Means ± SEM. From van Beresteijn *et al.* (1991).

appendicular and axial skeleton was slightly higher, in the women with a habitual dietary intake of less than 800 mg calcium per day (range 564-796 mg per day) than in those with an intake of more than 1350 mg per day (range 1354-2580 mg per day) ($P \leq 0.11$) (Figure 2). Body mass index was higher in the low calcium group and was significantly related to the rate of bone loss (van Beresteijn *et al.*, 1990a; van Beresteijn *et al.*, 1990b; van Beresteijn *et al.*, 1991). When women with similar body mass index from the low and high calcium intake groups were compared, no differences in bone parameters were found (Figure 3), indicating that habitual dietary calcium intake at the levels consumed by these subjects was not an important risk factor for postmenopausal bone loss or for postmenopausal bone mineral density.

Grouping the data according to the menopausal status of the women allowed a longitudinal evaluation of the changes in calcium and bone metabolism during the perimenopausal period (from 2 years before to 2 years after menopause), the early postmenopausal period (from 2 to 6 years after menopause), and

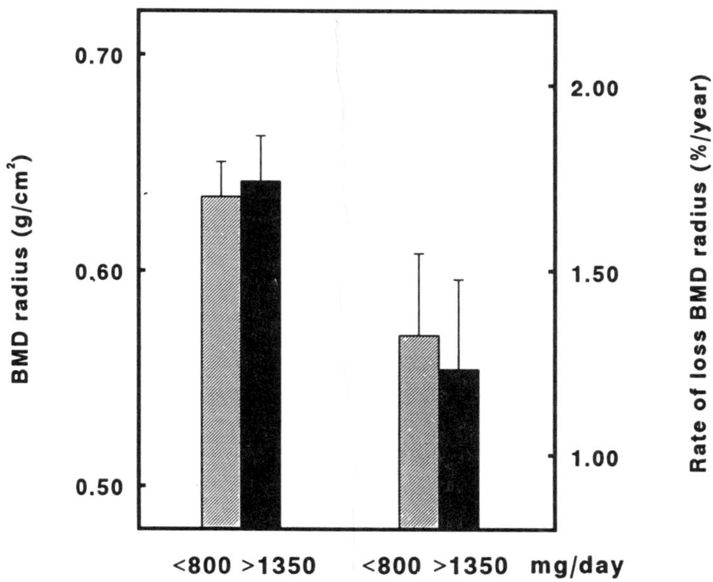


Figure 3. Bone parameters of the radius of postmenopausal women with low (< 800 mg/day, n = 10) and high (> 1350 mg/day, n = 10) habitual calcium intake, matched for body mass index (Means ± SEM).

Table 1. Changes of Body Mass Index and Variables of Bone and Calcium Metabolism in Women During the Perimenopausal, Early Postmenopausal and Late Postmenopausal Period (Mean ± SEM)

| | Menopausal period | | | | | | | | |
|---|---|---|---|---|---|---|---|---|---|
| | Peri[a] (n = 38) | | | Early Post[b] (n = 145) | | | Late Post[c] (n = 70) | | |
| Variable | b[d] | ± | SEM | b[d] | ± | SEM | b[d] | ± | SEM |
| Rate of loss BMD radius (%/4 yr)[e] | -4.0[f] | ± | 0.80 | -6.0[f] | ± | 0.10 | -2.4[f] | ± | 0.36 |
| BMI (kg/m$^2$/4 yr) | 0.36 | ± | 0.20 | 0.21 | ± | 0.14 | 0.50[f] | ± | 0.11 |
| Serum Ca (mmol/L/4 yr) | 0.084[f] | ± | 0.012 | 0.016[f] | ± | 0.006 | -0.002 | ± | 0.007 |
| Serum $Ca^{++}$ (mmol/L/4 yr) | 0.055[f] | ± | 0.006 | 0.030[f] | ± | 0.004 | -0.002 | ± | 0.003 |
| Serum P (mmol/L/4 yr) | 0.173[f] | ± | 0.024 | 0.024 | ± | 0.012 | -0.017 | ± | 0.011 |
| Serum alkaline phosphatase (U/L/4 yr) | 10.48[f] | ± | 0.71 | 2.28[f] | ± | 0.43 | -0.72 | ± | 0.57 |
| Serum iPTH (μEq bPTH/L/4 yr)[g] | -0.022 | ± | 0.027 | -0.012 | ± | 0.013 | 0.009 | ± | 0.018 |
| Urinary Ca/Cr (mmol/mmol/4 yr)[h] | 0.082[f] | ± | 0.033 | -0.014 | ± | 0.013 | 0.067[f] | ± | 0.014 |
| Urinary P/Cr (mmol/mmol/4 yr)[h] | 0.240[f] | ± | 0.117 | 0.062 | ± | 0.062 | 0.168[f] | ± | 0.050 |
| Urinary OHProl/Cr (μmol/mmol/4 yr)[h] | 2.16 | ± | 1.27 | -3.22[f] | ± | 0.56 | -2.02[f] | ± | 0.70 |

[a] 2 years before to 2 years after menopause
[b] 2 to 6 years after menopause
[c] 6 to 10 years after menopause
[d] Mean regression coefficient calculated from individual regression lines versus time multiplied by four
[e] Calculated from the individual regression lines versus time, as percentage of initial level
[f] Significantly different from zero, $P < 0.05$ (Student's t-test, two sided)
[g] Mid-region fragment assay
[h] Urinary excretion of calcium (Ca), phosphorus (P) and hydroxyproline (OHProl) expressed as their ratio to creatinine (Cr)

the late postmenopausal period (from 6 to 10 years after menopause). The biochemical data presented in Table 1 clearly show evidence of increased skeletal turnover following loss of ovarian function. Serum alkaline phosphatase, an indicator of bone formation, and urinary hydroxyproline, an indicator of bone resorption increase during the perimenopausal period. With increasing menopausal age there is a reduction in bone turnover and thereby in rate of bone loss. Bone loss starts before menopause, increases during the early postmenopausal period and decreases thereafter. During the first 10 years after menopause, the fall in bone mineral density was attributable to ovarian failure rather than to aging, the contribution of age was not significant; within most age groups years since menopause was significantly negatively related to BMD whereas age was not related to BMD in the various menopausal groups (Figures 4 and 5, van Beresteijn and van't Hof, 1990).

The changes in biochemical parameters of calcium metabolism during the different menopausal periods (Table 1) support the Theory III pathway (Figure 1) for the pathogenesis of peri- and early postmenopausal bone loss: serum calcium values increased significantly following the loss of ovarian function, paralleled

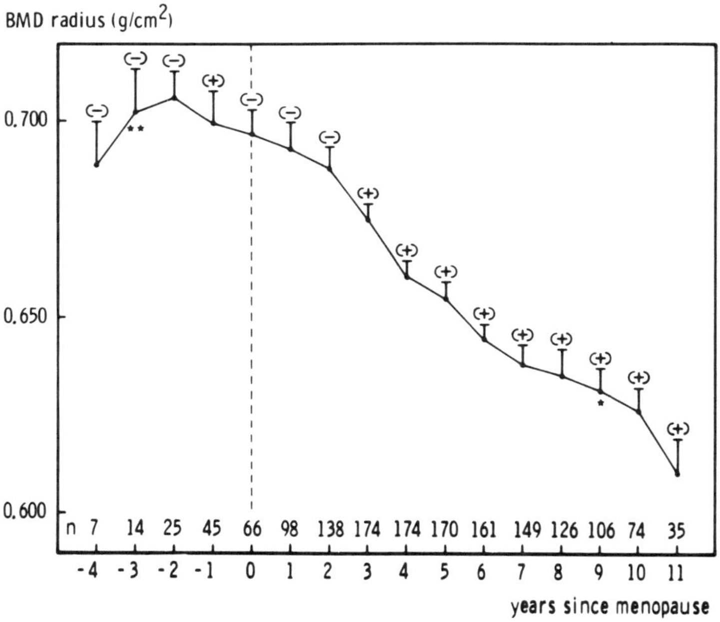


Figure 4. Mean BMD of the radius as a function of years since menopause. n = number of subjects in each menopausal group. The relation between age and BMD within menopausal groups is indicated in parentheses, * P <0.05, ** P < 0.01. Means ± SEM. From van Beresteijn and van't Hof (1990)

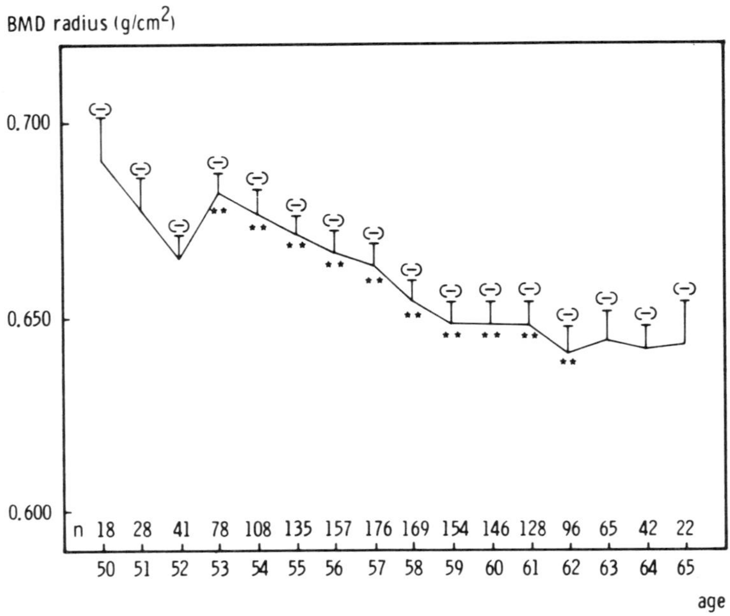


Figure 5. Mean BMD of the radius as a function of age. n = number of subjects in each age group. The relation between years since menopause and BMD within age groups is indicated in parentheses, *P = < 0.051, **P = < 0.01. Means ± SEM. From van Beresteijn and van't Hof (1990).

by a trend to decreasing serum iPTH levels. The increase in 24–hour urinary calcium excretion observed during the perimenopausal period was much less than could be anticipated from the observed yearly rate of bone loss, indicating that there was a reduction in intestinal calcium absorption as a result of the increase in serum calcium. In spite of increased bone loss observed during the early postmenopausal period, the data of Table 1 show that serum calcium and urinary calcium excretion do not increase further. Apparently, the women adapted to the putative systemic calcium excess by further reducing intestinal calcium absorption.

During the 10–year follow-up neither the initial values nor the changes in bone and calcium metabolism were influenced by habitual calcium intake (van Beresteijn *et al.,* 1990a). At the end of the follow-up period there were no significant differences between the subjects with the highest and those with the lowest habitual calcium intakes with respect to parameters of bone turnover except for their levels of serum intact hPTH 1-84, which were significantly higher in the low calcium intake group (Table 2). Serum intact hPTH 1-84 was not significantly related to habitual calcium intake and urinary calcium excretion was similar in both groups. The lower serum intact hPTH 1-84 values in the high calcium intake group may, therefore, have been compensatory for an increased rate of bone loss rather than being the result of an increased absorption rate. The difference in serum intact hPTH 1-84 values was not accompanied by differences in the serum concentration of 1,25-dihydroxyvitamin D or vitamin D-binding protein.

Table 2. Parameters of Calcium and Bone Metabolism in Late Postmenopausal Women with Low and High Habitual Calcium Intake*

| Variable | Habitual calcium intake (mg/day) < 800 (n = 23) | Habitual calcium intake (mg/day) > 1350 (n = 29) |
|---|---|---|
| Years since menopause | 9.4 ± 2.5 | 10.0 ± 1.7 |
| Serum Ca (mmol/L) | 2.39 ± 0.11 | 2.35 ± 0.10 |
| Serum $Ca^{++}$ (mmol/L) | 1.19 ± 0.05 | 1.16 ± 0.06 |
| Serum alkaline phosphatase (U/L) | 30.4 ± 8.4 | 30.6 ± 6.2 |
| Serum hPTH 1-84 (ng/L) | 20.5 ± 9.1[a] | 15.1 ± 5.5 |
| Serum 1,25 DHCC (pmol/L) | 63.5 ± 16.3 | 64.9 ± 12.7 |
| Serum vitamin D-binding protein (ng/mL) | 376 ± 31 | 373 ± 31 |
| Serum IGF-I (ng/L) | 138.1 ± 31.0 | 137.8 ± 38.7 |
| Serum osteocalcin (ng/ml) | 7.5 ± 2.4 | 8.4 ± 1.9 |
| Urinary Ca/Cr (mmol/mmol)[b] | 0.460 ± 0.198 | 0.474 ± 0.198 |
| Urinary OHProl/Cr (μmol/mmol)[b] | 22.0 ± 9.7 | 19.2 ± 6.9 |

*Mean ± SD
[a] Significantly different from high-calcium group, $P < 0.05$
[b] Urinary excretion of calcium (Ca) and hydroxyproline (OHProl) expressed as their ratio to creatinine (Cr)

## 6. Results of Previous Studies

Since 1979 a number of prospective as well as controlled intervention studies have been carried out on women within five years of the menopause. Most, but not all, of these studies have been included in recently published reviews (Kanis and Passmore, 1989a; Kanis and Passmore, 1989b; Nordin and

Heaney, 1990; Cumming, 1990; Dawson-Hughes, 1991; Heaney, 1992). Hansen *et al.* (1991) followed a group of Danish early postmenopausal women over a 12–year period. Although measurements were not carried out annually and only current calcium intake was estimated at follow-up, their results agreed with those of the above study on Dutch women: bone mineral density in the radius decreased significantly but calcium intake was not related to the rate of bone loss from the radius. As in the Dutch sample, calcium intake was high (mean 1184, range 600–2880 mg per day). Dawson-Hughes *et al.* (1990) demonstrated that calcium supplementation was of no benefit in preserving bone mass during the years immediately following menopause, even when dietary calcium intake was less than 400 mg per day. Elders *et al.* (1991) and Elders (1991) reported that daily supplementation with 1000 or 2000 mg calcium as carbonate to early postmenopausal women for three years retarded bone loss from the spine during the first year but not during the second and third year. At the end of the trial, bone mineral density did not differ among the three groups. This result may reflect the phenomenon of bone remodeling transients. A minor favorable effect of calcium supplementation on cortical bone loss was found and supplementation appeared to reduce bone turnover throughout the trial.

Bone loss slows down from about 6–7 years after menopause and it has been postulated that during the late postmenopausal period the age-related component of postmenopausal bone loss becomes more important, with the practical consequence that after the estrogen-withdrawal effect has waned, the calcium economy becomes more dependent upon calcium intake (Riggs and Melton, 1990). Although the contribution of an age-related component to postmenopausal bone loss is still controversial, the study of Dawson-Hughes *et al.* (1990) is in favor of this hypothesis; in contrast to peri- and early postmenopausal women, calcium supplementation to at least 800 mg per day reduced bone loss at all sites in women who were 6 years or more past menopause and had a habitual dietary calcium intake of less than 400 mg per day. However, women with a higher habitual intake did not benefit from calcium supplementation, suggesting that the relationship between calcium intake and bone mass displays a threshold effect (Heaney, 1986). It is possible that the threshold value for calcium increases with age. To answer this question long-term studies in the elderly, evaluating bone loss as well as the biochemical and hormonal mechanisms involved, are needed.

These studies corroborate the conclusion drawn by Cumming (1990) from his meta-analysis of published studies up to 1989: calcium supplementation is most effective in reducing postmenopausal bone loss when habitual calcium intake is low, the mean age of subjects is high, and/or the subjects have clinical evidence of osteoporosis. In all studies, the effect of calcium supplementation was found to be less effective than estrogen replacement therapy in decreasing bone loss.

Nordin *et al.* (1990) concluded from cross-sectional data that in women up to age 70 years the age-related component of the fall in forearm mineral density was highly significant and dominant over the menopause-related component. In contrast, the age-related component was not significant for vertebral bone loss (Nordin *et al.,* 1992). However, these results on forearm bone mineral density are completely at variance with other cross-sectional (Richelson *et al.,* 1984; Nilas and Christiansen, 1987) and longitudinal data (Falch and Sandvik, 1990).

The biochemical and hormonal changes in calcium and bone metabolism occurring before and after menopause have been reported earlier (Falch and Gautvik, 1988; Nilas and Christiansen, 1989), but not, apparently, in relation to habitual dietary calcium intake. Although it has been reported that 1,25-dihydroxyvitamin D values tend to be lower postmenopausally (Stevenson, 1984), no changes in vitamin D metabolite levels during the phase of rapid bone loss could be detected when measured longitudinally (Falch *et al.,* 1987). From experiments in rats, it is speculated that a reduced concentration of intestinal receptors for 1,25-dihydroxyvitamin D plays a role in the homeostatic decrease in intestinal calcium absorption following loss of ovarian function (Horst *et al.,* 1990).

Calcium supplements in the form of calcium carbonate or milk have been shown to suppress bone turnover, irrespective of menopausal status and habitual dietary calcium intake (Recker and Heaney, 1985; Elders, 1991). Pharmacological manipulation of the remodeling rate may, on the one hand, decrease remodeling-related skeletal losses and, on the other hand, decrease the turnover time of bone and delay skeletal repair. Despite the maintenance of skeletal mass appreciable decreases of bone turnover seem to increase the risk of fracture (Kanis, 1984).

## 7. Conclusion

The biochemical and hormonal changes in calcium and bone metabolism occurring within 10 years after menopause do not support the hypothesis that manipulating calcium intake by increasing dietary calcium intake, or by calcium supplementation, has a beneficial effect on bone loss during that period. Following ovarian failure, calcium becomes available from the downward revision of skeletal mass, leading to a homeostatic decrease in intestinal calcium absorption. As a consequence, the requirement for calcium from external sources is lower. The available data lead to the conclusion that habitual calcium intakes above 800 mg per day do not prevent or decrease postmenopausal bone loss up to the age of 65 years. Whether this conclusion holds true for women beyond the age of 65 years awaits further studies. One should be cautious in recommending an increase in the current RDA of 800 mg per day in the elderly before its beneficial effect has been clearly demonstrated, since decreases in bone turnover by calcium supplements may increase the risk of fracture.

ACKNOWLEDGMENT. The measurement of intact hPTH 1-84 and IGF-I in the serum samples of the low and high calcium intake groups at the end of the follow-up period by Professor dr. R. Bouillon, Laboratory for Experimental Medicine and Endocrinology, Katholieke Universiteit Leuven, Belgium, is highly appreciated. Without the cooperation of the women in Ede and the co-workers of the Department of Nutrition of NIZO the longitudinal study would not have been possible.

## References

Albright, F., Smith, P.H., and Richardson, A.M., 1941, Postmenopausal osteoporosis; its clinical features, *JAMA* **116**:2465.

Beresteijn, E.C.H. van, and Hof, M.A. van't, 1990, "Contributions of ovarian failure and aging to cortical bone loss in perimenopausal women: a mixed-longitudinal study" in: *Osteoporosis 1990* (Christiansen, C., and Overgaard, K., eds.), Handelsstrykkeriet Aalborg, Aalborg pp. 1023-1025.

Beresteijn, E.C.H. van, Hof, M.A. van't, Waard, H. de, Neeter, R., Winkeldermaat, H.J., Visser, R.M., Schaafsma, G., Schaik, M. van, and Duursma, S.A., 1986, Design and data quality of a mixed-longitudinal study to elucidate the role of dietary calcium and phosphorus on bone mineralization in pre-, peri- and postmenopausal women, *Am. J. Clin. Nutr.* **43**:538.

Beresteijn, E.C.H. van, Hof, M.A. van't, Schaafsma, G., Waard H. de, and Duursma, S.A., 1990a. Habitual dietary calcium intake and cortical bone loss in perimenopausal women: a longitudinal study, *Calcif. Tissue Int.* **47**:338.

Beresteijn, E.C.H. van, Hof, M.A. van't, Waard, H. de, Raymakers, J.A., and Duursma, S.A., 1990b, Relation of axial bone mass to habitual calcium intake and to cortical bone loss in healthy early postmenopausal women, *Bone* **11**:7.

Beresteijn, E.C.H. van, Dekker, P.R., Heiden-Winkeldermaat, H.J. van der, Schaik, M. van, Visser, R.M., and Waard, H. de, 1991, "The habitual calcium intake from milk products and its significance for bone health; a longitudinal study," in: *Nutritional Aspects of Osteoporosis* (P. Burkhardt and R.P. Heaney, eds.), Raven Press, New York pp. 206-212.

Cumming, R.G., 1990, Calcium intake and bone mass: A quantitative review of the evidence, *Calcif. Tissue Int.* **47**:194.

Davis, J.W., Ross, P.D., Wasnich, R.D., MacLean, C.J., and Vogel, J.M., 1991, Long-term precision of bone loss rate measurements among postmenopausal women, *Calcif. Tissue Int.* **48**:311.

Dawson-Hughes, B., 1991, Calcium supplementation and bone loss: a review of controlled clinical trials, *Am. J. Clin. Nutr.* **54**:274S.

Dawson-Hughes, B., and Dallal, G.E., 1991, Effect of radiographic abnormalities on rates of bone loss from the spine, *Calcif. Tissue Int.* **46**:280.

Dawson-Hughes, B., Dallal, G.E., Kral, E.A., Sadowski, L., Sahyoun, N., and Tannenbaum, S., 1990, A controlled trial of the effect of calcium supplementation on bone density in postmenopausal women, *N. Engl. J. Med.* **323**:878.

Elders, P.J.M., 1991, "The long-term effect of calcium supplementation on bone loss in perimenopausal women," in: *Perimenopausal Bone Loss*, Thesis VU, University Press, Amsterdam pp. 107-123.

Elders, P.J.M., Netelenbos, J.C., Lips, P., Ginkel, F.C. van, and Stelt, P.F. van der, 1988, Accelerated vertebral bone loss in relation to the menopause: a cross-sectional study on lumbar bone density in 286 women of 46 to 55 years of age, *Bone Miner.* **5**:11.

Elders, P.J.M., Netelenbos, J.C., Lips, P., Ginkel, F.C. van, Khoe, E., Leeuwenkamp, O.R., Hackeng, W.H.L., and Stelt, P.F. van der, 1991, Calcium supplementation reduces vertebral bone loss in perimenopausal women: a controlled trial in 248 women between 46 and 55 years of age, *J. Clin. Endocrinol. Metab.* **73**:533.

Ettinger, B., Genant, H.K., and Cann, C.E., 1985, Longterm estrogen therapy prevents bone loss and fracture, *Ann. Intern. Med.* **102**:319.

Falch, J.A., and Gautvik, K.M., 1988, A longitudinal study of pre- and postmenopausal changes in calcium metabolism, *Bone* **9**:15.

Falch J.A., and Sandvik, L., 1990, Perimenopausal bone loss: a 10-year prospective study, *Bone* **11**:425.

Falch, J.A., Oftebro, H., and Haug, E., 1987, Early postmenopausal bone loss is not associated with a decrease in circulating levels of 25-hydroxyvitamin D, 1,25-dihydroxyvitamin D, or vitamin D-binding protein, *J. Clin. Endocrinol. Metab.* **64**:836.

Geusens, P., Dequeker, J., Verstraeten, A., and Nijs, J., 1986, Age-, sex-, and menopausal-related changes in vertebral and peripheral bone: a population study using dual and single photonabsorptiometry and radiogrammetry, *J. Nucl. Med.* **27**:1540.

Hansen, T., Overgaard, K., Riis, B.J., and Christiansen, C., 1991, Potential risk factors for development of postmenopausal osteoporosis examined over a 12-year period, *Osteoporosis Int.* **1**:95.

Hansson. T., and Roos, B., 1986, Age changes in the bone mineral content of the lumbar spine in normal women, *Calcif. Tissue Int.* **38**:249.

Harris, S., and Dawson-Hughes, B., 1992, Rates of change in bone mineral density of the spine, heel, femoral neck and radius in healthy postmenopausal women, *Bone Miner.* **17**:87.

Heaney, R.P., 1986a, En recherche de la difference ($p < 0.05$), *Bone Miner.* **1**:99.

Heaney, R.P., 1986b, "Calcium, bone health, and osteoporosis," in: *Bone and Mineral Research 4* (W.A. Peck, ed.), Elsevier, Amsterdam-New York-Oxford, pp. 255-301.

Heaney, R.P., 1991, "Assessment and consistency of calcium intake," in: *Nutritional Aspects of Osteoporosis* (P. Burckhardt, and R. P. Heaney, eds.), pp. 99-104, Raven Press, New York.

Heaney, R.P., 1992, Calcium in the prevention and treatment of osteoporosis, *J. Int. Med.* **231**:169.

Heaney, R.P., Recker, R.R., and Saville, P.D., 1978, Menopausal changes in calcium performance balance, *J. Lab. Med.* **92**:953.

Horst, R.L., Goff, J.P., and Reinhardt, T.A., 1990, Advancing age results in reduction of intestinal and bone 1,25-dihydroxyvitamin D receptor, *Endocrinol.* **126**:1053.

Kanis, J.A., 1984, Treatment of osteoporotic fracture, *Lancet i*:27.

Kanis, J.A., 1991, Calcium requirements for optimal skeletal health in women, *Calcif. Tissue Int.* **49**:S33.

Kanis, J.A., and Passmore, R., 1989a, Calcium supplementation of the diet-I, *Br. Med. J.* **298**:137.

Kanis J.A., and Passmore, R., 1989b, Calcium supplementation of the diet-II, *Br. Med. J.* **298**:205.

Kreiger, N, Kelsey, J.L., Holford, T.R., and O'Connor, T., 1982, An epidemiologic study of hip fracture in postmenopausal women, *Am. J. Epidemiol.* **116**:141.

Krφlner, B., 1982, Seasonal variation of lumbar spine mineral content in normal women, *Calcif. Tissue Int.* **35:**145.

Matkovic, V., Kostial, K., Simonovic, I., Buzina, R., Broderec, A., and Nordin, B.E.C., 1979, Bone status and fracture rates in two regions of Yugoslavia, *Am. J. Clin. Nutr.* **32:**540.

Mazess, R.B., 1982, On aging bone loss, *Clin. Orthop.* **165:**239.

National Institutes of Health, 1984, Osteoporosis: consensus conference, *JAMA* **254:**799.

Nilas, L., and Christiansen, C., 1987, Bone mass and its relationship to age and the menopause, *J. Clin. Endocrinol. Metab.* **65:**697.

Nilas, L., and Christiansen, C., 1989, The pathophysiology of peri- and postmenopausal bone loss, *Br. J. Obst. Gynaec.* **96:**580.

Nordin, B.E.C., 1960, Osteomalacia, osteoporosis and calcium deficiency, *Clin. Orthop.* **17:**235.

Nordin, B.E.C., and Heaney, R.P., 1990, Calcium supplementation of the diet: justified by present evidence, *Br. Med. J.* **300:**1056.

Nordin, B.E.C., Need, A.G., Chatterton, B.E., Horowitz, M., and Morris, H.A., 1990, The relative contributions of age and years since menopause to postmenopausal bone loss, *J. Clin. Endocrinol. Metab.* **70:**83.

Nordin, B.E.C., Need, A.G., Bridges, A., and Horowitz, M., 1992, Relative contributions of years since menopause, age, and weight to vertebral density in postmenopausal women, *J. Clin. Endocrinol. Metab.* **74:**20.

Recker, R.R., and Heaney, R.P., 1985, The effect of milk supplements on calcium metabolism and calcium balance, *Am. J. Clin. Nutr.* **41:**254.

Richelson, L.S., Wahner, H.W., Melton, L.J., and Riggs, B.L., 1984, Relative contribution of aging and estrogen deficiency to postmenopausal bone loss, *N. Engl. J. Med.* **311:**1273.

Riggs, B.L., and Melton, L.J., 1986, Involutional osteoporosis, *N. Eng. J. Med.* **314:**1676.

Riggs, B.L., and Melton, L.J., 1990, Clinical heterogeneity of involutional osteoporosis: implications for preventive therapy, *J. Clin. Endocrinol. Metab.* **70:**1229.

Riggs, B.L., Wahner, H.W., Mazess, R.B., Offord, K.P., and Melton, L.J.,1981, Differential changes in bone mineral density of the appendicular and axial skeleton with aging and its relationship to spinal osteoporosis, *J. Clin. Invest.* **67:**328.

Schaadt, P., and Bohr, H., 1988, Differential trends of age-related diminution of bone mineral content in the lumbar spine, femoral neck, and femoral shaft in women, *Calcif. Tissue Int.* **42:**71.

Stevenson, J.C., 1984, "Interrelation of oestrogen and the major calcium regulating hormones," in: *The Climacteric–an Update* (H. van Herendael, B. van Herendael, F.E. Riphagen, K. Goessens and H. van de Pas, eds.), MTP Press Ltd, Lancaster, pp. 199-206.

Stevenson, J.C., Lees, B., Devenport, M., Cust, M.P., and Ganger, K.F., 1989, Determinants of bone density in normal women: risk factors for future osteoporosis? *Br. Med. J.* **298:**924.

Weiss, N.S., Ure, C.I., Ballard, J.H., Williams, A.R., and Daling, J.R., 1980, Decreased risk of fractures of the hip and lower forearm with postmenopausal use of estrogen, *N. Engl. J. Med.* **303:**1195.

Chapter 5

# Osteoporosis in Japan: Factors Contributing to the Low Incidence of Hip Fracture

Takuo Fujita

## 1. Introduction

Hip fracture or fracture of the proximal part of the femur is the most serious complication of osteoporosis, progressively increasing with age, especially in females, to become an urgent public health problem all over the world. Among elderly subjects confined to bed, or so-called bedridden patients, osteoporosis and resultant hip or vertebral fractures represent the second major cause responsible for such a miserable state, next only to cerebrovascular accidents, in Japan

The hip fracture incidence in Japan, however, is only one-half to one-third that reported in the U.S.A. and northern Europe. Similar low figures are also reported from other Asian countries. Since the Japanese population is almost uniformly Asiatic and the study populations of the U.S.A. and northern Europe consist mainly of Caucasians, a racial and genetic difference should first be taken into consideration. Patterns of nutritional intake are also sufficiently different between the two groups to justify detailed analysis of the effects of dietary habits on the development of osteoporotic hip fractures. Finally, the lifestyle, including

Takuo Fujita • Calcium Research Institute, 250 Makamicho, Kishiwada, Osaka 596, Japan

*Advances in Nutritional Research,* Vol. 9
Edited by Harold H. Draper
Plenum Press, New York, 1994

the work and exercise patterns, is also quite different between Japan and Western countries and its influence on the incidence and patterns of fracture should also be evaluated.

Factors contributing to hip fracture are complex. The decrease in bone mass or density in the elderly is no doubt the major cause of the bone fragility leading to fracture, but bone quality should also be taken into account. Chances of falling are also important, as they provide the impact to the bone resulting in fracture; hence the agility and motor functions of the subjects should also be evaluated. The mode of falling, the property of the floor on which the subjects fall, and soft tissue protection have also been discussed as additional elements influencing the occurrence, site and severity of the fracture. In order to explain the difference in the hip fracture incidence between the two populations, therefore, a comparative review of all these factors will be necessary before weighing the contribution of each factor to the difference in incidence.

## 2. Is Hip Fracture Incidence Truly Low in Japan?

According to a nationwide survey on hip fractures in Japan, about 50,000 people sustain a fracture each year out of a population of 125,000,000. This figure certainly appears low compared to the occurrence of hip fractures in 250,000 subjects in the United States, which has a population approximately twice as large. The annual incidence of hip fracture in the United States thus appears to be about 2.5 times that in Japan (Fujita, 1992; Cummings *et al.,* 1985). Local studies in Niigata and Tottori provided similar figures (Kawashima, 1989; Yamamoto *et al.,* 1991) and a correspondingly low incidence has also been reported from other Asian countries, notably from Singapore and Hong Kong (Wong, 1966; Lau *et al.,* 1990).

Cultural differences influencing the attitude to patient care should be taken into consideration. In Japan, about 400,000 patients are in a bedridden state and the majority of these are elderly patients. The way to take the best care of a patient is to attend to his or her every need, according to the Japanese way of thinking with reference to filial duty, even if the patient is confined to bed and cannot do anything. In consequence, the number of such patients is alarmingly increasing and even sons and daughters faithful to their parents are finding it difficult to take complete care of them in advanced age at home. Hence, a 10-year campaign to prevent the bedridden state was recently launched in Japan. In the United States, on the contrary, the best way to take care of patients is to let them take care of themselves, keeping them moving until the last moment with the full use of wheel chairs. This is possible only in the spacious American home, since the narrow Japanese house does not allow the use of wheel chairs. It may, therefore, be assumed that there are more bedridden patients in Japan than in the United States. Bedridden patients are latent hip fracture patients, since no doubts remain as to the consequences of getting up patients with extreme immobilization osteoporosis after a prolonged bedrest. Fractures of the hip and other sites occur even without a fall. Fractures caused by a simple shift of position to change sheets are a common daily occurrence in Japanese geriatric hospitals. Assuming that there are 50,000 long-term bedridden elderly in Japan who can be classified as latent hip fracture patients, and only 25,000 such patients in the United States, the total of manifest and latent hip fracture patients would be 100,000 in Japan and 275,000 in the United States. The age-adjusted incidence then becomes only 1.3 times higher in the United States than in Japan.

Hip fracture incidence is constantly changing with changes in medical care and life- and work style. The incidence is rapidly increasing in Hong Kong (Lau, *et al.,* 1990) and Niigata (Kawashima, 1989). Since hip fracture is a serious condition with severe disturbances of motor function, it would hardly be overlooked once it occurred, though a faint possibility remains that insufficient medical records and statistics may provide falsely low figures for its incidence. Improvements in medical care and treatment in some developing countries may be increasing the figures related to fracture incidence and surgical repair, in addition to the true increase in incidence. As a whole, however, hip fracture incidence still appears to be lower in Japan than in the United States and other Western countries, though the actual difference may not turn out to be as wide as it seems.

## 3. Factors Leading to Hip Fracture

Osteoporosis, severe osteopenia (decrease of bone mass) giving rise to difficulties in supporting body weight, is no doubt the most important factor in hip fracture, though osteomalacia has also been reported to be responsible for fractures in a small proportion of patients (Shimizu *et al.,* 1991). Bone mass may be measured and expressed in several different ways. The size of bones should be related to the size of the body they support. Bone mineral content (BMC) measured by single or dual photon absorptiometry or dual energy X-ray absorptiometry and expressed as g/cm, and even so-called bone mineral density (BMD) calculated by dividing it again with the distance of scanning expressed as $g/cm^2$ is still influenced by the bone size or body size. These measurements overestimate bone density, which should actually be expressed as $g/cm^3$ to express the amount of mineral per unit volume of the bone. Many reports of higher bone density in subjects with larger body size may be explained on this basis (Ross, 1991a). Since Caucasians are generally larger in body frame than their Asian counterparts, BMC or BMD expressed in this way consequently may be higher in Caucasians. Quantitative computed tomography (QCT), on the other hand, expresses the bone density as $g/cm^3$ presumably without interference by body size factors. Fujii *et al.* (1989) compared the spinal trabecular bone density of Japanese and American whites using QCT. Japanese had a significantly lower density, contrary to the blacks who had a higher bone density with lower fracture incidence. Thus Japanese apparently have a lower bone density than Caucasians and this makes the lower hip fracture incidence in Japan rather paradoxical.

## 4. The Influence of Bone Quality on Fracture Incidence

In addition to bone quantity, another property of bone may also influence the occurrence of fractures. This factor may tentatively be called bone quality, but its actual definition has never been established. In addition to the chemical properties of bone itself in terms of the proportion of osteoid tissue, morphological findings such as discontinuity of the trabeculae may cause a difference in bone quality influencing susceptibility to fracture. Ultrasound transmission has been proposed as one method to assess bone quality in addition to bone quantity.

Signet apparatus for the measurement of ultrasonic transmission through the patella measures apparent ultrasound conduction velocity (AUC), expressed as $k(E/d)^{0.5}$, where E = Young's modulus of elasticity or $k'd^2$. AUC may therefore be expressed as $Kd^{0.5}$. This would indicate that AUC is a function of the product of d, the density or quantity of bone, and K, a complex value expressing the quality of bone. Measurement of the AUC through the patella in American white women and Japanese women of the same age, using the identical apparatus and technique, revealed exactly the same range of values in both groups (Heaney et al., 1989). Since the quantity of bone or "d" has already been shown to be less in Japanese than in American whites, it follows that K, which expresses bone quality, must be higher in Japanese women than in Caucasians. Though the relationship between ultrasonic transmission and fracture risk is still unknown and the significance of K or bone quality is even more obscure, a possible difference in the quality of bone between Caucasians and Japanese should be taken into account when factors contributing to the difference in fracture rate are discussed.

The patella, like the vertebral body, consists mainly of trabecular bone. The characteristics of the trabecular bone that influence ultrasound transmission and possibly fracture rate include the continuity of the trabecular network, as demonstrated by strut analysis. Loss of trabecular bone occurs in two modes. One is a localized loss leading to disconnection of the trabeculae, occurring immediately after menopause and during immobilization, while the other is a diffuse narrowing of the trabeculae as seen in aging, hyperparathyroidism and renal osteodystrophy. Since endocrine factors should be much the same in different groups of people, a difference in lifestyle affecting the degree of immobilization is the most probable factor contributing to difference in bone quality. This was indicated by an archeological study on Germans who lived 900 years ago (Vogel *et al.,* 1990). Unlike modern man, working people 900 years ago did not lose the trabecular connectivity of their bone despite the universal age-related decrease in bone quantity. People of nobility, who did not work as much, developed the disconnectivity of trabeculae seen in modern people. Thus the quality of the bone that possibly contributes to a low incidence of hip fracture may be influenced by life- and work-style among other factors.

## 5. Lifestyle Differences

Since the skeletal system is a complex weight-bearing structure, each part responds differently to weight and force exerted with various intensities and from various directions. Bone needs external physical stress to maintain its quantity and quality. The spine, for example, is always compressed by gravitation, regardless of position: standing, sitting or walking. The hip, however, may be exposed to different kinds of physical force depending on the work load. Lower hip fracture incidence in Japanese than in American whites, despite an almost equal vertebral fracture rate, strongly suggests the influence of lifestyle on bone quality as well as quantity.

Japanese people have been living on a small island for several thousand years. Due to constant overpopulation and lack of habitable space, their residences have always been very small and crowded. In consequence, an average family used a single room for living, dining and sleeping. By sitting directly on the floor with the knees completely flexed, space for chairs was saved. The beds were made directly on the floor, simply by spreading a mat, the removal of which immediately converted the room into a living room. Kitchen space was also very limited. Women living in such a household for many years automatically exercise and strengthen their hip and pelvic muscles, because they have to sit directly on the mattress to greet a guest and stand up to go to the kitchen to get some tea the next moment. They would then quickly return from the kitchen to serve the tea to the guest. The repetition of such daily routine, together with child-raising, cooking and washing strengthened the muscles and bones of Japanese women. The squatting type toilet still commonly used in Japan would be intolerable to Western women, who are used to the comfortable sitting position. A similar situation is common in most Asian countries. Most Japanese women of osteoporotic age never have driven a car. It was their responsibility to buy and bring home on foot all the food and materials for living for the family. Therefore, they had to be quite strong. Such a lifestyle is naturally not shared by American white women. They live in a big house, sit on chairs and sleep in beds. Though their household work is also hard, at least their home space is more abundant and there seems to be less opportunity to exercise the hip muscles. Cars have been available to most of them since their youth. Such a difference in lifestyle causes some difference in motor function and agility. Japanese elderly women who are free of neurological conditions and adverse effects of drug treatment may not fall as readily as their American counterparts.

Things are, however, rapidly changing all over the world. Japan and other Asian nations are being rapidly "Americanized." There is nothing more reasonable than to live as comfortably as possible within one's means. Though home space is still quite limited in Japan, Western style houses are becoming more and more popular. Girls no longer want to sit directly on a mattress or work as hard as their mothers did. Many of them drive cars. It is of profound concern to see the changes in hip fracture rate occurring along with such Americanization of lifestyle.

## 6. Nutritional Intake of the Japanese

The diet of the Japanese has been characterized by low calcium, low fat, low protein and high sodium intakes. Drinking water is almost exclusively soft, containing only a small amount of calcium. This may be explained by the volcanic soil which contains little calcium and retains little water. Abundant rain accelerates the leaching of soil minerals, so that river and well water is quite low in mineral content. As an exception, the inhabitants of the Island of Okinawa, which is surrounded by a coral ridge, have a higher calcium intake from water than those in other parts of Japan. In addition, the narrow space between the mountains and the sea in Japan was used almost exclusively to grow rice and little space was left for raising cattle. Their protein intake thus depended mainly on fish; little meat and dairy products, including milk, butter and cheese, were consumed. The nutrient intake of the Japanese thus heavily depends on marine products and calcium is no exception. The main source of calcium in Japan is small fish with bones, seaweeds and some soy bean products, unlike the typical Western diet, which supplies calcium mainly through milk and cheese. According to the results of the National Nutritional Survey conducted by the Ministry of Health and Welfare through extensive sampling of representative households all over Japan and direct interviews, the mean daily calcium intake of Japanese adults stayed between 500 and 600 mg/day over the last 15 years. Some increase occurred above the very low level of 300-400 mg/day during the immediate postwar period (Table 1). The Recommended Daily Allowance therefore has been set at 600 mg/day for adults. This is lower than in most Western Countries, probably reflecting the consistently low calcium intake by the Japanese people, and the influence of a strong dairy industry in most Western countries. WHO recommends 400-500 mg/day because this is the amount usually available on cereal-based diet without significant supplement by dairy products. Although a superficial analysis of the global epidemiological data on hip fracture (Hegsted, 1986; Gallagher *et al.*, 1980) may suggest that the hip fracture rate is positively correlated with calcium intake, the fracture rate is evidently influenced by multiple factors such as racial differences and variations in lifestyle, in addition to nutritional factors. High calcium intake is thus far from being a risk factor for an increased hip fracture rate in Western countries and low calcium intake should hardly be conceived of as a protecting factor against hip fracture in non-Western countries. Blacks are apparently not as susceptible to fracture as Caucasians and an active lifestyle in the non-Western countries may help to develop strong hip musculature and bones. Finally, the vast difference in life expectancy between developed and developing countries does not justify a simple comparison of fracture incidence among different nations. In developing countries where the mean life expectancy of women is as low as 50 years, an osteoporotic fracture would be an exceptional occurrence with an extremely low incidence.

Table 1. Mean Daily Nutritional Intake by Japanese Adults 1975-1991[a]

| Item | 1975 | 1980 | 1985 | 1987 | 1989 | 1991 |
|---|---|---|---|---|---|---|
| Energy (kcal) | 2226 | 2119 | 2088 | 2053 | 2067 | 2061 |
| Protein (g) | 81.0 | 78.7 | 79.0 | 78.5 | 79.2 | 80.2 |
| Animal protein (g) | 38.9 | 39.2 | 40.1 | 40.1 | 41.7 | 42.4 |
| Fat (g) | 55.2 | 55.6 | 56.9 | 56.6 | 58.3 | 58.9 |
| Animal fat (g) | 26.2 | 26.9 | 22.6 | 27.6 | 28.0 | 28.3 |
| Carbohydrate (g) | 335 | 309 | 298 | 291 | 289 | 290 |
| Calcium (mg) | 552 | 539 | 553 | 551 | 524 | 540 |
| NaCl (g) | 13.5 | 12.9 | 12.1 | 11.7 | 12.2 | 12.2 |

[a] Japan Ministry of Health and Welfare, 1991, *Current Status of the Nutritional Intake in Japan,* Daiichi Publishers, Tokyo, p. 33.

In Japan, a higher calcium intake of up to 1100 mg/day is recommended for pregnant and lactating women. The RDA for the young is still not much higher than that for adults, i.e., less than 700 mg/day. Eating habits cannot be changed easily despite a rise in the living standard, which gives the freedom to choose foods of one's preference. Most people eat what they like, and not what they should eat. In addition, there is low solar exposure in the northern half of Japan, restricting the ultraviolet irradiation to the skin necessary to the conversion of provitamin D to vitamin D to assist intestinal calcium absorption.

As shown in Table 1, sodium intake has been high in Japan, reflecting the frequent use of salt as a preservative for fish and vegetables. Despite a continuing effort to decrease salt intake, it still remains considerably above 10 g/day. High salt intake increases urinary calcium excretion, thereby contributing to the risk of negative calcium balance. Protein and phosphorus intakes are generally low. These nutrients tend to increase and decrease urinary calcium excretion, respectively, with mutually counteracting effects that are important in maintaining calcium balance. Animal protein, especially fish intake, is especially low with a consequent low vitamin D intake that becomes more pronounced with advance in age.

Table 2 shows the mean calcium intake of healthy male and female Japanese living at home by age groups. Older people tend to take more calcium, exceeding 600 mg only in the seventh decade. Elderly people in geriatric institutions consume less calcium (probably around 300 mg/day). The low calcium intake of younger people is also conspicuous, partly because of excessive and unjustified dieting. General malnutrition apparently contributes to the development of hip fractures and osteoporosis during prolonged illness. As shown in Table 3, a comparison between hip fracture patients and age-matched controls revealed lower body weight, lower serum albumin, lower serum total calcium and lower milk intake in the patients (Hayashi, 1991).

Table 2. Mean Calcium Intake by Age in Normal Japanese Volunteers Living at Home (mg/day)[a].

| Age | Ca Intake | Male | Female |
|---|---|---|---|
| –19 | 386 | 368 | 424 |
| 20–29 | 374 | 362 | 418 |
| 30–39 | 488 | 474 | 515 |
| 40–49 | 556 | 556 | 555 |
| 50–59 | 581 | 362 | 611 |
| 60–69 | 624 | 474 | 618 |
| 70– | 537 | 541 | 536 |
| Mean | 514 | 451 | 561 |

[a]Japan, Ministry of Health and Welfare, 1991, *Current Status of the Nutritional Intake in Japan*, Daiichi Publishers, Tokyo, p. 70.

Table 3. Data for Elderly Patients with Hip Fracture and Age-matched Controls (Mean ± SD)

| Item | Patients (N = 41) | Controls (N = 57) | P value[a] |
|---|---|---|---|
| Age (years) | 79 ± 9 | 79 ± 9 | n.s. |
| Body weight (kg) | 42 ± 6 | 45 ± 8 | 0.05 |
| Total serum Ca (mg/dl) | 7.0 ± 0.8 | 7.6 ± 1.6 | 0.001 |
| Serum albumin (g/dl) | 2.1 ± 0.3 | 2.7 ± 0.5 | 0.1 |
| Balance (sec)* | | | |
| > 4 | 10 | 21 | |
| < 4 | 4 | 11 | 0.1 |
| Milk Intake | | | |
| > 200 ml/day | 20 | 36 | |
| < 200 ml/day | 21 | 21 | 0.1 |

[a] Significance of the differences between the two groups was tested by Student's t-test.
* Standing on one foot, an indicator of balance sensation and coordination of muscle strength.

In contrast to hip fracture, spinal compression fracture apparently occurs at a frequency as high in Japanese as in American whites, according to the cross-cultural comparison made by Ross (1991b). In fact, the prevalence of spinal compression fractures among Japanese may be even higher than that in their Western counterparts. This phenomenon is also substantiated by simple observation of elderly women walking on the street. Roundback or kyphotic women are quite common in Japan. This suggests a continued high activity despite osteoporosis or poor working conditions, in a narrow kitchen, or manual farming that forces them to flex the back throughout their lives. If spinal compression fractures are at least as common in Japan as in the U.S.A. and other Western countries, the paradox of a lower hip fracture incidence in spite of a lower calcium intake may be easier to understand, in terms of a difference in lifestyle.

In a nutritional survey on an institution for the elderly in Tokyo, an extremely low calcium intake of 300 to 400 mg/day was found without milk ingestion, while the use of milk raised it to 500 mg/day. The yearly shortening of body height was significantly greater than in those who did not drink milk regularly. Thus, nutritional intake, especially that of calcium, no doubt remains important in the prevention of spinal compression fractures even for Japanese, but hip fracture incidence is apparently kept low by some other factors.

Low calcium intake and poor general nutritional status in Japanese do not explain their lower incidence of fractures. In fact, a higher incidence of hip fractures would be expected in Japan, where low calcium intake and poor general nutritional intake are still prevalent, especially in the elderly population. Hip fracture requires an urgent explanation, since it certainly is the most serious complication of osteoporosis.

## 7. Summary and Conclusion

Hip fracture incidence seems to be lower in Japan than in many Western countries, but the difference is apparently becoming smaller with progressive Westernization of the Japanese lifestyle and nutritional habits. Nutrition cannot explain the lower incidence of hip fracture. A lower calcium intake prevails in Japan. Genetic differences in body build, including a lower center of gravity, better motor function and agility, well developed hip musculature and small but more fracture-resistant bones secondary to a difference in life- and work-style may contribute to fewer falls and a lower fracture rate among Japanese than among their Western counterparts. Such traditional lifestyle habits as sitting directly on the floor are rapidly decreasing, and time will tell how much of the low incidence of hip fracture in Japan can be explained by lifestyle and how much by genetic and other factors. The Japanese women who now enjoy a low hip fracture incidence led a hard physical life when they were young. This may be a lesson to the young of future generations in how to avoid bone fractures when they are old. Bone health may be achieved by enjoying life through sports or even the tea ceremony in place of the hard physical work of their ancestors, which is gradually disappearing.

## References

Cummings, S.R., Kelsey, J.L. Newitt, M.C., and O'Doud, K.J., 1985, Epidemiology of osteoporosis and osteoporotic fracture, *Epidemiol. Rev.* **7**:178.

Fujii, Y., Tsutsumi, M., and Tsunenari, T., 1989, Quantitative computed tomography of lumbar vertebrae in Japanese patients with osteoporosis, *Bone and Mineral* **6**:87.

Fujita, T., 1992, Comparison of osteoporosis and calcium intake between Japan and the United States, *Proc. Soc. Exp. Biol. Med.* **200**:149.

Gallagher, T.C., Melton, L.C., Riggs, B.L., and Bergstrath, E., 1980, Epidemiology of fractures of the proximal femur in Rochester, Minnesota, *Clin. Orthoped. Rel. Res.* **150**:163.

Hayashi, Y., 1991. "Abnormalities of the musculoskeletal system," in: *Falls and Fractures of the Elderly* (Committee for Lake Riwa Symposium on Science of Aging, eds.), pp. 59-72, Ishiyaku Publ., Tokyo.

Heaney, R.P., Avioli, L.V., and Chestnut, C.H. III, 1989, Osteoporotic bone fragility: detection by ultrasound transmission velocity, *JAMA* **261**:2976.

Hegsted, D.M., 1986, Calcium and osteoporosis, *J. Nutr.* **116**:2316.

Kawashima, T., 1989, Epidemiology of the femoral neck fracture in 1985, Niigata Prefecture, Japan, *J. Bone Min. Metab.* **7**:118.

Lau, E.M.C., Cooper, C., Donnan, S., and Barker, D.J.P., 1990, "Incidence and risk factors for hip fractures in Hong Kong Chinese," in: *Osteoporosis 1990: Proceedings Third International Symposium on Osteoporosis* (C. Christiansen and K. Overgaard, eds.), pp. 66-70, Osteopress, Copenhagen.

Ross, P.D., 1991a, Bone mass and other risk factors for fracture, *J. Bone Min. Metab.* **9**:146.

Ross, P.D., 1991b, "Comparison of vertebral fracture between Caucasians and Japanese," in: *Proceedings of the International Conference on Osteoporosis, 1991* (T. Fujita, ed.), p. 56 (in press).

Shimizu, S., Kushida, K., Sumi, Y., Denda, M., Yamazalo. K., Inoue, T., 1991, Osteomalacia in elderly patients with hip fractures, *J. Bone Min. Metab.* **9**:259.

Vogel, M., Hahn, M., Caselitz, P., Woggan, J., Pompesuis-Kempa, M., and Delling, G., 1990, "Comparison of trabecular bone structure in man today and an ancient population in western Germany," in: *Bone Morphometry* (H.E. Takahashi, ed.), pp. 220-224, Nishimura Publishers.

Wong, P.C.N., 1966, Fracture epidemiology in a mixed southeastern Asian community (Singapore), *Clin. Orthop. Rel. Res.* **45**:55.

Yamamoto, K., Hagino, H., and Nakamura, T., 1991. Incidence and risk factors for hip fractures in Japanese elderly women, *J. Bone Min. Metab.* **9**:168.

Chapter 6

# Osteoporosis in Asia

E.M.C. Lau and J. Woo

## 1. Osteoporosis in Asia—a Modern Epidemic

During the last two decades, osteoporosis has evolved from a relatively rare condition to an epidemic one in urbanized parts of Asia. In the past, there was a definite geographical pattern to osteoporosis and osteoporotic fractures. The age-adjusted incidence of hip fracture among Caucasians from Europe and North America was two- to threefold higher than the rates observed in Chinese, Bantu and Maori in the late 1960s (Table 1) (Maggi *et al.*, 1991).

With urbanization, the incidence of hip fractures has increased dramatically in several Asian countries. In Hong Kong, the age-adjusted rates increased more than twofold from 1966 to 1989 (Table 2) (Lau *et al.,* 1990). The incidence increased mainly in elderly men and postmenopausal women. Similarly, in Singapore the incidence of hip fracture increased from 0.7 per 1,000 in women who were 60 years and over in 1957-62 (Wong, 1966) to 1.5 per 1,000 in the same group in 1980 (Lee *et al.*, 1988).

---

E.M.C. Lau • Department of Community and Family Medicine, Chinese University of Hong Kong, Lek Yuen Health Centre, Shatin, N. T., Hong Kong. J. Woo • Department of Medicine, Chinese University, Hong Kong.

*Advances in Nutritional Research,* Vol. 9
Edited by Harold H. Draper
Plenum Press, New York, 1994

Table 1. Age-adjusted incidence rates (per 100,000) of hip fracture by sex in the population over 50 years of age, by geographical area.

| Geographic area, years of survey | Age-adjusted rates | | Age and sex-adjusted rates: total |
|---|---|---|---|
| | Women | Men | |
| Norway, 1983-84 | 1293 | 551 | 968 |
| Oslo, 1978-79 | 701 | 310 | 530 |
| Stockholm, 1972-81 | 622 | 291 | 477 |
| Denmark, 1973-79 | 620 | 203 | 437 |
| New Zealand Whites, 1973-76 | 620 | 151 | 414 |
| California Whites, 1983-84 | 559 | 207 | 402 |
| Rochester, NY, 1965-74 | 510 | 174 | 364 |
| Texas Whites, 1980 | 530 | 205 | 384 |
| Hong Kong, 1985 | 353 | 181 | 277 |
| Hong Kong, 1965-67 | 153 | 96 | 128 |
| California Asians, 1983-84 | 338 | 104 | 235 |
| Texas Hispanics, 1980 | 263 | 118 | 197 |
| Yorkshire, UK, 1973-77 | 275 | 96 | 196 |
| California Blacks, 1983-84 | 219 | 144 | 185 |
| Kuopio, Finland, 1968 | 100 | 249 | 183 |
| California Hispanics, 1983-84 | 197 | 90 | 151 |
| New Zealand Maori, 1973-76 | 107 | 182 | 149 |
| Singapore, 1955-62 | 75 | 100 | 86 |
| Johannesburg Bantu, 1950-64 | 26 | 38 | 31 |

Source: Maggi *et al.*, 1991.

There is no information on the time trends in the incidence of hip fracture in China. However, a survey conducted in Chengdu in 1989 (Huang, unpublished results) showed that the crude incidence rate for hip fracture was 30/100,000 in both men and women. In Hong Kong the corresponding crude incidence rate was 30/100,000 in men and 77/100,000 in women (Lau *et al.*, 1990). This seems to suggest that while the incidence of hip fracture in men in Chengdu is as high as in Hong Kong, the rate in women is lower. It is necessary to compare age-adjusted rates before further conclusions can be made.

There has also been an increase in the incidence of hip fracture in Japan. The age-specific incidence rate in Japan in 1989 (Orimo, 1990) was very similar to that in Hong Kong. Moreover, a 20% increase in the age-adjusted incidence of hip fracture was reported in Niigata, Japan, from 1985 to 1989 (Orimo, 1990). The age-adjusted incidence rate in Japan remains lower than that observed in Europe and North America.

Table 2. Age-specific hip fracture rates in Hong Kong per 100,000 population in 1966, 1985 and 1989.

| | Men | | | Women | | |
|---|---|---|---|---|---|---|
| Age Group | 1966 | 1985 | 1989 | 1966 | 1985 | 1989 |
| 40-49 | 6 | 13 | 8 | 7 | 11 | 3 |
| 50-59 | 16 | 28 | 21 | 23 | 32 | 25 |
| 60-69 | 71 | 54 | 77 | 57 | 135 | 113 |
| 70-79 | 224 | 339 | 307 | 173 | 501 | 505 |
| $\geq$ 80 | 321 | 1156 | 898 | 716 | 1521 | 1327 |

Source: Lau *et al.,* 1990.

Hence the incidence of hip fracture in many Asian countries has increased in recent years, particularly in men. Moreover, the population in Asia is aging rapidly. The number of elderly subjects in Asia will increase fourfold from 1990 to 2050, so that there will be a total of 532,548,000 men and women who are 65 years and over in 2050. This implies that the absolute number of subjects with osteoporosis and hip fracture will be very large. What will be the cause of this epidemic of osteoporosis in Asia and how can it be prevented?

## 2. Calcium

### 2.1. Calcium and Osteoporosis

Nutritional factors are very important in the etiology of osteoporosis. Dietary calcium intake has a particularly large effect, which may be modified by other factors such as physical activity, protein intake and sodium intake.

Most of the controversy over the relationship between dietary calcium intake and osteoporosis has been resolved. A meta-analysis (Cumming, 1990) was conducted to study this relationship and the results of randomized controlled trials on the effects of calcium supplementation are summarized in Table 3. In postmenopausal women, the mean increase in bone mineral content (BMD) with calcium supplements ranged from 0 to 4%. BMD increased by 1% or less in subjects whose dietary calcium intake was more than 548 mg per day. In contrast to postmenopausal women, the role of supplementing the diet of premenopausal women and of men with calcium tablets or food rich in calcium remains unclear.

Table 3. Selected characteristics and mean effects in intervention studies of the effect of calcium supplements on bone mass

| Study | Mean age (yr) | Calcium dose (mg) | Baseline calcium intake (mg) | Mean effect of calcium[a] (%) |
|---|---|---|---|---|
| Postmenopausal women | | | | |
| Albanese et al. (1975)* | 81 | 750 | 450 | 4.27 |
| Ettinger et al. (1987)* | 52 | 1,000 | 792 | 0.00 |
| Hansson et al. (1987) | 66 | 1,000 | - | 0.85 |
| Horsman et al. (1977)* | 50 | 800 | - | 1.69 |
| Lamke et al. (1978) | 60 | 800 | - | 4.35 |
| Polley et al. (1987) | 57 | 1,000 | 716 | 0.82 |
| Polley et al. (1987) | 57 | Diet | 692 | 0.12 |
| Recker et al. (1977) | 57 | 1,040 | 548 | 1.01 |
| Recker et al. (1985) | 59 | Diet | 737 | -0.01 |
| Riis et al. (1987) | 50 | 2,000 | - | 0.76 |
| Smith et al. (1981) | 82 | 750 | - | 2.03 |
| Smith et al. (1989) | 55 | 1,500 | 710 | 0.75 |
| Premenopausal women | | | | |
| Smith et al. (1989) | 42 | 1,500 | 652 | 0.02 |

[a] Percent difference per year (percent bone lost per year in the group given calcium supplement minus percent bone lost per year in control group). The mean effect is the average of the "percent difference per year" for all bone sites measured in that study. A positive figure indicates a protective effect of calcium.

* Not a randomized trial.

Source: Cumming, 1990

The results of a large randomized controlled trial conducted in the USA (Dawson-Hughes *et al.*, 1990) showed that the effects of calcium supplementation vary with menopausal status and skeletal sites. A calcium supplement of one gram per day reduced the rate of bone loss at the radius in both early and late postmenopausal women. However, this regime prevented bone loss at the spine in late but not in early postmenopausal women. As described by Dawson-Hughes (1991) in a review of controlled clinical trials, little is known about the effects of calcium supplementation on bone density at the hip.

Calcium is a threshold element, in the sense that if the intake is above a certain level, further supplementation will have little benefit. The threshold of calcium intake is likely to be 500–600 mg per day. As the mean dietary calcium intake in many populations in Asia is at or below this level, calcium supplementation is potentially very important for the prevention of osteoporosis.

## 2.2. Calcium Intake in Asia

A low calcium content is a unique feature of Asian diets. The results of a survey on calcium consumption in Chinese men and women living in Hong Kong showed that the mean calcium intake was lower than 500 mg per day in all age groups, and calcium intake decreased with age (Pun *et al.,* 1989). The low calcium intake in the elderly was confirmed by other studies (Ho *et al.*, 1988; Lau *et al*, 1989). The calcium content of the Japanese diet is slightly higher than that of the Chinese. In a survey in which calcium intake was assessed by a 3-day food diary, the mean calcium intake of premenopausal Japanese women was 458 mg (SD = 171 mg; n = 89) and of postmenopausal women was 601 mg (SD = 209 mg; n = 89) (Lacey *et al.*, 1991).

Another unique feature of the Asian diet is the non-dairy source of calcium. The percentages of dietary calcium from different food groups in Chinese, Japanese and American diets are presented in Table 4 (data compiled from Lacey *et al.*, 1991 and Pun *et al.*, 1989). In Chinese, vegetables and soya products accounted for about 41% of the calcium intake of all age groups, while dairy products accounted for only about 23% of the total intake. The Japanese diet was also characterized by infrequent consumption of dairy products, which accounted for only 24% of calcium intake in both premenopausal and post-menopausal women. The other major sources of calcium included vegetables, fish and soybean products (to-fu) (Table 4).

Table 4. Contributions of food groups to daily calcium intake of elderly Chinese, Japanese and American women.

| | Percentage contribution | | |
|---|---|---|---|
| Food groups | Hong Kong Chinese | Japanese | Americans |
| Dairy Products | 23 | 24 | 42 |
| Vegetables | 32 | 18 | 10 |
| Soya Bean Products | 9 | 15 | - |
| Grain | 10 | 12 | 15 |
| Fish | 8 | 15 | - |
| Others | 18 | 16 | 33 |
| Total | 100 | 100 | 100 |

In the past, it was commonly believed that calcium absorption from non-dairy sources was unsatisfactory. Calcium absorption from 10 different foods has been studied (Weaver, 1992). Calcium absorption from vegetables (with the exception of spinach, which is high in oxalate), and from soya products was as good as from milk (which averages 25–35%). However, it would be difficult to increase calcium intake significantly by increasing the dietary intake of soya and vegetables, as their calcium content is lower than that of dairy products and a huge amount would have to be consumed.

## 2.3. Calcium and Osteoporosis in Asia

The results of case-control studies and randomized trials strongly suggest that a low dietary calcium intake is one of the most important etiological factors in osteoporosis and osteoporotic fractures in Chinese. A case-control study was conducted to investigate the role of load-bearing activities and calcium intake in hip fracture in Hong Kong (Lau *et al.*, 1988). Four hundred consecutive patients (280 women and 120 men) with hip fracture were compared to 400 surgical inpatients and 400 noninstitutionalized elderly subjects. Calcium intake was assessed by the food frequency method. The weekly frequency of consumption of nine food items identified as important sources of calcium in the Chinese diet was recorded. The results of multiple logistic regression showed that the risk of hip fracture increased significantly with a low calcium intake, a statistically significant trend being demonstrated in both sexes (Table 5).

A randomized controlled trial was conducted to investigate the effects of calcium supplementation and load-bearing exercise on bone density (Lau et al., 1992). Fifty Hong Kong Chinese women who were 62 to 92 years old were randomized to enter one of four treatment groups: (1) Calcium supplement group—800 mg of calcium as lactate-gluconate per day; (2) Exercise group—stepping up and down on a 23 cm block 100 times plus 15 minutes muscle-strengthening exercise 4 times a week, plus placebo tablet daily; (3) Calcium supplement and exercise group; (4) Placebo group.

Bone mineral density was measured at the hip and spine by dual X-ray densitometry and the change in BMD after 10 months in the four treatment groups was compared. The results are shown in Table 6. Calcium supplements led to an increase in bone mineral density in two sites at the hip, and its effects were enhanced by exercise. These results imply that calcium supplementation and exercise may be important in the prevention of osteoporosis in elderly Chinese women.

Table 5. Dietary calcium intake and the risk of hip fracture in 400 patients with hip fracture and 800 controls.

| Fifths of the distribution of calcium intake (mg/day) | No. of patients | No. of controls | Adjusted* Relative Risk | 95% Confidence interval |
|---|---|---|---|---|
| | | Women | | |
| <75 | 93 | 137 | 1.9 | 1.2 to 2.9 |
| 75- | 47 | 72 | 1.9 | 1.1 to 3.1 |
| 83- | 42 | 105 | 1.1 | 0.7 to 1.9 |
| 129- | 57 | 126 | 1.2 | 0.8 to 2.0 |
| ≥244 | 41 | 120 | 1.0 | |
| | | Men | | |
| <75 | 44 | 67 | 2.1 | 1.1 to 4.2 |
| 75- | 14 | 30 | 1.4 | 0.6 to 3.4 |
| 83- | 23 | 44 | 1.7 | 0.8 to 3.7 |
| 129- | 20 | 40 | 1.5 | 0.7 to 3.2 |
| ≥244 | 19 | 59 | 1.0 | |

*Adjusted for cigarette smoking and alcohol consumption.
Source: Lau *et al.*, 1988

Table 6 Percentage changes in bone mineral density for the four groups (mean and 95% confidence interval)

| | Group I: calcium supplements ($n$ = 12) | Group II: exercise and placebo ($n$ = 11) | Group III: calcium supplements and exercise ($n$ = 15) | Group IV: placebo ($n$ = 12) |
|---|---|---|---|---|
| Spine (L2-4) | -0.08(-5.2-5.1) | -1.9(-6.7-2.8) | -1.1(-3.7-1.4) | -2.5(-6.5-1.4) |
| Femoral neck | -3.5(-9-1.8) | -6.6(-12-0.8) | 5.0(-0.77-10) | -1.1(-7.4-5.3) |
| Ward's triangle | 2.5(-5.9-11) | -6.0(-15-3.2) | 17(3-31) | -2.4(-10-5.9) |
| Intertrochanteric area | 2(-1.6-5.7) | 0.1(-6.5-6.7) | 11(1.3-22) | 0.25(-3.3-3.8) |

The percentage changes at Ward's triangle and the intertrochanteric area for the calcium supplements group were statistically significant ($P<0.05$) by two-way analysis of variance. The interaction between calcium supplements and exercise was significant ($P<0.05$) at the femoral neck by two-way analysis of variance.
Source: Lau *et al.*, 1992

In Japan, the results of an ecological cross-sectional study (Orimo, 1990) showed that the average calcium intake was negatively correlated with the incidence of hip fracture in various prefectures in Japanese women. This relationship was not demonstrated in men. In another cross-sectional survey, current calcium intake was not found to be significantly associated with forearm bone mineral density in premenopausal or postmenopausal women (Lacey *et al.*, 1991). However, the current intakes of milk and vegetables were found to be positively correlated with forearm BMD ($P = 0.05$ and $P < 0.01$, respectively). The current calcium intake is a cross-sectional measurement and is a poor representation of calcium intake in the past. As it is a low calcium intake over long periods that may result in a low BMD, the lack of association between current calcium intake and BMD is plausible. On the other hand, the current intakes of milk and vegetables might have been good measures of dietary habits; hence the positive correlation between BMD and current intake. Such ecological and correlation studies were weak in design and do not provide conclusive evidence on the relationship between calcium intake and osteoporosis in Japanese.

# 3. Physical Activity

## 3.1 Physical Activity and Osteoporosis

The relationship between weight-bearing exercise and bone density is well known. Several cross-sectional and longitudinal studies have shown a direct relationship between weight-bearing exercise and bone mass (Pocock *et al.*, 1986; Simkin *et al.*, 1987; Dalsky *et al.*, 1988; Kelly *et al.*, 1990). The results of a randomized trial provided direct evidence that exercise prevented bone loss at the upper trunk and thigh in women who were 50–62 years old (Chow *et al.*, 1987). In a more recent study by Prince *et al.* (1991), bone loss in the distal forearm in an exercise group resembled that in a control group. However, bone loss was significantly reduced in an exercise and calcium group and a group on estrogen and exercise showed significant bone gain. These observations imply that the effects of exercise on bone density are less than those of calcium and estrogen. However, as load-bearing exercise can enhance the effects of calcium supplementation and estrogen replacement therapy in preventing bone loss, it is a very important modality for the prevention of osteoporosis.

## 3.2. Exercise and Osteoporosis in Asia

Urbanization and a reduction in load-bearing activity may have resulted in the recent increase in the prevalence of osteoporosis and in the incidence of hip fracture in several developed countries in Asia. As mentioned previously, the incidence of hip fracture increased by 100% in both Hong Kong and Singapore in the last 20 years. In these countries, the effects of a low calcium intake may have been offset by a high level of load-bearing activity in the past. With urbanization, hard physical labour becomes much less necessary in everyday life. Eventually, the effect of a low calcium intake may be manifested as an epidemic of osteoporosis in urbanized parts of Asia.

The relationship between load-bearing activities and the risk of hip fracture in Hong Kong Chinese was studied in the case-control study described previously (Lau, *et al.*, 1988). The results for physical activity are presented in Table 7. The relative risk for hip fracture was twofold among women who reported walking outdoors, upstairs, uphill, or with a load less than once a day, as compared to those who performed these activities every day. Among men, the relative risk of hip fracture was greater for subjects who walked outdoors, uphill, or with a load less than once a day, but the increase was not statistically significant.

The results of the randomized trial in Hong Kong Chinese described previously (Lau *et al.*, 1992) showed that exercise *per se* cannot increase bone density in elderly Chinese women. However, exercise enhanced the effect of calcium supplements. This is in keeping with observations in Caucasians, and confirms evidence that exercise alone cannot increase BMD in situations where calcium intake is inadequate. However, exercise is important in preventing osteoporosis in Asians due to its interaction with calcium supplementation.

# 4. Vitamin D

## 4.1. Low Vitamin D Status—a Risk Factor for Osteoporosis?

Low vitamin D status may be a risk factor for osteoporosis. In Caucasians living in the UK, the plasma level of 25-hydroxyvitamin D (25-OHD) decreases with age and is low in the institutionalized elderly (Baker *et al.*, 1979). In two cross-sectional studies conducted in the UK (Baker *et al.*, 1979; Cooper *et al.*, 1989), the 25-OHD level of patients with hip fracture was found to be lower than that of controls.

The beneficial effects of vitamin D supplementation in North Americans were demonstrated in a randomized controlled trial conducted by Dawson-Hughes (1991). While receiving a calcium supplement of 377 mg/day, women who increased their vitamin D intake from 100 IU to 500 IU per day had less loss of spinal bone mineral than a placebo group in six winter months (loss in the treatment group was -0.54% compared with -1.22% in the controls, $P < 0.05$). These results may not be directly applicable to Asians, as the effect of vitamin D supplementation depends both on the vitamin D status and the calcium intake of the population.

Table 7. Current physical activity and the risk of hip fracture in 400 patients with hip fracture and 800 controls. Adjusted relative risk and 95% confidence intervals (C.I.) *

| | Women. | | | |
|---|---|---|---|---|
| Frequency of activity | No. of patients | No. of controls | *Adjusted relative risk | 95% C.I. |
| Walking outdoors: | | | | |
| - Less than once a day | 105 | 151 | 1.7 | 1.2 to 2.3 |
| - Once daily or more | 175 | 409 | 1.0 | |
| Walking upstairs: | | | | |
| - Less than once a day | 182 | 323 | 1.4 | 1.0 to 1.9 |
| - Once daily or more | 97 | 235 | 1.0 | |
| Walking uphill: | | | | |
| - Less than once a day | 267 | 523 | 1.6 | 0.8 to 3.1 |
| - Once daily or more | 13 | 37 | 1.0 | |
| Walking with a load: | | | | |
| - Less than once a day | 268 | 513 | 2.3 | 1.2 to 4.7 |
| - Once daily or more | 11 | 45 | 1.0 | |
| | Men | | | |
| Walking outdoors: | | | | |
| - Less than once a day | 25 | 40 | 1.3 | 0.7 to 2.4 |
| - Once daily or more | 95 | 200 | 1.0 | |
| Walking upstairs: | | | | |
| - Less than once a day | 71 | 133 | 1.0 | 0.6 to 1.7 |
| - Once daily or more | 49 | 107 | 1.0 | |
| Walking uphill: | | | | |
| - Less than once a day | 111 | 216 | 1.9 | 0.7 to 4.7 |
| - Once daily or more | 9 | 24 | 1.0 | |
| Walking with a load: | | | | |
| - Less than once a day | 111 | 218 | 1.4 | 0.6 to 3.3 |
| - Once daily or more | 9 | 22 | 1.0 | |

* Adjusted for cigarette smoking and alcohol consumption
Source: Lau *et al.*, 1988

## 4.2. Vitamin D Status in Chinese and Japanese

As there is adequate sunshine in Southeast Asia throughout the year, the average vitamin D level among the elderly in these areas is much higher than that of subjects living in Europe and North America. A cross-sectional survey was conducted to measure the 25-OHD level of a cluster sample of noninstitutionalized elderly men and women living in Hong Kong (Woo *et al.*, 1989b). In subjects who were older than 60 years, the mean 25-OHD level was 33 μg/L (SD = 2.8 μg/L; n = 157) in men and 27 μg/L in women (SD = 0.4 μg/L; n = 242).

Table 8. Serum 25-OHD levels in temperate countries, Australia and Hong Kong

| Author | Country | Mean 25-OHD level, $\mu g/L$ |
|---|---|---|
| Von Knorring *et al.* (1982) | Finland | 6.8 |
| Schmidt-Gayk *et al.* (1977) | FRG | 8.4 |
| McKenna *et al.* (1985) | Ireland | 8.4 |
| Lester *et al.* (1977) | UK | 8.6 |
| Rapin *et al.* (1982) | Switzerland | 9.2 |
| Chapuy *et al.* (1983) | France | 11.9 |
| Omdahl *et al.* (1982) | USA | 12.4 |
| Lund *et al.* (1975) | Denmark | 22.0 |
| Tassie *et al.* (1985) | Australia | 27.2 |
| Lau *et al.* (1989) | Hong Kong | 27.0 |

The values for temperate countries were quoted by McKenna *et al.* (1985). Original values presented as nmol/L were converted to $\mu g/L$ after multiplying by a factor of 0.4.

This was much higher than the 25-OHD levels observed in Europe and North America (Table 8). Moreover, the level in this elderly group was not lower than that of 55 medical students living in Hong Kong (Mean = 26.9 µg/L; SD = 5.6 µg/L) (Woo *et al.*, 1989).

A low vitamin D level is much more prevalent in elderly Hong Kong Chinese living in institutions. In a survey conducted in four chronic care institutions in Hong Kong (Woo *et al.*, 1989), the mean 25-OHD level was 18.8 µg/L in men (SD = 9.5 µg/L; n = 44) and 15.7 µg/L in women (SD = 6.9 µg/L; n = 121). This was much lower than the level observed in community-dwelling individuals.

Table 9. Plasma 25-OHD levels in patients and controls ($\mu g/L$).

| | Patients | Controls |
|---|---|---|
| Men younger than 70 | 22.5±7.2 (28) | 33.9±10.2 (63) |
| Men 70 years and over | 18.5±6.9 (32) | 32.2±8.6 (90) |
| Women younger than 70 | 17.8±5.5 (31) | 29.0±6.2 (81) |
| Women 70 years and over | 17.1±6.2 (107) | 26.0±6.8 (134) |

Values are mean ± SD with the number of subjects given in parentheses. The 25-OHD level was lower in patients than in controls for all groups ($P<0.001$).

## 4.3. Vitamin D Level and Osteoporosis in Asia

The mean 25-OHD level was lower in Chinese patients with hip fracture than in controls. In the case-control study of hip fracture in Hong Kong Chinese (Lau *et al.*, 1988) the 25-OHD level was measured and was compared to the levels in community-dwelling individuals (Lau *et al.*, 1989). The results are presented in Table 9. The 25-OHD levels were lower in both men and women with hip fractures than in the controls ($P < 0.01$). Moreover, among patients, the 25-OHD levels were lower in women than in men ($P < 0.01$). There was no significant difference in serum 25-OHD concentration between young and old patients. Twenty percent of the male patients and 30% of the female patients had a 25-OHD level below the 95% confidence limits of the controls (14.6 μg/L in men and 13.7 μg/L in women). Among subjects with a low 25-OHD level, 78% of the men and 68% of the women were 80 years and over; they were less ambulant and tended to stay indoors most of the time. None of the patients with a low 25-OHD level had a biochemical bone profile resembling osteomalacia. Although a low 25-OHD level is a risk factor for hip fracture, one cannot conclude from these results that a low level causes osteoporosis, but it may increase the risk of hip fracture by causing muscle weakness and the susceptibility to falls. Moreover, the relationship between a low serum 25-OHD level and osteoporosis is confounded by other risk factors such as lack of activity. A recent case-control study of hip fracture patients in Hong Kong Chinese suggests that there is no direct relationship between vitamin D status and osteoporosis, since the levels of parathyroid hormone and $1,25(OH)_2D$ were the same among cases and controls (MacDonald *et al.*, in press). The effect of vitamin D supplementation in the prevention of osteoporosis has not been studied in Chinese.

The effect of vitamin D supplementation has been studied extensively in Japan. The results of two controlled trials showed that a vitamin D supplement given as 1 μg of 1-α-OHD per day maintained spinal bone mineral density and prevented spinal fracture (as quoted in Fujita, 1990); Similar effects were observed for 0.5 μg of $1,25\text{-}(OH)_2D_3$ given in divided doses. Nevertheless, as calcium deficiency is the main underlying factor for osteoporosis in Asia, supplementing the diet with calcium should be used for primary prevention, while vitamin D supplementation should be secondary.

# 5. Protein

## 5.1. Protein Intake and Osteoporosis

Most of the evidence for a relationship between protein intake and bone density has come from cross-sectional and correlation studies. Although the results of these studies suggest that both protein deprivation and excessive protein intake may lead to an increase in bone loss, they do not provide conclusive evidence that a change in protein intake would prevent osteoporosis.

Extreme protein deprivation may result in osteoporosis (Garn and Kangas, 1981), but this may not be relevant in populations on moderate protein intakes. In an ecological cross-sectional study of fracture rates in Yugoslavia (Matkovic *et al.,* 1979), the average protein intake was higher in areas with lower fracture rates. However, there were possible confounding effects of other nutrients. In two cross-sectional studies conducted in elderly women, dietary protein intake (Tylavsky and Anderson, 1988) and serum albumin level (Orwell, 1992) were positively associated with BMD. The results of all other cross-sectional studies on the relationship between protein intake and bone density were negative (Angus *et al.*, 1988; Freudenheim *et al.*, 1986; Mazess and Barden, 1991; Yano and Heilbrun, 1985). In a longitudinal study on the relationship between protein intake and bone density, Freudenheim *et al.* (1986) found that a high protein intake was associated with lower rates of appendicular bone loss in pre- and postmenopausal women, but there were possible confounding effects of other nutrients.

There is also evidence to suggest that a high protein intake is associated with a negative calcium balance. A strong correlation existed between animal protein intake consumption and incidence of hip fractures in women from various countries ($r = 0.9$) (Swaminathan, 1989). Heaney and Recker (1982) reported that among noninstitutionalized subjects, the level of dietary protein intake was negatively correlated with calcium balance. In experiments in which high levels of purified proteins were fed to human adults, calciuria was observed (Hegsted and Linkswiler, 1981; Schuette and Linkswiler, 1982; Licata *et al.*, 1981). However, when diets containing a similar amount of protein in the form of normal foods were consumed, no calciuria was observed (Spencer *et al.*, 1988; Cummings *et al.*, 1979). The difference in response is due to the phosphate naturally associated with protein in the diet, which increases the synthesis of PTH by depressing serum calcium, and thereby increasing the parathyroid hormone dependent reabsorption of calcium by the renal tubules. The dietary protein-phosphorus relationship is critical to the maintenance of calcium homeostasis on a high protein diet.

### 5.2. Protein Intake and Osteoporosis in Asia

Westernization may have caused an increase in protein intake among the inhabitants of Asia without a concurrent increase in calcium intake. For instance, the protein intake of community-dwelling elderly men and women in Hong Kong in the 1990s was high: 60 ± 21.8 g (Mean ± SD, n = 102) in elderly men and 60 ± 21.4 g, n = 151 in women (Woo *et al.*, 1988). This was equivalent to an intake of 1.2 g per Kg of body weight, and was high with respect to the recommended dietary intake of 0.8 g protein/Kg body weight (WHO/FAO/UNU).

Protein intake is even higher in Japanese women. The average intake found by Lacey *et al.* (1988) was 70.6 g (SD = 19.6; n = 89) in premenopausal women and was 72.6 g (SD = 15.6; n = 89) in postmenopausal women. In the same study, protein intake was found to be positively correlated with forearm BMD in both premenopausal and postmenopausal women ($r = 0.2$; $P < 0.05$).

As calcium intake is low in Asia and a high protein intake may lead to excessive excretion of calcium, a high protein intake is likely to aggravate calcium deficiency. However, this may not apply to Chinese subjects living in institutions due to their very low protein intake. Mean protein intake was 40 g (SD = 13; n = 43) in men and 44 g (SD = 23; n = 125) in women living in long-term care institutions in Hong Kong (Woo *et al.*, 1989a). Although the protein intake was not expressed per body weight, the protein calorie percentage was 9% in men and 12% in women, values significantly lower than those observed in elderly subjects living at home in Hong Kong. The very low protein intake may result in a low bone mineral density and muscle mass, and may increase the risk of osteoporotic fractures in this group.

## 6. Sodium

### 6.1. Sodium Intake and Osteoporosis

There is no direct evidence that a high sodium intake is an important cause of osteoporosis in human beings. It is known that sodium reabsorption and calcium reabsorption in the renal tubules are linked, and a strong positive correlation between sodium excretion and calcium excretion has been shown in many studies (Goulding, 1981; Goulding *et al.*, 1986; Law *et al.*, 1988). The increase in calcium excretion is accompanied by an increase in hydroxyproline excretion suggesting that there may be an increase in bone resorption as a result of high sodium intake (Goulding, 1981; Goulding and Campbell, 1984). In experimental animals it has been clearly shown that increasing salt intake leads to reduced bone mass (Goulding and Campbell, 1984), but this has not been demonstrated in humans.

## 6.2. Sodium Intake in Asia

The high sodium content of the Chinese and the Japanese diet is reflected in high urinary sodium/creatinine levels. The average sodium/creatinine ratio was 26.2 (SD unknown) (Kesteloot *et al.*, 1987) in Northern Chinese women and 16.8 (SD = 8.8; n = 584) (Woo *et al.*, 1991) in Hong Kong Chinese women. These values are higher than those observed in Caucasians (Intersalt Group, 1988). Salt intake is very high in Japan (over 12 g per day) (Lacey *et al.*, 1991). Their high salt intake may aggravate the low calcium intake in these oriental populations by increasing calcium excretion and thereby inducing osteoporosis. Nevertheless, further studies on the relationship between salt intake and bone density would be required to justify salt restriction as a means of preventing osteoporosis in Asians.

# 7. Conclusions and Recommendations

Osteoporosis and osteoporotic fractures are major public health problems in Asia. With an increasing prevalence of fractures and an aging population, the magnitude of the problem will be tremendous in the coming decades. Nutritional intervention is an important measure in the prevention of osteoporosis in these populations. Calcium supplementation and regular exercise are most effective measures in preventing bone loss in elderly men and women. Other effective dietary interventions may include maintaining an optimal protein intake, vitamin D supplementation of the deficient, and avoidance of an extremely high sodium intake.

# References

Albanese, A.A., Edelson, A.H., Lorenze, A.H., Woodhull, M.C., and Wein, E.H., 1975, Problems of bone health in elderly, *NY State J. Med.* **75**:326.

Angus, R.M., Sambrook, P.N., Pocock, N.A., and Eisman, J.A., 1988, Dietary intake and bone mineral density, *Bone and Mineral* **4**:265.

Baker, M.R., McDonnell, H., Peacock, M., Nordin, B.E.C., 1979, Plasma 25-hydroxyvitamin D concentrations in patients with fractures of the femoral neck, *Br. Med. J.* **1**:589.

Chow, R., Harrison, J.E., and Notarius, C., 1987, Effect of two randomised exercise programmes on bone mass of healthy postmenopausal women, *Br. Med. J.* **295**:1441.

Cooper, C., McLaren, M., Wood, P.J., Coulton, L., and Kanis, J.A.,1989, Indices of calcium metabolism in women with hip fractures, *Bone and Mineral* **5**:193.

Cumming, R.G., 1990, Calcium intake and bone mass: a quantitative review of the evidence, *Calcif. Tissue Int.* **47**:194.

Cummings, J.H., Hill, M.J., Jivrai, T., Houston, J., Branch, W.J., and Jenkins, D.J.A., 1979, The effect of meat protein and dietary fiber on colonic function and metabolism:

1. Changes in bowel habit, bile acid excretion and calcium absorption. *Am. J. Clin. Nutr.* **32**:2086.

Dalsky, G.P., Stocke, K.S., Ehsani, A.A., Slatopolsky, E., Lee, W.C., and Birge, S.J. Jr., 1988, Weight-bearing exercise training and lumbar bone mineral content in postmenopausal women, *Ann. Intern. Med.* **108**:824.

Dawson-Hughes, B., 1991, Calcium supplementation and bone loss: a review of controlled clinical trials, *Am. J. Clin. Nutr.* **54**:274S.

Dawson-Hughes, B., Dallal, G.E., Krall, E.A., Harris, S., Sokoll, L.J., and Falconer, G., 1991, Effect of vitamin D supplementation on wintertime and overall bone loss in healthy postmenopausal women, *Ann. Intern. Med.* **115**:505.

Dawson-Hughes, B., Dallal, G.E., Krall, E.A., Sadowski, L., Nadine Sahyoun, R.D., and Tannenbaum, S., 1990, A controlled trial of the effect of calcium supplementation on bone density in postmenopausal women, *N. Engl. J. Med.* **323**:878.

Ettinger, B., Genant, H.K., and Cann, C.E., 1987, Postmenopausal bone loss is prevented by treatment with low-dosage estrogen with calcium, *Ann. Intern. Med.* **106**:40.

Freudenheim, L.L., Johnson, N.E., Smith, E.L., 1986, Relationships between usual nutrient intake and bone-mineral content of women 35-65 years of age: Longitudinal and cross-sectional analysis. *Am. J. Clin. Nutr.* **44**:863.

Fujita, T., 1990, Studies of osteoporosis in Japan, *Metabolism* **39**:38.

Garn, S.M., and Kangas, J., 1981, "Protein intake, bone mass, and bone loss," in: *Osteoporosis: Recent Advances in Pathogenesis and Treatment* (H.F. DeLuca, H.M. Frost, W.S.S. Jee, C.C. Johnston, and A.M. Parfitt, eds.), pp. 257-267, University Park Press, Baltimore MD.

Goulding, A., 1981, Fasting urinary sodium/creatinine in relation to calcium/creatinine and hydroxyproline/creatinine in a general population of women, *N.Z. Med. J.* **93**:294.

Goulding, A., and Campbell, D.R., 1984, Effects of oral loads of sodium chloride on bone composition in growing rats consuming ample dietary calcium. *Min. Electrolyte Metab.* **10**:58.

Goulding, A., Everitt, H.E., Cooney, J.M., and Spears, G.F.S., 1986, "Sodium and osteoporosis," in: *Recent Advances in Clinical Nutrition* (M.L. Wahlquist and A.S. Truswell, eds.), pp. 99-108, John Libby, London.

Hansson, T., and Roos, B., 1987, The effect of fluoride and calcium on spinal bone content: a controlled, prospective (3 years) study, *Calcif. Tissue Int.* **40**:315.

Heaney, R.P., and Recker, R.R., 1982, Effects of nitrogen, phosphorus, and caffeine on calcium balance in women, *J. Lab. Clin. Med.* **99**:46.

Hegsted, M., and Linkswiler, H.M., 1981, Long-term effects of level of protein intake on calcium metabolism in young adult women, *J. Nutr.* **111**:244.

Ho, S.C., Donnan, S., and Sham, A., 1988, Dietary intake among elderly Chinese in Hong Kong, *J. Hum. Nutr. and Dietetics* **1**:205.

Horsman, A., Gallagher, J.C., Simpson, M., and Nordin B.E.C., 1977, Prospective trial of estrogen and calcium in postmenopausal women. *Br. Med. J.* **2**:789.

Intersalt Cooperative Research Group, 1988, Intersalt: an international study of electrolyte excretion and blood pressure. Results for 24 hour urinary sodium and potassium excretion. *Br. Med. J.* **297**:319.

Kelly, P.J., Eisman, J.A., and Sambrook, P.N., 1990, Interaction of genetic and environmental influences on peak bone density, *Osteoporosis Int.* **1**:56.

Kesteloot, H., Huang D.X., Li, Y.L., Gebsers, J., and Joosens, J.V., 1987, The relationship between cations and blood pressure in the People's Republic of China, *Hypertension* **9**:654.

Lacey, J.M., and Anderson, J.J.B., 1988, "Older women in Japan and the United States: physical and nutritional comparisons," in: *Proceedings of the Fifth International Congress on Bone Morphometry* (H.E. Takahashi, ed.), pp. 562-565, Nishimura, Japan.

Lacey, J., Anderson, J.J.B., Fujita, T., Yoshimoto, Y., Fukase, M., Tsuchie, S., and Koch, G.G., 1991, Correlates of cortical bone mass among premenopausal and postmenopausal Japanese women, *J. Bone Min. Res.* **6:**651.

Lamke, B., Sjoberg, H-E., and Sylvan, M., 1978, Bone mineral content in women with Colles' fracture: effect of calcium supplement, *Acta Orthop. Scand.* **49:**143.

Lau, E., Donnan, D., Barker D., and Cooper, C., 1988, Physical activity and calcium intake in fracture of the proximal femur in Hong Kong, *Br. Med. J.* **296:**1441.

Lau, E.M.C., Woo, J., Swaminathan, R., MacDonald, D., and Donnan, S.P.B., 1989, Plasma 25-hydroxyvitamin D concentration in patients with hip fracture in Hong Kong, *Gerontology* **35:**198.

Lau, E., Cooper, C., Donnan, S., and Barker, D.J.P., 1990, "Incidence and risk factors for hip fractures in Hong Kong Chinese," in: *Proceedings of the Third International Symposium on Osteoporosis* (G. Christiansen and K. Overgaard, eds.), pp. 66-70, Osteopress, Copenhagen.

Lau, E.M.C., Woo, J., Leung, P.C., Swaminathan, R., and Leung, D., 1992, The effects of calcium supplementation and exercise on bone density in elderly Chinese women, *Osteoporosis Int.* **2:**168.

Law, L.K., Swaminathan, R., and Donnan, S.P.B., 1988, Relationship between sodium excretion and calcium excretion in healthy subjects, *Med. Sci. Res.* **16:**643.

Lee, S.T., Lee, K.O., and Bose, K., 1988, Osteoporosis in elderly Chinese, *Br. Med. J.* **296:**1402.

Licata, A.A., Bou, E., Bartter, F.C., and West, F., 1981, Acute effects of dietary protein in calcium metabolism in patients with osteoporosis, *J. Gerontol.* **36:**14.

MacDonald, D., Lau, E., Chan, E.L.P., Mak, T., Woo, J., Leung, P.C., and Swaminathan, R., Serum intact parathyroid hormone levels in hip fracture, *Calcif. Tissue Int.* (in press).

Maggi, S., Kelsey, J.L., Litvak, J., and Heyse, S.P., 1991, Incidence of hip fractures in the elderly: A cross-national analysis, *Osteoporosis Int.* **1:**232.

Matkovic, V., Kostial, K., Simonovic, I., Buzina, R., Brodare, A., and Nordin, B.E.C., 1979, Bone status and fracture rates in two regions of Yugoslavia, *Am. J. Clin. Nutr.* **32:**540.

Mazess, R.B., and Barden, H.S., 1991, Bone density in premenopausal women: Effects of age, dietary intake, physical activity, smoking, and birth-control pills, *Am. J. Clin. Nutr.* **53:**132.

Orimo, H., 1990, "Epidemiology of fractures in Asia," in: *Proceedings of the Third International Symposium on Osteoporosis* (C. Christiansen and K. Overgaard, eds.), pp. 66-70, Osteopress, Copenhagen.

Orwell, E.S., 1992, The effects of dietary protein insufficiency and excess on skeletal health, *Bone* **13:**343.

Pocock, N.A., Eisman, J.A., Yeates, M.G., Sambrook, P.N., and Eberl, S., 1986, Physical fitness is a major determinant of femoral neck and lumbar spine bone mineral density, *J. Clin. Invest.* **78:**618.

Polley, K.J., Nordin, B.E.C., Baghurst, P.A., Walker, C.J., and Chatterton, B.E., 1987, Effect of calcium supplementation on forearm bone mineral content in postmenopausal women: a prospective, sequential controlled trial, *J. Nutr.* **117:**1929.

Prince, R.L., Smith, M., Dick, I.M., Roger, I.P., Webb, P.G., Henderson, N.K., and Harris, M.M., 1991, Prevention of postmenopausal osteoporosis, *N. Engl. J. Med.* **325:**1192.

Pun, K.K., Chan, L.W.L., and Chung, V., 1989, The problem of calcium deficiency in Hong Kong, *The Hong Kong Practitioner* **11**:287.

Recker, R.R., and Heaney, R.P., 1985, The effect of milk supplements on calcium metabolism, bone metabolism and calcium balance, *Am. J. Clin. Nutr.* **41**:254.

Recker, R.R., Saville, P.D., and Heaney, R.P., 1977, Effect of estrogens and calcium carbonate on bone loss in postmenopausal women, *Ann. Intern. Med.* **87**:649.

Riis, B., Thomsen, K., and Christiansen, C., 1987, Does calcium supplementation prevent postmenopausal bone loss? *N. Engl. J. Med.* **316**:173.

Schuette, S.A., and Linkswiler, H.M., 1982, Effects on Ca and P metabolism in humans by adding meat, meat plus milk, or purified proteins plus Ca and P to a low protein diet, *J. Nutr.* **112**:338.

Simkin, A., Ayalon, J., and Leichter, I., 1987, Increased trabecular bone density due to bone-loading exercises in postmenopausal osteoporotic women, *Calcif. Tissue Int.* **40**:59.

Smith, E.L., Reddan, W., and Smith, P.E., 1981, Physical activity and calcium modalities for bone mineral content in aged women, *Med. Sci. Sports. Exer.* **13**:60.

Smith, E.L., Gilligan, C., Smith, P.E., and Sempos, C.T., 1989, Calcium supplementation and bone loss in middle-aged women, *Am. J. Clin. Nutr.* **50**:833.

Spencer, H., Kramer, L., and Osis, D., 1988, Do protein and phosphorus cause calcium loss? *J. Nutr.* **118**:657.

Swaminathan, R., 1989, Nutritional factors in osteoporosis, *Journal of AFES.* **10**:3.

Tylavsky, F.A., and Anderson, J.J.B., 1988, Dietary factors in bone health of elderly lactoovo-vegetarian and omnivorous women, *Am. J. Clin. Nutr.* **48**:842.

Weaver, C.M., 1992, Calcium bioavailability and its relation to osteoporosis, PSEBM **200**:157.

Wong, P.C.N., 1966, Fracture epidemiology in a mixed Southeastern Asian Community (Singapore), *Clinical Orthopaedics* **45**:55.

Woo, J., Cheung, C.K., Ho, S.C., Mak, Y.T., and Swaminathan, R., 1988, Protein nutritional status in elderly Chinese in Hong Kong, *Eur. J. Clin. Nutr.* 42:903.

Woo, J., Ho, S.C., Cheung, C.K., Mak, Y.T., and Swaminathan, R., 1989a, Protein calorie malnutrition in elderly chronic care institutions in Hong Kong, *Nutr. Rep. Int.* **40**:1011.

Woo, J., Ho, S.C., Mak, Y.T., MacDonald, P., and Swaminathan, R., 1989b, Vitamin nutritional status in elderly Chinese subjects living in chronic care institutions, *Nutr. Res.* **9**:1071.

Woo, J., Lau, E., Chan, A., Cockram, C., and Swaminathan, R., 1991, Blood pressure and urinary cations in a Chinese population, *J. Hum. Hypertension* **6**:299.

Yano, K., and Heilbrun, L.K., 1985, The relationship between diet and bone mineral content of multiple skeletal sites in elderly Japanese-American men and women living in Hawaii, *Am. J. Clin. Nutr.* **42**:877.

Chapter 7

# Calcium and Osteoporosis?

D. M. Hegsted

## 1. Introduction

> One farmer says to me, "You cannot live on vegetable food solely, for it furnishes nothing to make bones with," and so he religiously devotes a part of his day to supplying his system with the raw material of bones, walking behind his oxen, which, with vegetable–made bones, jerk him and his lumbering plough along in spite of every obstacle.
>
> Henry David Thoreau
> 1817–1862

It seems we are about back to square one and the question that still needs an answer is whether diet, or calcium intake, plays a really significant role in the etiology of osteoporosis, especially hip fractures. Logic may tell us that calcium intake ought to be important but the evidence is weak.

## 2. International Data

It has long been apparent that international data on food consumption patterns provide no support for the idea that osteoporosis represents calcium deficiency. A large proportion of the world's population consumes low calcium

D. M. Hegsted • New England Regional Primate Research Center, Harvard Medical School, Southborough, MA 01772-9102.

*Advances in Nutritional Research,* Vol. 9
Edited by Harold H. Draper
Plenum Press, New York, 1994

diets—at least as judged by recommendations in the U. S. Although there are few quantitative data on the incidence of fractures in the elderly in such populations, it is certain that it is relatively low when compared to the U. S. and northern European countries. Indeed, populations consuming low calcium diets appear to be protected against hip fractures. Although there has been little study of bone mineral or calcium metabolism in those areas where calcium intakes are low, we are now beginning to see data from Japan and a few other populations. They show that bone mineral content is low, as might be expected from the calcium intake, but, paradoxically, hip fracture rates are also low. Until this inconsistency is resolved, it is apparent that there must be a large gap in our understanding of the etiology of osteoporosis.

Loss of bone mineral increases the risk of fractures and it would seem certain that a true deficiency of calcium would result in loss of bone mineral. The availability of more accurate measures of bone mineral content has made it possible to estimate rates of bone loss in individuals and groups over time. Loss of bone or reduction in mineral content has been equated with osteoporosis. However, these are quantitative issues. The questions are: what level of intake is inadequate, and how much clinical significance can be attached to modest differences in bone mineral content? Japanese women have fewer hip fractures than expected from similar densitometry levels in the U. S. population (Fujita and Fukase, 1992; Ross *et al.*, 1991). Similarly, it is reported that Gambian women (Prentice *et al.*, 1991) have lower bone mineral content than English women and show about the same rate of age-related bone loss, yet have far fewer fractures than would be expected when compared to data obtained from English women, in whom fractures are common. Blacks in the U. S. have fewer fractures than do whites, presumably because of increased bone mass, yet it appears that Gambian women have less bone than do whites. Hip fractures are also reported to be lower in Mexican Americans than in Texan non-Hispanic whites (Bauer, 1988) although those of Mexican origin consume less calcium. As data accumulate from low incidence areas, it seems certain that many countries with low fracture incidence will fall at the lower end of the distribution shown in Figure 1 (Hegsted, 1986). Thus, even if calcium intake has statistically significant effects upon bone mineral content, densitometry readings alone may not provide a good indication of the relative risk of fractures. At least in the cross-cultural comparisons, there must be other, apparently more important, factors involved than simply the amount of bone mineral.

## 3. Western Diet

It is apparent (Figure 1) that the data on hip fractures are reminiscent of data on coronary heart disease, certain cancers, etc., which implicated the

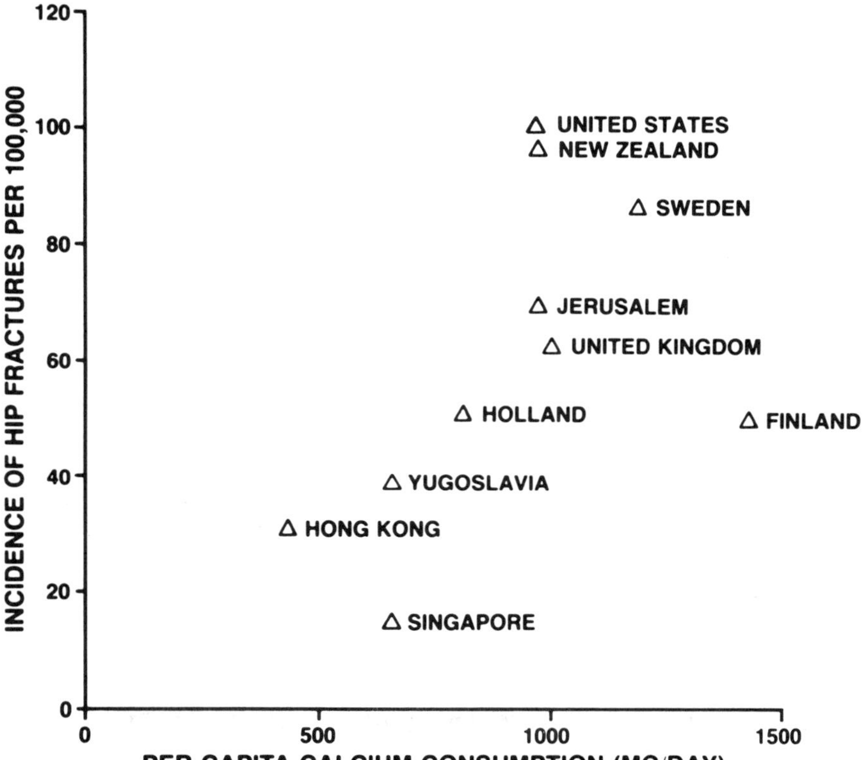


Figure 1. Available calcium in the food supply compared to the incidence of hip fractures in females of several nations (Hegsted, 1986).

"Western type diet" as a causal factor. There are too few data on fracture rates to determine how well these correlate with other chronic diseases common in the Western world. High fat intake, particularly animal fat and cholesterol, has been implicated in heart disease and cancer, but many other dietary factors may be involved. One is dairy products, which are high in saturated fatty acids. High calcium intakes in Western populations are due largely to high levels of consumption of milk and other dairy products; meat consumption is also high. Although the data are not entirely consistent, many studies indicate that increased protein consumption, especially animal protein, increases urinary calcium excretion. Thus, it is possible that high protein intakes increase calcium requirements. While this may sound like an attractive possibility to explain the epidemiologic data, it would only indicate that "calcium deficiency" would occur at somewhat higher levels of intake. High intakes of calcium should be protective and calcium supplementation should be effective. Since this does not appear to be true, if excessive protein intake is detrimental, it is not likely explained by an effect upon calcium excretion or requirements.

## 4. Risk of Fractures

A Lancet editorial (Anonymous, 1982) concluded that "a reduction in fracture rate remains the only true test of treatment regimens" and this should apply to prevention, as well. While measures of bone mineral content are important, the primary interest is in how these relate to fractures, especially fractures of the hip. We may soon have a better understanding of the real clinical significance of moderate changes or differences in bone mineral content.

While there is no doubt that loss of bone increases the risk of fractures and that there is a generalized loss of bone with aging, there may be different implications from measurements that have been reported with a variety of techniques and at various sites. For example, Colles' fracture is apparently "not typically a low mass fracture" (Heaney, 1989). And apparently spinal compression fractures are at least as frequent in Chinese women as in Americans (Tsai *et al.*, 1991). Thus generalizations about "osteoporosis" from measurements at one site may not be appropriate even though measurements at several sites may correlate rather well (Tsai *et al.*, 1991).

Although the data are not entirely consistent, it is likely that, at least within populations, bone mineral content does correlate positively with calcium intake and that calcium supplements do increase bone density (Dawson-Hughes *et al.*, 1987; Anderson, 1990; Cumming, 1990). Whether this can be confirmed probably depends upon the usual intake and the range of calcium intakes (less likely if the general intake is high and the range of intakes small), the age and time after menopause of the subjects investigated, and the reliability of the dietary and measurement data. Calcium supplements may appear to slow bone loss but it now seems apparent that they do not provide effective treatment (Riis *et al.*, 1987; Kanis and Passmore, 1989; Christiansen, 1992). Thus, modest differences in bone mineral content may have little clinical significance.

Measurement of bone mineral has been criticized in that it may not provide adequate identification of individual risk, i. e., identify those who will develop fractures. This seems to be an unjust criticism. It is well known that elevated serum cholesterol levels are a major risk factor of coronary heart disease yet measurements of serum cholesterol are rather poor indicators of individual risk. Indeed, although the Cholesterol Education Campaign established a serum cholesterol level of 200 mg/dl as desirable, most heart attacks occur in individuals with serum cholesterol levels near or below this level. Selection of 200 mg/dl represents a pragmatic judgment—an attempt to balance the feasibility of obtaining a more desirable serum cholesterol level in the population with the improvement in risk as serum cholesterol levels decline. Eventually, quantitative data on risk of fractures versus bone mineral content should allow some decisions to be made about the feasibility and utility of whatever can be done to increase bone mineral content. It does not appear to me that pharmacological levels of calcium intake can be recommended for the general public. The cost of calcium supplements in the U. S. was thought to be about $160 million in 1985 (Kolata, 1986) and presumably is climbing, even though the benefits are tenuous.

## 5. Adaptation

The most provocative question one can ask is whether high calcium intakes cause or increase susceptibility to osteoporosis. It is certain that the body has protective mechanisms that prevent excessive accumulation of body calcium when intakes are high and resist depletion of body calcium when intakes are low. Such mechanisms exist for all nutrients with varying degrees of effectiveness. Adaptation to various intakes of calcium must occur (Nicholls and Nimalasuriya, 1939; Mitchell, 1949; Hegsted *et al.*, 1952). Young animals have a great ability to adapt to a low calcium diet (Henry and Kon, 1953; Gershoff *et al.*, 1958). Urinary calcium can drop to near zero and retention of dietary calcium can be nearly 100% under these conditions. Malm (1958) demonstrated adaptation to a lowering of calcium intakes in adult men but this can be a rather slow process and many studies are not long enough or accurate enough to detect it. Although the inhabitants of the low calcium consuming countries face a variety of nutritional problems, they reproduce, develop their skeletons, and many survive until old age. Such populations must be efficient in metabolizing dietary calcium, and they are the ones that appear to be protected against fractures, while high calcium consumers must be inefficient. Whether adaptation to a low calcium intake during growth or adulthood provides protection against osteoporosis in elderly or postmenopausal women is unknown and would be difficult to determine. It would, however, be consistent with the epidemiologic data on fractures. The possibility that diet or environmental factors in early life may affect the development of chronic disease many years later is suggested by the possible effects of infant feeding on susceptibility to heart disease (Fall *et al.*, 1992) and by evidence that early patterns of development affect susceptibility to breast cancer (de Waard, 1986).

The populations of Central America are of special interest. Calcium intakes in the north are relatively high from the consumption of lime-treated corn rather than from dairy products. Thus, the diet approximates that of other developing countries, except that it is high in calcium. In the south, where rice replaces corn as the major cereal, calcium intakes are lower. Although there appear to be no data on fracture rates, bone size or loss of bone with aging (estimated from cortical measurements) appears not to be related to calcium intake (Garn *et al.*, 1969; Garn, 1970). Thus, these data suggest that there is no particular advantage or disadvantage related to high or low calcium intake alone with respect to the development of osteoporosis. Whether this would be true of other dietary sources of calcium, such as dairy products, is not known.

It is often recommended that calcium intake be high prior to the menopause so that maximal bone size will be achieved. While a true calcium deficiency would no doubt reduce bone size, the level of intake required is undefined. The Central American data provide no evidence that the ultimate amount of bone depends upon calcium intake. Nor, apparently, are there other data to support the belief that insufficient calcium is a primary determinant of bone size in most populations.

## 6. Dietary Data

A word should be said about dietary data collected in the field. These always rely upon the recollection and cooperation of the subjects interviewed. There are no independent methods for assessing the reliability of such data with the exception of the double-labeled water technique for estimating energy expenditure, which can be compared to reported energy intakes. The data available are discouraging since they indicate gross biases in the estimates of total food intake (Livingston *et al.*, 1990; Schoeller, 1990). Even 7-day weighed intakes underestimate energy expenditure—in some individuals by as much as 50%. The reliability of reported calcium intakes is unknown but is likely low if estimates of total food intake are low. Individuals vary in the reliability of the data they provide. Hence, classification of individuals on the basis of such data is hazardous. Since physical activity should modify energy consumption and also affect susceptibility to osteoporosis, errors in reported intakes may be particularly important. All data collected on individuals represent current intake, of course, and this may not be particularly relevant.

Chapuy *et al.* (1992) have reported that supplements of tricalcium phosphate and vitamin D reduced nonvertebral fractures by 30–40% in elderly French women confined to nursing homes. This appears impressive but leaves many questions unanswered. The diets were assessed by food records and the average calcium intake was 511±172 mg/day and the vitamin D intake 123±45 IU/day. If these are realistic estimates, the large standard deviations indicate that many were consuming less than 300 mg of calcium and the vitamin D intakes were extremely low. If the low consumers were those who responded, the data may not be surprising. The data would seem to be reasonably consistent with those of Dawson–Hughes *et al.* (1987, 1990, 1991) that only women with calcium intakes of less than 400 mg/day respond to calcium and that vitamin D supplements were helpful in maintaining bone mineral in Boston women. In prior studies, Chapuy *et al.* (1987) concluded that "statistical analysis indicates that in our patients the mild secondary hyperparathyroidism is due mainly to vitamin D deficiency and to a lesser extent to low-calcium intake." Studies are needed to clearly define the roles of calcium and vitamin D.

## 7. Genetics

Could the population differences in fracture rates be of genetic origin? Are the northern European races uniquely susceptible to osteoporosis? Within any population genetics plays a large role (Pocock *et al.*, 1987; Seeman *et al.*, 1989), as it does in susceptibility to many diseases, but the question is whether average susceptibility differs greatly among various populations. Unless specific genes can be identified, the answer is likely to come from the study of migrants. Such studies of heart disease, cancer, etc., provided support for the importance of diet, rather than differential susceptibility in various countries, as a causal factor. Migrants from low prevalence areas developed disease patterns typical of their new area of residence. Whether epidemiologic studies will reveal similar changes in fracture incidence remains to be seen, but the limited data so far available on Japanese populations seem discouraging. Fracture rates in elderly Hawaiian and Okinawan Japanese are apparently similar and are much below the rates in Caucasians in the U. S. (Ross *et al.*, 1991). Additional data of this kind are greatly needed. As noted before, fractures or other disease-relevant rates are needed, rather than (or in addition to) bone mineral measurements.

It may be difficult to interpret data from migrant populations. The relevant dietary data are what people ate in the past—not current intakes—and these are virtually impossible to estimate, particularly in a population with a gradually changing diet. In addition, the time required to develop the disease pattern of the new area of residence, if it occurs, may vary greatly. Changes in blood lipid levels, that are relevant to atherosclerotic heart disease, occur rapidly with dietary change, but it requires decades to develop atherosclerotic heart disease. Early diet and environment may affect susceptibility to the chronic diseases of adults and the elderly. Several generations may be needed to fully adapt to a new dietary pattern and some groups, like the Chinese, may maintain their usual diets even in a new environment

## 8. Calcium Balance Studies

All estimates of calcium requirements have been based on the results of studies on calcium balance. It has long been clear that high intakes tend to yield excessively high estimates of "apparent retention"—often levels that are not biologically reasonable (Hegsted, 1976). If retention of calcium in elderly women with high intakes of calcium were as high as has often been reported, osteoporosis should be cured with calcium supplements. The reasons for this have been debated over the years. Analytical errors tend to be greater when intakes are high. The most likely errors—overestimate of actual intake and underestimate of excretion—tend to yield positive values. Most balance trials are relatively short and adaptation to the new level of intake may not occur. There may be a temporary gain in body calcium when the intake is raised, or a loss when it is lowered, that cannot be sustained. Thus, all of the likely errors in balance trials tend to yield apparent but erroneous retention; these are viewed as desirable, while losses are viewed as unsatisfactory. Most of the data from balance trials may represent no more than a statistical artifact. As emphasized recently by Kanis (1991), even if randomly selected values for intake and excretion are used to calculate balances, a fairly impressive dose-response curve is obtained. This is simply because intake appears on both sides of the equation. While it should be possible to obtain valid balance data under strictly controlled metabolic ward conditions, it is apparent that calcium balance data, such as those recently summarized by Matkovic and Heaney (1992), show a rather poor correlation between intake and balance. How much of the trend lines is real and how much artifactual is not known, but the correlation coefficients are not much better than those obtained with random numbers.

If we cannot rely on estimated requirements obtained from balance studies, we are left with only observational data on populations or individuals. The primary issue is fractures, not bone mineral content. Other than population data, which are now beginning to accumulate, there is little information on fractures. Data on bone mineral content are difficult to interpret (Riggs *et al.*, 1987; Stevenson *et al.*, 1988; Dawson-Hughes *et al.*, 1990). Data on other factors that need consideration, such as fluoride (Pak *et al.*, 1989), vitamin D (Dawson-Hughes *et al.*, 1991) calcitriol (Riggs *et al.*, 1985), etc., are also fragmentary.

## 9. National Policy

All policy decisions are made, and necessarily must be made, with inadequate data. Policy decisions with regard to calcium have been dominated by dubious data obtained from calcium balance studies and the apparent logic that high intakes may do some good and should do no harm. Even most treatment regimens, such as estrogens and fluoride, are usually combined with high calcium intakes; there are few studies at more modest dietary levels. The recommendation that elderly women consume 1 to 1.5 g/day would require consumption of calcium supplements, an extra quart of milk per day or widespread fortification of foods. The evidence that any of these strategies is worthwhile is weak and the costs are not negligible. The urgent question that needs an answer is whether modification of calcium intake produces worthwhile clinical results.

## References


Anderson, J.J.B., 1990, Dietary calcium and bone mass through the lifecycle, *Nutrition Today,* March–April, p. 9.

Anonymous, 1982, An overview of osteoporosis, *Lancet* **2**:423.

Bauer, R.L., 1988, Ethnic differences in hip fractures: a reduced incidence in Mexican Americans, *Am. J. Epidemiol.* **127**:145.

Chapuy, M.C., Chapuy, P., and Meunier, P.J., 1987, Calcium and vitamin D supplements: effect on calcium metabolism in elderly people, *Am. J. Clin. Nutr.* **46**:324.

Chapuy, M.C., Arlot, M.E., Duboeuf, F., Brun, J., Crouzet, B., Arnaud, S., Delmas P.D., and Meunier, P.J., 1992, Vitamin $D_3$ and calcium to prevent hip fractures in elderly women, *N. Engl. J. Med.* **327**:1637.

Christiansen, C., 1992, Prevention and treatment of osteoporosis, *Bone* **13**:S35.

Cumming, R.G., 1990, Calcium intake and bone mass: a quantitative review of the evidence, *Calcif. Tissue Int.*, **47**:194.

Dawson-Hughes, B., Jacques, P., and Shipp, C., 1987, Dietary calcium intake and bone loss from the spine in healthy postmenopausal women, *Am. J. Clin. Nutr.* **46**:685.

Dawson-Hughes, B., Dallal, G.E., Krall, E.A., Sadowski, L., Sahyoun, N., and Tannenbaum, S., 1990, A controlled trial of the effect of calcium supplementation on bone density in postmenopausal women, *N. Engl. J. Med.* **323**:878.

Dawson-Hughes, B., Dallal, G.E., Krall, E.A., Harris, S., Sokoll, L.J., and Falconer, G., 1991, Effect of vitamin D supplementation on wintertime and overall bone loss in healthy postmenopausal women, *Ann. Int. Med.* **115**:505.

de Waard, F., 1986, "Body size, body mass and cancer of the breast," in: *Dietary Fat and Cancer,* Progress in Clinical and Biological Research, vol. 222 (C. Ip, D.F. Birt, A.E. Rogers and C. Mattlin, eds.), Liss, New York, pp. 33-41.

Fall, C.H.D., Barker, D.J.P., Osmond, C., Winter, P.D., Clark, P.M.S., and Hales, C.N., 1992, Relation of infant feeding to adult serum cholesterol concentration and death from ischaemic heart disease, *Br. Med. J.* **304**:801.

Fujita, I., and Fukase, M., 1992, Comparison of osteoporosis and calcium intake between Japan and the United States, *Proc. Soc. Exp. Biol. Med.* **200**:149.

Garn, S. M., 1970, *The Earlier Gain and Later Loss of Cortical Bone*, Thomas, Springfield IL. 146 pp.

Garn, S.M., Rohman, C.G., and Wagner, B., 1969, Population similarities in the onset and rate of endothelial bone loss, *Clin. Orthop.* **65**:51.

Gershoff, S.N., Legg, M.A., and Hegsted, D.M., 1958, Adaptation to different calcium intakes in dogs, *J. Nutr.* **64**:303.

Heaney, R.P., 1989, Osteoporotic fracture space: an hypothesis, *Bone and Mineral* **6**:1.

Hegsted, D.M., 1976, Balance studies, *J. Nutr.* **106**:307.

Hegsted, D.M., 1986, Calcium and osteoporosis, *J. Nutr.* **116**:2316.

Hegsted, D.M., Moscoso, I., and Collazos, C., 1952, A study of the minimum calcium requirement of adult men, *J. Nutr.* **46**:181.

Henry, K.M., and Kon, S.K., 1953, The relationship between calcium retention and body stores of calcium in the rat: effect of age and of vitamin D, *Br. J. Nutr.* **7**:147.

Kanis, J.A., 1991, Calcium requirements for optimal skeletal growth, *Calcif. Tissue Int.* **49**:S33.

Kanis, J.A., and Passmore, R., 1989, Calcium supplementation of the diet, *Br. Med. J.* **298**:137.

Kolata, G., 1986, How important is dietary calcium in preventing osteoporosis? *Science* **233**:519.

Livingstone, M.B.E., Prentice, A.M., Strain, J.J., Coward, W.A., Black, A.E., Barker, M.E., McKenna, P.G., and Whitehead, R.G., 1990, Accuracy of weighed dietary records in studies of diet and health, *Br. Med. J.* **300**:708.

Malm, O.J., 1958, Calcium requirement and adaptation in adult men, *Scand. J. Clin. Lab. Invest.* **10**:Suppl. 36.

Matkovic, V., and Heaney, R.P., 1992, Calcium balance during human growth: evidence for threshold behavior, *Am. J. Clin. Nutr.* **55**:992.

Mitchell, H.H., 1949, Adaptation to undernutrition, *Am. J. Diet. Assoc.* **20**:511.

Nicholls, L., and Nimalasuriya, A., 1939, Adaptation to a low calcium intake in reference to the calcium requirements of a tropical population. *J. Nutr.* **18**:563.

Pak, C.Y.C., Sakhaee, K., Zerwekh, J.E., Parcel, C., Peterson, R., and Johnson, K., 1989, Safe and effective treatment of osteoporosis with intermittent slow release sodium fluoride: augmentation of vertebral bone mass and inhibition of fractures, *J. Clin. Endocrinol. Met.* **68**:150.

Pocock, N.A., Eisman, J.A., Hopper, J.L., Yeates, M.G., Sambrook, P.N., and Ebert, S., 1987, Genetic determinants of bone mass in adults: a twin study, *J. Clin. Invest.* **80**:706.

Prentice, A., Shaw, J., Laskey, M.A., Cole, T.J., and Fraser, D.R., 1991, Bone mineral content of British and rural Gambian women aged 18–80+ years, *Bone and Mineral* **12**:201.

Riggs, B.L., and Nelson, K., 1985, Effect of long term treatment with calcitriol on calcium absorption and mineral metabolism in postmenopausal osteoporosis, *J. Clin. Endocrinol. Met.* **61**:457.

Riggs, B.L., Wahner, H.H., Melton, L.J., Richelson, L.S., Judd, H.L., and O'Fallon, W.M., 1987, Dietary calcium intake and rates of bone loss in women, *J. Clin. Invest.* **80**: 979.

Riis, B., Thomsen, K., and Christiansen, C., 1987, Does calcium supplementation prevent postmenopausal bone loss? A double-blind controlled clinical study, *N. Engl. J. Med.* **316**:173.

Ross, P.D., Norimatsu, H., Davis, J.W., Yano, K., Wasnich, R.D., Fujiwara, S.K., Hosodas, Y., and Melton, J., 1991, A comparison of hip fracture incidence among native Japanese, Japanese Americans, and American Caucasians, *Am. J. Epidemiol.* **133**:801.

Schoeller, D.A., 1990, How accurate is self-reported dietary energy intake? *Nutr. Rev.* **48**:373.

Seeman, E., Hopper, J.L., Bach, L.A., Cooper, M.E., Parkinson, E., McKay, J., and Jerums, G., 1989, Reduced bone mass in daughters of women with osteoporosis, *N. Engl. J. Med.* **320**:554.

Stevenson, J.C., Whitehead, M.I., Padwick, M., Endacott, J.A., Sutton, C., Banks, L.M., Freemantle, C., Spinks, T.J., and Hesp, R., 1988, Dietary intake of calcium and postmenopausal bone loss, *Br. Med. J.* **97**:15.

Tsai, K-S., Huang, K-M., Chieng, P-U., and Su, C-T., 1991, Bone mineral density of normal Chinese women in Taiwan, *Calcif. Tissue Int.* **48**:161.

van Beresteijn, E.C.H., van't Hof, M.A., Schaafsma, G., de Waard, H., and Duursma, S.A., 1990, Habitual dietary calcium intake and cortical bone loss in perimenopausal women: a longitudinal study, *Calcif. Tissue Int.* **47**:338.

Chapter 8

# Ethnic and Genetic Differences in Susceptibility to Osteoporotic Fractures

John J. B. Anderson and William S. Pollitzer

## 1. Introduction

This chapter contains a review of current information relative to the genetic and environmental determinants of ethnic and racial differences in bone mass and susceptibility to osteoporotic fractures. Evidence for differences in bone mass at all stages of the life cycle is documented, with particular reference to blacks vs. whites and whites vs. Asians. The potential of genetic markers to explain these differences is discussed.

## 2. Ethnic and Racial Differences in Adult Bone Mass

### 2.1. Genetic Determinants

Genetic factors are responsible for an estimated 70–80 % of bone mass during the first 20 years of life; the remainder is determined by environmental and lifestyle factors, including nutrition. These estimates are based on different lines of research, mostly on white female subjects (Slemenda *et al.,* 1991). No known estimates have been published on black, Asian and other ethnic groups.

John J. B. Anderson • Department of Nutrition, Schools of Public Health and Medicine,
William S. Pollitzer • Department of Cell Biology and Anatomy, School of Medicine,
University of North Carolina, Chapel Hill NC 27599-7400

*Advances in Nutritional Research,* Vol. 9
Edited by Harold H. Draper
Plenum Press, New York, 1994

Virtually all comparative black-white studies in North America show that black females have 5–10% greater bone mass (mineral content) and 5% greater bone density than white females at the same measurement sites when matched for age, the differences appearing as early as 1 year of age ( Cohn *et al.,* 1977; Goldsmith *et al.,* 1973; Wakefield *et al.,* 1980; Mayor *et al.,* 1980; Lohman *et al.,* 1984; Liel *et al.,* 1988; DeSimone *et al.,* 1989; Luckey *et al.,* 1989; Nelson *et al.,* 1988; Li *et al.,* 1989; Bell *et al.,* 1991, 1992; McCormick *et al.,* 1991; Venkataraman and Duke, 1991; Ortiz *et al.,* 1992). Black males also exhibit greater bone mass than white males at practically every stage of the life cycle. In contrast, white females have greater bone mass than Asian females, although adjustments for body weight, height and body mass index tend to eliminate these differences.

Table 1 provides a summary of data on comparative single-photon absorptiometric (SPA) measurements of bone mineral content of the mid-shaft of the radius, a non-weight-bearing bone of the forearm, across the life cycle of females. Note that bone mass changes in the two ethnic groups occur in parallel across the life cycle; radial bone mineral content increases from birth until some time between 20 and 40 years, and falls off significantly after approximately age 50 in both ethnic groups.

The mechanisms responsible for black-white differences in bone mass have not yet been elucidated, but two possible explanations have been put forth. One deals with calcium and bone metabolism and the other with reproductive hormones. Bell and coworkers (1991, 1992) support the concept of a difference in calcium-regulating hormones that favors retention of calcium by blacks. These researchers showed that black children 7–13 years old had a lower urinary calcium excretion than whites. Bell *et al.* (1985) found a lower plasma concentration of 25-hydroxyvitamin D, a higher concentration of parathyroid hormone, and a lower urinary calcium excretion in older black women than in whites. Meier *et al.* (1991) confirmed a lower calcium excretion in adult black women but found no difference in plasma 25-hydroxyvitamin D or parathyroid hormone concentration when the women were matched for weight, age and dietary intakes of calcium and vitamin D. Dawson-Hughes and colleagues (1993) observed that adult black women had higher levels of 1,25-dihydroxyvitamin D than white subjects of similar age at the same levels of calcium intake. Also, there were indications that blacks have a higher threshold for 1,25-dihydroxyvitamin D action on the intestine, resulting in a lower efficiency of calcium absorption. Although these studies on the control of calcium metabolism do not fully explain black-white differences in bone mass, they support the concept, based on evidence for a lower blood concentration of osteocalcin and a lower urinary calcium excretion, that lower bone turnover by blacks contributes to the accumulation of a greater bone mass (Bell *et al.,* 1985, 1991, 1992; Weinstein and Bell, 1988; Luckey *et al.,* 1989; Meier *et al.,* 1991; and Dawson-Hughes *et al.,* 1993).

Table 1. Comparative Mid-Radial Bone Mineral Content (g/cm) of Normal Black and White Females Across the Lifecycle Measured by Single-Photon Absorptiometry (Mean ± SD)

| Reference | Age[a] | N | Black | N | White | $P < 0.05$ |
|---|---|---|---|---|---|---|
| Venkataraman & Duke (1991) | 1-7 | 23 | 0.091 ± 0.028 | 51 | 0.094 ± 0.024 | No |
| Li *et al.* (1989) | 1-3 | 19 | 0.186 ± 0.051 | 9 | 0.160 ± 0.042 | Yes |
| Wakefield *et al.* (1980) | 10 | 58 | 0.57 --[b] | 56 | 0.52 --[b] | Yes |
| Tylavsky and Anderson (unpubl) | 20 | 170 | 0.907 ± 0.117 | 795 | 0.840 ± 0.260 | Yes |
| Adeleke (1993) | 21 | 25 | 0.908 ± 0.125 | 42 | 0.821 ± 0.114 | Yes |
| Cohn *et al.* (1977) | 40-49 | 6 | 0.946 --[b] | 11 | 0.875 --[b] | --[b] |
| Nelson *et al.* (1988) | 57 | 36 | 0.89 ± 0.10 | 99 | 0.76 ± 0.14 | Yes |
| Cohn *et al.* (1977) | 60-69 | 5 | 0.840 --[b] | 7 | 0.714 --[b] | --[b] |
| Cohn *et al.* (1977) | 70-79 | 8 | 0.764--[b] | 6 | 0.558--[b] | --[b] |

[a]Age in years, except for days in Venkataraman & Duke (1991)
[b]Data not reported

In support of the hormone concept, Richards *et al.* (1992) found greater serum concentrations of estradiol and lower levels of androstenedione in pubertal black males and females than in whites. Buchanan and coworkers (1988) observed that higher levels of circulating androgens in females were associated with increased trabecular, but not cortical, bone density. The implications of these differences in steroid hormone levels for bone mass are not yet clear, but higher estrogen levels are considered to favor the retention of mineral by the skeleton of either sex, and especially of females because of almost a lifetime priming of bone estrogen receptors.

Studies on twins and parent-offspring pairs have provided much of our current knowledge of the contribution of hereditary and environmental factors to bone mass and density. Smith and co-workers (1973), Christian and colleagues (1989), Pocock *et al.* (1987), Eisman *et al.* (1991) and Slemenda *et al.* (1992) have found significantly higher correlations between bone density at different skeletal sites in monozygotic twins than in dizygotic twins and less intrapair variance in bone measurements.

In a continuation of a 16-year study on men by Christian *et al.* (1989) the value for radial bone mass correlation between monozygotic twins continued to be much greater than that for dizygotic twins. Furthermore, these researchers found that midlife bone loss in these men was associated with several specific environmental factors, including cigarette smoking, alcohol intake, and physical activity (Slemenda *et al.*, 1992).

Other approaches to assessing genetic contributions to bone mass have made use of mother-daughter (offspring) intrapair correlational analyses. In general, these reports have demonstrated high intrapair correlations between mothers and daughters. Mothers with low bone mass typically have daughters with low bone mass ( Lutz, 1986; Sowers *et al.*, 1986; Seeman *et al.*, 1989; Tylavsky *et al.*, 1989; Lutz and Tesar, 1990; Matkovic *et al.*, 1990; Hansen *et al.*, 1992). In the study of Tylavsky *et al.* (1989), it was estimated that white daughters at 18 years of age would need, on average, to accumulate 5–10% additional bone mass to achieve the same mean value as their mothers at age 44.

Bone mass changes during the life cycle are especially affected by hereditary factors. For example, in women the age of menopause is determined by genetic factors, though it can be affected by environmental variables, especially diet. In the last 50 years the mean age of menopause in American women has increased from the last half of the fifth decade of life to the first half of the sixth decade, primarily because of an improvement in the overall nutritional status of women. Genetic factors probably contribute to the rapid loss of bone in some women, i.e., the "fast losers," in the early postmenopausal years. Loss of ovarian estrogen production is considered the major contributor to bone loss in women following the menopause, but it is possible that changes in other hormonal factors, such as adrenal corticosteroids and androgens affect bone turnover. Little information is available on genetic determinants of hormonal changes at the menopause (Eichner *et al.,* 1992), but they must be important for bone loss during the climacteric transition.

Little information exists on the genes that determine differences in bone mass, density and loss among individuals and ethnic and racial groups. To date, only one gene has been identified for a specific bone molecule, namely the one for osteocalcin or bone GLA protein (BGP), which behaves as an autosomal dominant gene (McKusick, 1992). The occurrence of a specific gene for osteocalcin was predicted by Kelly and coinvestigators (1991) on the basis of measurements of blood concentrations of osteocalcin in monozygotic and dizygotic twins. Another bone protein, $\alpha_2$-HS-glycoprotein, has also been reported to be under strong genetic influence (Eichner *et al.,* 1990). Other bone markers that may be under genetic control are bone-specific alkaline phosphatase, collagen cross-link pyrilinodine peptides (remnants), tartrate-resistant acid phosphatase, and other bone proteins such as osteonectin and bone sialoprotein II (Delmas, 1991). Over the next several years it can be anticipated that new genes related to bone tissue will be identified that will enhance our understanding of the role of hereditary factors in bone development.

A report by Peacock *et al.* (1992) suggests that genetic factors are operative in the regulation of several aspects of calcium metabolism, including intestinal calcium absorption and parathyroid hormone secretion, and thereby may have an effect on bone density. The genes that govern these processes, however, have no known serum markers. The genetic regulation of vitamin D-binding proteins and the calcium-binding proteins or calbindins, especially those found in the small intestinal mucosa, recently has been studied by Eichner *et al.* (1992).

## 2.2. Environmental Determinants

Environmental determinants of bone mass, including dietary and other lifestyle factors, increase in importance prior to puberty in both boys and girls, perhaps most significantly during the two years immediately preceding puberty. Johnston *et al.* (1992) observed a greater increase in bone mass at three different skeletal sites in prepubertal twins receiving a calcium supplement of 1000 g per day for two years than in similarly treated postpubertal twins. This increase was independent of sex and age of puberty, but not independent of Tanner developmental stage. Radial bone density was more responsive to calcium supplementation than that of the vertebrae and hips.

Clearly, by age 18 virtually all females and by age 22 practically all males in nations with adequate food supplies have completed their growth in height. Maximum growth is dependent on sufficient amounts of energy and protein in the diet (Anderson and Henderson, 1991; Anderson, 1992). Some investigators support the concept that dietary calcium intakes and exercise patterns play important roles in the development of peak bone mass during the early to mid-periods of adolescence (Tylavsky *et al.,* 1992; Sandler *et al.,* 1985). It has been concluded that growth ceases in most girls by age 16 to 17 in Western nations, approximately four years after menarche (Bonjour *et al.,* 1991). Hence, calcium intake may have an important effect on bone development prior to this age, and on the achievement of peak bone mass during the last four years of the second decade of life (Johnston *et al.,* 1992; Matkovic *et al.,* 1990; Sentipal *et al.,* 1991; Anderson and Metz, 1993). Calcium is now generally considered to be a threshold nutrient (Matkovic and Heaney, 1992).

In young adult males, environmental factors appear to influence bone development as well as bone maintenance after completion of peak bone mass. During the early years of the third decade of life in males, the level of dietary calcium has a significant impact on radial bone mass (Anderson *et al.,* unpublished results). During middle adult life other environmental factors may affect the loss of bone mass in males (Slemenda *et al.,* 1992; Meier *et al.,* 1987; Kelly *et al.,* 1990).

Although modest differences in nutrient intakes exist between blacks and whites in the U.S.A. (Mettlin, 1980), these differences cannot explain the greater bone mass of blacks. It is especially noteworthy that blacks have a significantly lower calcium intake throughout the lifecycle. However, among both young black and white females, low calcium consumers have lower bone mass than high calcium consumers (Anderson *et al.,* 1987; Adeleke *et al.,* 1993; Tylavsky *et al.,* 1992).

Table 2. Comparative Anthropometric Data for 20-year-old Females: American Blacks and Whites and Japanese Living in Japan (Mean $\pm$ SE).

| Variable | White (n=790) | Black (n=190) | Japanese (n=19) |
|---|---|---|---|
| Age, years | 20.4 $\pm$ 0.1 | 20.3 $\pm$ 0.1 | 20.5 $\pm$ 0.2 |
| Height, cm | 165.0 $\pm$ 0.2 | 163.8 $\pm$ 0.5 | 156.1 $\pm$ 1.3* |
| Weight, kg | 59.5 $\pm$ 0.3 | 63.9 $\pm$ 1.0 | 48.5 $\pm$ 1.1* |
| BMI, kg/m $^{1.5}$ | 28.1 $\pm$ 0.1 | 30.4 $\pm$ 0.5 | 24.8 $\pm$ 0.5* |
| BMI, kg/m$^2$ | 18.0 $\pm$ 0.1 | 19.5 $\pm$ 0.5 | 15.5 $\pm$ 0.5* |

* Statistically different from the other two groups, $p < 0.05$.
BMI = body mass index, either to the power of 1.5 or 2
Anderson, Tylavsky and Lacey (Unpublished data).

# 3. Ethnic and Racial Differences in Bone Development

Table 2 illustrates some of the anthropometric differences among three different ethnic groups of 20-year-old females: black and white North Carolinians and Japanese living in Kobe, Japan. Besides dietary factors, several other lifestyle factors influence skeletal development and growth during the first 20 years of life.

## 3.1. BMC of African and U.S. Black and White Children

Recent reports confirm the greater bone mineral content (BMC) of blacks compared to whites in the U.S.A., but not in Africa. The pioneer work of Trotter and colleagues showed that the difference between U.S. blacks and whites appears early in life (Trotter and Peterson, 1970) and continues in adulthood (Trotter *et al.,* 1960). These investigators used Archimedes' principle to determine that black children had greater bone density than whites, starting within the first year of life. They also found slight but nonsignificant differences in the density of late fetal bones using this method (Choi and Trotter, 1970). More direct measurements of BMC and bone mineral density (BMD, a linear measurement of bone density as opposed to volumetric density), using current absorptiometric technology revealed no differences in BMC between black and white newborns (Venkataraman and Duke, 1991).

Investigations on African children have provided additional information about the temporal sequence of skeletal growth among blacks. These children have not received as much general nutritional support of skeletal development as have their white counterparts living in Africa, the United Kingdom, or North America. Inadequate nutritional status is indicated by studies of bone mass in Gambia. The BMC of the radius in 134 British and 243 rural Gambian children from birth to 36 months showed marked differences both in growth rate and calcium intake. Radial BMC increased with age in both groups and there was no difference between groups in the rate of increase after adjustment for body size (height, weight, and bone width). Gambian children had lower mean BMC values than British children of the same age. The difference was 11% at birth and increased to 31% by 36 months of age. A reduced but significant differential remained after adjustment for body size (Prentice *et al.,* 1990).

The observations of Patel *et al.* (1992) on black and white South African children also revealed no difference in appendicular bone mass. The contrast in the results of these studies on African children in South Africa and Gambia versus those on U.S. black and white children has not been entirely explained, but one possible factor in the higher bone mass of American children and young adults is their higher intakes of protein, energy and calcium. The earlier growth acceleration, menarche and overall physical development of North American black children must reflect at least in part the influence of dietary adequacy, although other factors such as medical care cannot be ruled out. This nutritional hypothesis was first formulated by Garn *et al.* (1972) and it has been further refined by others (Heaney, 1986; Anderson 1992).

Further information on the reasons for racial differences in BMD between black and white children aged 7–12 years comes from research performed in Charleston, South Carolina. BMD was determined at four sites on 20 black boys, 18 black girls, 33 white boys, and 35 white girls. BMD increased with age and body weight in both races, but the mean values were greater in black than in white children in practically all comparisons at all sites (Bell *et al.,* 1991). Another study was designed to obtain information on the underlying biological mechanisms responsible for the greater BMD of the black children in Charleston (Bell *et al.,* 1992). A controlled daily intake of calcium was given to a subset of the children: 6 black boys, 8 black girls, 5 white boys, and 6 white girls. Serum 25-hydroxyvitamin D was lower; serum 1,25-dihydroxyvitamin D was higher, and urinary calcium was lower in black than in white children. The lower urinary loss of calcium by black children denotes greater calcium retention by a faster growing skeleton.

The study of Gilsanz *et al.* (1991) showed that puberty is the crucial time of bone formation in both black and white children. Vertebral BMD, which is representative of nearly all cancellous bone tissue in the body, was measured using quantitative computed tomography (QCT) in 75 black and 75 white females between 2 and 20 years of age. The peripubertal girls were matched for stage of sexual development. There was no difference in BMD in the two groups up to the time of puberty. The magnitude of the pubertal increase in BMD, however, was much greater in black females than in white females (34% vs. 11%). This study, in particular, supports the strong contribution of genetic determinants of increases in BMD during pubertal development, independent of dietary intake. Genetic control of osteoid synthesis, rather than intestinal calcium absorption and/or of renal reabsorption of calcium, must be considered as the dominating factor governing adolescent bone development.

If nutritional intake is insufficient to support adequate growth by children, they will remain stunted and often will have disproportionate growth, i.e., shorter legs and arms relative to trunk. In many cases, catchup growth and reasonable recovery are possible when adequate nutrition is provided. In an interesting historical investigation of this phenomenon, growth suppression and recovery has been reported to be widespread among black slaves living in the southern U.S.A. in the 19th century (Steckel, 1986, 1987). Steckel provides data indicating that poorly nourished black children underwent remarkable recoveries in growth, when they were adequately fed. The high ratio of adult height to child height in these blacks is considered to be unique. Thus, the catchup phenomenon appears to be greater in blacks than in any other ethnic groups examined.

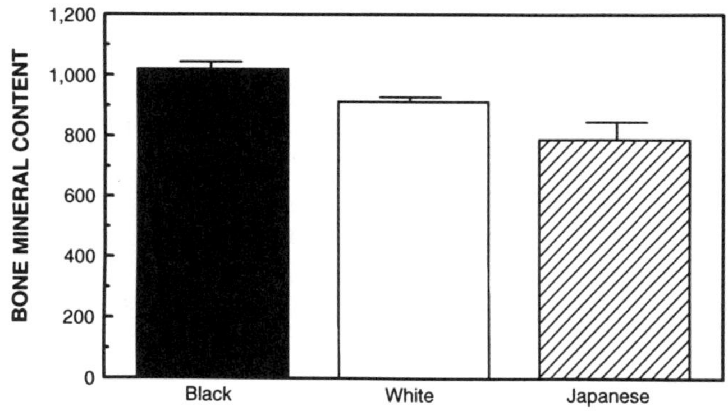


Figure 1. Comparative radial (midshaft) bone mineral content (BMC) of 20-year-old- black, white and Japanese females. Means ± SEM are unadjusted. Anderson, Tylavsky and Lacey (unpubl. data)

## 3.2. BMC of Asian Children

The bone mineral status of Asian populations has been studied only recently. While the BMC and BMD of Japanese, Chinese, and Korean children are lower than that of comparably aged American children and adults (Sugimoto *et al.,* 1992), adjustment for height, weight or body mass index brings the bone measurements on Asian populations closer to those on whites. Nevertheless, hereditary differences between Asians and whites likely exist and contribute to the differences in BMC and BMD. An example of the differences in bone parameters in 20-year-old females of three different racial groups is illustrated in Figure 1. This figure includes data from the same subjects as in Table 2 (Anderson, Tylavsky and Lacey, unpublished). Russell-Aulet and coinvestigators (1993) also found that statistical adjustments for height, weight and other factors brought the bone mass measurements of Asian and white women much closer.

## 3.3. Lifestyle Factors

Aside from diet and exposure to sunlight, the major factors influencing bone development early in life are environmental conditions including housing, water supply, medical care, and involvement in diverse physical activities at work and at play. Factors that negatively affect bone mass are excessive use of alcoholic beverages, excessive cigarette smoking, and chronic high-dose therapy with glucocorticoid hormones.

Physical activities that provide strains (loads) on the skeleton, either directly or through muscle insertions, improve both the mass and the structural integrity of bone tissue at the sites affected (Lanyon, 1984). Thus, the regular activities of daily living or physical work and sports are beneficial in bone development so long as they are not so excessive as to interfere with nutrient utilization, hormonal regulation of growth processes, or, in the case of menstruating adolescent or young adult females, with ovarian production of estrogens. In contrast to adults, documentation of these effects in children is limited.

Johnston *et al.* (1992) and Tylavsky *et al.* (1992) obtained evidence that regular physical activity is an important determinant of gain in bone mass during early bone development. Other studies also indicate that there is a beneficial effect of exercise on bone mass, but further longitudinal investigations are needed to identify the most important determinants.

## 4. Adult Changes in Bone Mass: Ethnic Differences

### 4.1. Bone Loss and Fracture Rates

Fractures are more common among whites than blacks (Farmer *et al.*, 1984). The explanation offered for this racial difference is that blacks have, on average, a greater bone mass at any decade of life than whites, including the elderly decades (Goldsmith *et al.*, 1973; Cohn *et al.*, 1977). Current bone mass of postmenopausal women predicts the risk of future fractures (Wasnich *et al.*, 1985; Hui *et al.*, 1988; Gardsell *et al.*, 1989, 1991; Cummings *et al.*, 1990). Stated another way, fractures are more likely to occur in women (and men) with a bone mass or bone density below some bone-specific threshold. For example, in elderly women, fractures occur at a significantly greater rate when distal radial bone density is below a given fracture threshold (Lester *et al.*, 1990). Thus, low bone mass is probably the most significant risk factor for fracture, independent of any specific event, such as a fall (Cummings and Nevitt, 1989).

Despite a lower calcium intake and a lower body weight and body mass index among elderly Japanese women, hip fracture rates remain significantly lower than for Caucasians living in the United States and Europe (Fujita and Fukase, 1992). This paradox of a lower bone mass coupled with a lower hip fracture rate remains unresolved, but Lacey *et al.* (1991) and other investigators have suggested that the traditional lifestyle of Japanese women, which involves many daily functions at floor level, has forced them to get up and down many times each day. This form of musculo-skeletal activity associated with the daily activities of living may be effective in protecting the bone mineral content of the proximal femur.

Amenorrhea is a major risk factor for low bone mass among all ethnic groups. However, the expected benefit of oral contraceptive agents on the bone mass of amenorrheic women has not been verified by experimental data. Lindsay (1992) found no difference in bone homeostasis between hypo- and eumenorrheic ovulatory premenopausal women attributable to the use of oral contraceptives. Irregular menstrual cycles in mid-adult premenopausal Japanese women were found to be negatively correlated with bone mass (Lacey *et al.,* 1991). Low bone mass was also seen in women with inadequate progesterone production in the luteal phase of the menstrual cycle (Prior *et al.,* 1990). A rapid loss of bone mass following the menopause has been found in Japanese women living in Japan (Hagino *et al.,* 1992). Estrogen replacement therapy, initiated as early as possible during the menopause, is the most effective modality used to maintain bone mass (Lindsay, 1976; Christiansen *et al.,* 1980; Lindsay, 1991).

## 4.2. Physical Activity

Physical activity remains the lifestyle factor most often observed to impact positively on bone mass late in life. However, activity must be continuously maintained in order to be effective over long periods of time. If exercise is decreased, the rate of bone loss will increase for a period of time, perhaps as long as a year, until a new steady state is achieved. The same phenomenon occurs after the menopause, either natural or surgical, and after the cessation of estrogen therapy in postmenopausal women. In this case, a relatively short period of rapid bone loss occurs, lasting approximately six months, then a period of slower loss occurs for a few years until the new steady state sets in.

An example of gain and loss of vertebral bone mass in postmenopausal women on a controlled exercise program has been described by Dalsky *et al.* (1988). These investigators observed a significant increase in the density of the lumbar vertebrae after one year of exercise, but when women stopped exercising they returned to their pre-exercise bone values within six months. Only those women who continued exercising on their own during the year after the program was stopped were able to retain much of the bone mass gained during the period of programmed exercise. This well-designed work, as well as other investigations, illustrate the importance of maintaining physical activity for an indefinite period..

Late postmenopausal women continue to lose bone mass at a fairly steady or linear decay rate until ambulatory difficulties occur (Smith *et al.,* 1975; Anderson *et al.,* 1990; Reed *et al.,* 1994). When an ambulatory condition or immobility occurs, elderly individuals undergo a new phase of exponential bone loss, which lasts for a long period of time and is associated with a rapid increase in the risk of bone fracture. Very little information is available on this immobility-related loss in the elderly. Hospitalized patients and bedridden subjects are known to undergo a period of rapid bone loss until they begin to reambulate.

Cross-sectional data on elderly Japanese and Hong Kong Chinese women confirm the effect of physical activity on bone mass (Lau *et al.,* 1988, 1992; Lacey *et al.,* 1991). The modern lifestyle in Western countries, which has been rapidly accepted by Asian countries, in all probability has contributed to their increasing rate of hip fractures by fostering physical inactivity. Modernization has been cited as the main reason for the epidemic of hip fractures in Hong Kong and other Asian countries in recent years (Lau *et al.,* 1990). This increase is also associated with a low calcium intake (300–400 mg per day) and a low bone mass in a significant fraction of Chinese women (Lau *et al.,* 1993).

No studies of the effects of physical activity on the bone mass of adult black populations have been reported. Based on studies of college age and young adult women, it would appear, however, that the bone mass of black women responds to external loads in a manner similar to that in white females. This area is ripe for new experimental efforts.

### 4.3. Dietary Factors

An investigation of the linkages between dietary factors and bone mass among young adult black women yielded results similar to those reported for white females (Anderson *et al.,* 1992; Adeleke, 1993). The relationships between dietary variables and bone mass in Japanese Americans living in Hawaii were explained by Yano *et al.* (1985) in terms of positive associations, after adjusting for other variables, between milk calcium intake and bone mineral content in both men and women.

Dietary factors have been shown not to have much effect in maintaining bone mass during the late decades of life (Reed *et al.,* 1994), so that even elderly white women consuming calcium at or above the RDA on a regular basis continue to lose bone mass at the same rate of 1% per year as do the low-calcium subjects. Data on the calcium-bone relationship are not available for older black women, but data from the study of Goldsmith *et al.* (1973) suggest that postmenopausal black women may lose bone mass at a greater rate than similarly aged white females. Corroborative findings are not yet available to support this finding, which may have been due to small numbers of women in the study. It is not generally accepted that elderly black women have greater rates of bone loss than white women in the U.S.A. (Farmer *et al.,* 1984). Supplementation of elderly women, such as those studied by Chapuy *et al.* (1992) using vitamin D and calcium, support the notion that some individuals may benefit from these specific nutrient supplements.

## 4.4. Anthropometric Factors

Many researchers have found that high body mass index or body weight are protective of bone mass (Ribot *et al.,* 1988; Bevier *et al.,* 1989; Hassager and Christiansen, 1989; Slemenda *et al.,* 1990). In cross-sectional studies on white women between 20 and 50 years of age, analyses showed consistently positive associations between body mass index, body fat mass or estimated lean body mass and bone measurements (Halioua and Anderson, 1990; Lindsay *et al.,* 1992). Reports by Liel and co-authors (1988), Luckey *et al.* (1989) and Meier *et al.* (1991) document the positive effect of body weight on bone measurements in both black and white women. Lacey *et al.* (1991) and Tsunerari and coworkers (1993) also found a positive association between bone mass and fat mass (or body weight) of Japanese women. Bell and coworkers (1985) found that obese adults, both males and females, exhibited alterations in the metabolism of vitamin D leading to secondary hyperparathyroidism, elevation in the serum concentration of 1,25-dihydroxyvitamin D and a lower urinary calcium excretion. Additional studies are needed to verify these findings among other ethnic groups.

The correlation between fat mass and bone mass is greater for females than males, an observation made for Caucasians (Reid *et al.,* 1992), Asian Chinese (Lau *et al.,* 1993) and Japanese (Tsunenari *et al.,* 1993). The explanation for this difference is not clear, but Felson and colleagues (1993) speculate that the peripheral conversion of androstenedione to estrogens is more extensive in females. Also, because of a greater number of estrogen receptors in the bone cells of females, they may conserve bone mass more effectively than males. Thus, obese postmenopausal women with higher circulating estrogen will likely maintain bone mass more effectively than lean women, who have basal levels of blood estrogen.

In addition to the benefits conferred by fatty tissue on the hormonal maintenance of bone mass in postmenopausal women (Felson *et al.,* 1993), body fat is protective against hip fractures from falls, especially backward falls (Cummings and Nevitt, 1989). An upper body fat mass provides a greater gravitational load at the most fracture-susceptible bone site, i.e., the hip.

Blacks differ from whites in body composition, especially muscle (lean tissue) mass and skeletal mass, though not in fat mass (Ortiz *et al.,* 1992). The report on elderly women in the NHANES Epidemiological Follow-up Study (Farmer *et al.,* 1989) indicates that any anthropometric indicator of a large body mass (body weight, skinfold thickness, body mass index, midarm circumference), was inversely related to the rate of fractures. Low body weight was demonstrated by Pruzansky *et al.* (1989) and low fat mass by Lindsay *et al.* (1992) to be important risk factors for hip fractures in both black and white women. Low body fat in elderly Asian female populations (Lau *et al.,* 1993; Tsunenari *et al.,* 1993) has also been associated with an increased rate of hip fractures.

Asian women with a high body mass index, body weight or fat mass as estimated by skinfolds also have greater bone mass (Lacey *et al.,* 1991; Lau *et al.,* 1993). Thus, body mass index and other variables in body composition must be considered in evaluating the significance of differences in bone mass in diverse ethnic groups. In a comparative study of bone measurements of adult and elderly Japanese subjects living in Japan and Japanese descendants living in Hawaii, however, adjusting BMC by dividing by bone width (i.e., BMD) increased the mean difference between the two groups (Ross *et al.,* 1989). Lifestyle and dietary variables have been offered as the major determinants of differences in bone mass among Japanese migrants to mainland U.S.A. (Kin *et al.,* 1993).

### 4.5. Co-Morbidity and Immobility

The frequent presence of one or more chronic diseases in elderly individuals can increase the rate of bone loss and the risk of an osteoporotic fracture. The increased frequency of fractures that follows a long period of bed rest is well documented. Disease episodes that require bed rest enhance bone loss with little subsequent regain of the excess loss. Elderly subjects who become ill and bedridden, even for periods of a few days, incur an increased risk of fracture. Falls under these conditions are especially contributory to hip fractures. Impaired mobility, in general, is an important risk factor for osteoporotic fractures (Anderson *et al.,* 1990; Ooms *et al.,* 1993).

The lower hip fracture rates for Asian, Hispanic and black women than for whites discharged from California hospitals suggest a strong influence of genetic factors, but other factors such as greater mobility caused by regular exercise cannot be ruled out (Silverman and Madison, 1988). Postmenopausal bone loss has been shown to decline after 60 years of age, but the fractional loss, calculated on the basis of residual bone mass, may actually be higher (Davis *et al.,* 1989). Any sustained loss of bone by elderly women, as well as men, therefore increases their risk of a fracture.

## 5. Summary

A plethora of investigations in recent years has demonstrated the occurrence of ethnic differences in bone mineral content, bone density and fracture rates. These findings indicate that genetic determinants exist both for bone development during growth and for bone loss during aging. Twin and parent-offspring studies have corroborated the existence of a hereditary component. It is most evident in the greater bone mass and lower fracture rate in blacks than in whites. Differences in bone mass between Asians and whites are less clear than between blacks and whites because of disparities in body size and other confounding factors. Black children and adults excrete less urinary calcium than whites on essentially the same diets and consequently retain more calcium in their skeletons. Better calcium retention is commensurate with the faster rate of bone growth of black children.

## References

Adeleke, V.M., 1993, "Determinants of Bone Mass In Young Adult Black And White Women." Master's Thesis, University of North Carolina, Chapel Hill, NC, 120 pp.

Adeleke, V.M, Anderson, J.J.B., Gallagher, P.N, Jr., 1993, Risk factors for low bone mass among young adult black women. *FASEB J.* 7 (no. 4, pt. II): A613.

Anderson, J.J.B., 1992, The role of nutrition in the functioning of skeletal tissue, *Nutr. Rev.* **50**:388.

Anderson, J.J.B., and Henderson, R.C., 1991, "Dietary factors in the development of peak bone mass," in: *Nutritional Aspects of Osteoporosis* (P. Burckhardt and R.P. Heaney, eds.), pp. 3-19, Serono Symposium Publication 85, Raven Press, New York.

Anderson, J.J.B., and Metz, J.A., 1993, Contributions of dietary calcium and physical activity to primary prevention of osteoporosis in females, *J. Am. Coll. Nutr.* **12**:378.

Anderson, J.J.B., Tylavsky, F.A., Lacey, J.M., and Talmage, R.V., 1987, "Black-white differences in bone status of young women according to calcium intake," in: *Calcium Regulating Hormones and Bone Metabolism 9* (D.V. Cohn, T.J. Martin, and P.J. Meunier, eds.), p. 696, Excerpta Medica, Amsterdam, (abstract).

Anderson, J.J.B., Reed, J.A., Tylavsky, F.A., Lester, G.E., Talmage, R.V., and Taft, T.N., 1990, "Lack of an effect of dietary calcium in preventing the loss of radial bone mass in high-calcium consuming elderly white women," in: *Osteoporosis 1990* (C. Christiansen and K. Overgaard, eds.) pp. 981-84, Osteopress ApS, Copenhagen, Denmark .

Anderson, J.J.B., Tylavsky, F.A., Adeleke, V.M., Wu, L-y, Talmage, R.V., and Taft, T.N., 1992, Black-white differences of college-age women in radial bone mass and anthropometry: Dietary and lifestyle variables, *J. Bone Min. Res.* 7 (supp. 1): S191 (abstract).

Bell, N.H., Greene, A., Epstein, S., Oexmann, M.J., Shaw, S., and Shary, J,. 1985, Evidence for alteration of the vitamin D-endocrine system in blacks, *J. Clin. Invest.* **76**:470.

Bell, N.H., Shary, J., Stevens, J., Garza, M., Gordon, L., and Edwards, J., 1991, Demonstration that bone mass is greater in black than in white children, *J. Bone Min. Res.* **6**:719.

Bell, N.H., Yeargey, A.L., Vieira, N., and Shary, J.R., 1992, Calcium absorption is the same and urinary calcium is lower in black than white children: Mechanism for calcium retention and greater bone mass in blacks, *J. Bone Min. Res.* 7 (supp. 1): S150 (abstract).

Bevier, W.C., Wiswell, R.A., Pyka, G., Kozak, K.C., Newhall, K.M., and Marcus, R., 1989, Relationship of body composition, muscle strength, and aerobic capacity to bone mineral density in older men and women, *J. Bone Min. Res.* **4**:421.

Bonjour, J.P., Theintz, G., Buchs, B., Slosman, D., and Rizzoli, R., 1991, Critical years and stages of puberty for spinal and femoral bone mass accumulation during adolescence, *J. Clin. Endocrinol. Metab.* **73**:555.

Buchanan, J.R., Myers, C., Lloyd, T., Leuenberger, P., and Demers, L.M., 1988, Determinants of peak trabecular bone density in women: The role of androgens, estrogen, and exercise, *J. Bone Min. Res.* **3**:673.

Chapuy, M.C., Arlot, M.E., Duboeuf, F., Brun, J., Crouzet, B., Arnaud, S., Delmas, P.D., and Meunier, P.J., 1992. Vitamin $D_3$ and calcium prevent hip fractures in elderly women. *N. Engl. J. Med.* **327**:1637.

Choi, S.C., and Trotter, M., 1970, A statistical study of the multivariate structure and race-sex differences of American white and negro fetal skeletons, *Am. J. Phys. Anthropol.* **33**:313.

Christian, J.C., Yu, P-L., Slemenda, C.W., and Johnston, C.C., Jr., 1989, Heritability of bone mass: A longitudinal study in aging male twins, *Am. J. Human Gen.* **44**:429.

Christiansen, C., Christensen, M.S., McNair, P., Hagen, C., Stocklund, K.E, and Transbol, I., 1980, Prevention of early postmenopausal bone loss: Controlled 2-year study in 315 normal females, *Eur. J. Clin. Invest.* **10**:273.

Cohn, S.H., Abesamis, C., Yasamura, S., Aloia, J., Zanzi, J., and Ellis, K.J., 1977, Comparative skeletal mass and radial bone mineral content in black and white women, *Metabolism* **26**:171.

Cummings, S.R., and Nevitt, M.C., 1989, A hypothesis: The causes of hip fractures, *J. Gerontol.* **44**:M107.

Cummings, S.R., Black, D.M., Nevitt, M.C., Browner, W.S., Cauley, J.A., Genant, H.K., Mascioli, S.R., Scott, J.C., Seeley, D.G., Steiger, P., Vogt, T.M., and the Study of Osteoporotic Fractures Research Group, 1990, Appendicular bone density and age predict hip fracture in women, *JAMA* **263**:665.

Dalsky, G.P., Stocke, K.S., Ehsani, A.A., Slatopolsky, E., Lee, W.C., and Birge, S.J., Jr., 1988, Weight-bearing exercise training and lumbar bone mineral content in post-menopausal women, *Ann. Int. Med.* **108**:824.

Davis, J.W., Ross, P.D., Wasnich, R.D., MacLean, C.J., and Vogel, J.M., 1989, Comparison of cross-sectional and longitudinal measurements of age-related changes in bone mineral content, *J. Bone Min. Res.* **4**:351.

Dawson-Hughes, B., Harris, S., Kramich, C., Dallal, G., and Rasmussen, H.M., 1993, Calcium retention and hormone levels in black and white women on high- and low-calcium diets, *J. Bone Min. Res.* **8**:779.

Delmas, P., 1991, Biochemical markers of bone turnover: Methodology and clinical use in osteoporosis, *Am. J. Med.* **91** (supp. 5B): 59S.

DeSimone, D.P., Stevens, J., Edwards, J., Shary, J., Gordon, L., and Bell, N.H., 1989, Influence of body habitus and race on bone mineral density of the midradius, hip, and spine in aging women, *J. Bone Min. Res.* **4**:827.

Eichner, J.E., Friedrich, C.A., Cauley, J.A., Kamboh, M.I., Gutai, J.P., Kuller, L.H., and Ferrell, R.E., 1990, Alpha$_2$-HS glycoprotein phenotypes and quantitative hormone and bone measures in postmenopausal women, *Calcif. Tissue Int.* **47**:345.

Eichner, J.E., Cauley, J.A., Ferrell, R.E., Cummings, S.R., and Kuller, L.H., 1992, Genetic variation in two bone-related proteins: Is there an association with bone mineral density or skeletal size in postmenopausal women? *Gen. Epidemiol.* **9**:177.

Eisman, J.A., Sambrook, P.N., Kelly, P.J., and Pocock, N.A., 1991, Exercise and its interaction with genetic influences in the determination of bone mineral density, *Am. J. Med.* **91** (supp. 5B): 5S.

Farmer, M.E., Harris, T., Madans, J.H., Wallace, R.B., Cornoni-Huntley, J.R., and White, L.R., 1989, Anthropometric indicators and hip fractures: The NHANES I Epidemiological Follow-up Study, *J. Am. Geriatr. Soc.* **37**:9.

Farmer, M.E., White, L.R., and Brody, J.A., 1984, Race and sex differences in hip fracture incidence, *Am. J. Public Health* **74**:1374.

Felson, D.T., Zhang, Y., Hannan, M.T., and Anderson, J.J., 1993, Effects of weight and body mass index on bone mineral density in men and women: The Framingham Study, *J. Bone Min. Res.* **8**:567.

Fujita, T., and Fukase, M., 1992, Comparison of osteoporosis and calcium intake between Japan and the United States, *Proc. Soc. Exp. Biol. Med.* **200**:149.

Gardsell, P., Johnell, O., and Nilsson, B.E., 1989, Predicting fractures in women by using forearm bone densitometry, *Calcif. Tissue Int.* **44**:235.

Gardsell, P., Johnell, O., and Nilsson, B.E., 1991, The predictive value of bone loss for fragility fractures in women: A longitudinal study over 15 years, *Calcif. Tissue Int.* **49**:90.

Garn, S.M., Sandusky, S.T., Nagy, J.M., and McCann, M.B., 1972, Advanced skeletal development in low-income Negro children, *J. Pediatr.* **80**:965.

Gilsanz, V., Roe, T.F., Mora, S., Costin, G., and Goodman, W.G., 1991, Changes in vertebral bone density in black girls and white girls during childhood and puberty, *N. Eng. J. Med.* **325**:1597.

Goldsmith, N.F., Johnston, J.O., Picetti, G., and Garcia, C., 1973, Bone mineral in the radius and vertebral osteoporosis in an insured population, *J. Bone Joint Surg.* **55A**:1276.

Hagino, H., Yamamoto, K., Teshima, R., Kishimoto, H., and Kagawa, T., 1992, Radial bone mineral changes in pre- and postmenopausal healthy women: Cross-sectional and longitudinal studies, *J. Bone Min. Res.* **7**:147.

Halioua, L., and Anderson, J.J.B., 1990, Age and anthropometric determinants of radial bone mass in premenopausal Caucasian women: A cross-sectional study, *Osteoporosis Int.* **1**:50.

Hansen, M.A., Hassager, C., Jensen, S.B., and Christiansen, C., 1992, Is heritability a risk factor for postmenopausal osteoporosis? *J. Bone Min. Res.* **7**:1037.

Hassager, C., and Christiansen, C., 1989, Influence of soft tissue body composition on bone mass and metabolism, *Bone* **10**:415.

Heaney, R.P., 1986, Calcium, bone health and osteoporosis, *Bone Min. Res. 4* (W.A.Peck, ed.), pp. 255-301, Excerpta Medica, Amsterdam.

Hui, S., Slemenda, C.W., and Johnston, C.C., Jr., 1988, Age and bone mass as predictors of fracture in a prospective study, *J. Clin. Invest.* **81**:1804.

Johnston, C.C., Jr., Miller, J.Z., Slemenda, C.W., Reister, T.K., Christian, J.C., and Peacock, M., 1992, Calcium supplementation and increases in bone mineral density in children, *N. Engl. J. Med.* **327**:82.

Kelly, P.J., Pocock, N.A., Sambrook, P.N., and Eisman, J.A., 1990, Dietary calcium, sex hormones, and bone mineral density in men, *Br. Med. J.* **300**:1361.

Kelly, P.J., Hopper, J.L., Macaskill, G.T., Pocock, N.A., Sambrook, P.N., and Eisman, J.A.. 1991, Genetic factors in bone turnover, *J. Clin. Endocrinol. Metab.* **72**:808.

Kin, K., Lee, J.H.E., Kushida, K., Sartoris, D.J., Ohmura, A., Clopton, P.L., and Inoue, T., 1993, Bone density and body composition on the Pacific rim: A comparison between Japan-born and U.S.-born Japanese-American women, *J. Bone Min. Res.* **8**: 861.

Lacey, J.M., Anderson, J.J.B., Fujita, T., Fukase, M., Tsuchie, S., and Koch, G.G., 1991, Correlates of cortical bone mass among premenopausal and postmenopausal Japanese women. *J. Bone Min. Res.* **6**:651.

Lanyon, L., 1984, Functional strain as a determinant for bone remodelling, *Calcif. Tissue Int.* **36** (supp. 1): S56.

Lau, E., Donnan, S., Barker, D.B.P., and Cooper, C.. 1988, Physical activity and calcium intake in fracture of the proximal femur in Hong Kong, *Br. Med. J.* **297**:1443.

Lau, E.M.C., Cooper, C., Wickham, C., Donnan, S., and Barker, D.J.P., 1990, Hip fracture in Hong Kong and Britain, *Int. J. Epidemiol.* **19:**1119.

Lau, E.M.C., Woo, J., Leung, P.C., Swaminathan, R., and Leung, D., 1992. The effects of calcium supplementation and exercise on bone density in elderly Chinese women, *Osteoporosis Int.* **2:**168.

Lau, E.M.C., Woo, J., Leung, P.C., and Swaminathan, R., 1993, Low bone mineral density, grip strength and skinfold thickness are important risk factors for hip fracture in Hong Kong Chinese, *Osteoporosis Int.* **3:** 66.

Lester, G.E., Anderson, J.J.B., Tylavsky, F.A., Sutton, W.R., Stinnett, S.A., Demasi, R.A., and Talmage, R.V., 1990, Update on the use of distal radial bone measurements in prediction of hip and Colles' fracture risk, *J. Orthop. Res.* **8:** 220.

Li, J-Y., Specker, B.L., Ho, M.L., and Tsang, R.C., 1989, Bone mineral content in black and white children 1 to 6 years of age: Early appearance of race and sex differences, *Am. J. Dis. Child.* **143:**1346.

Liel, Y., Edwards, J., Shary, J., Spicer, K.M., Gordon, L., and Bell, N.H., 1988, The effects of race and body habitus on bone mineral density of the radius, hip and spine in premenopausal women, *J. Clin. Endocrinol. Metab.* **66:**1247.

Lindsay, R., 1991, Estrogens, bone mass, and osteoporotic fractures, *Am. J. Med.* **91** (supp. 5B):10S.

Lindsay, R., 1992. The effect of sex steroids on the skeleton in premenopausal women, *Am. J. Obstet. Gynecol.* **166:**1993.

Lindsay, R., Hart, D.M., Aitken, J.M., MacDonald, E.B., Anderson, J.B., and Clarke, A.C., 1976, Long-term prevention of postmenopausal osteoporosis by oestrogen. *Lancet i:*1038.

Lindsay, R., Cosman, F., Herrington, B.S., and Himmelstein, S., 1992, Bone mass and body composition in normal women, *J. Bone Min. Res.* **7:**55.

Lohman, T.G., Slaughter, M.H., Boileau, R.A., Bunt, J., and Lussier, L., 1984, Bone mineral measurements and their relation to body density in children, youth and adults, *Human Biol.* **56:**667.

Luckey, M.M., Meier, D.E., Mandeli, J.P., DaCosta, M.C., Hubbard, M.L., and Goldsmith, S.J., 1989, Radial and vertebral bone density in white and black women: Evidence for racial differences in premenopausal bone homeostasis, *J. Clin. Endocrinol. Metab.* **69:**762.

Lutz, J., 1986, Bone mineral, serum calcium, and dietary intakes of mother/daughter pairs, *Am. J. Clin. Nutr.* **44:**99.

Lutz, J., and Tesar, R., 1990, Mother-daughter pairs: Spinal and femoral bone densities and dietary intakes, *Am. J. Clin. Nutr.* **52:**872.

Matkovic, V., and Heaney, R.P., 1992, Calcium balance during human growth: Evidence for threshold behavior, *Am. J. Clin. Nutr.* **55:**992.

Matkovic, V., Fontana, D., Tomineac, C., Goel, P., and Chesnut, C.H. III, 1990, Factors that influence peak bone mass formation: A study of calcium balance and the inheritance of bone mass in adolescent females, *Am. J. Clin. Nutr.* **52:**878.

Mayor, G.H., Sanchez, T.V., and Garn, S.M., 1980, " Adjusting photon-absorptiometry norms for whites to the black subject," in: *Fourth International Conference on Bone Measurement* (R.B. Mazess, ed.), pp. 99-106, NIH Publ. No. 80-1938.

McCormick, D.P., Ponder, S.W., Fawcett, H.D., and Palmer, J.L., 1991, Spinal bone mineral density in 335 normal and obese children and adolescents: Evidence for ethnic and sex differences, *J. Bone Min. Res.* **6:**507.

McKusick, V.A., 1992, *Mendelian Inheritance in Man,* 10th ed., 2 vol., Johns Hopkins University Press, Baltimore.

Meier, D.E., Orwoll, E.S., Keenan, E.J., and Fagerstrom, R.M., 1987, Marked decline in trabecular bone mineral content in healthy men with age. Lack of association with sex steroid levels, *J. Am. Geriatr. Soc.* **35:**189.

Meier, D.E., Luckey, M.M., Wallenstein, S., Clemens, T.L., Orwoll, E.S., and Waslien, C.I., 1991, Calcium, vitamin D, and parathyroid hormone status in young white and black women: Association with racial differences in bone mass, *J. Clin. Endocrinol. Metab.* **72:**703.

Mettlin, C., 1980, Nutritional habits of blacks and whites, *Prev. Med.* **9:**601.

Nelson, D.A., Kleerekoper, M., and Parfitt, A.M., 1988, Bone mass, skin color and body size among black and white women, *Bone and Mineral* **4:**257.

Ooms, M.E., Lips, L., Van Lingen, A., and Valkenburg, H.A., 1993, Determinants of bone mineral density and risk factors for osteoporosis in healthy elderly women, *J. Bone Min. Res.* **8:**669.

Ortiz, O., Russell, M., Daley, T.L., Baumgartner, R.N., Waki, M., Lichtman, S., Wang, J., Pierson, R.N., Jr., and Heymsfield, S.B., 1992, Differences in skeletal muscle and bone mineral mass between black and white females and their relevance to estimates of body composition, *Am. J. Clin. Nutr.* **55:**8.

Patel, D.P., Pettifor, J.M., Becker, P.J., Grieve, C., and Leschner, K., 1992, The effect of ethnic group on appendicular bone mass in children, *J. Bone Min. Res.* **7:**263.

Peacock, M., Johnston, C.C., Jr., and Christian, J.C., 1992, "Inheritance of calcium absorption, calcium-regulating hormones and bone turnover," in: *Calcium Regulating Hormones and Bone Metabolism* (D.V. Cohn, C. Gennari, and A.H. Tashjian, Jr., eds.), pp. 451-455, Elsevier Science Publ., Excerpta Medica, Amsterdam.

Pocock, N.A., Eisman, J.A., Hopper, L., Yeates, M.G., Sambrook, P.N., and Eberl, S., 1987, Genetic determinants of bone mass in adults: A twin study, *J. Clin. Invest.* **80:**706.

Pollitzer, W.A., and Anderson, J.J.B., 1989, Ethnic and genetic differences in bone mass: A review with an hereditary vs. environmental prespective, *Am. J. Clin. Nutr.* **50:**1244.

Prentice, A., Laskey, M.A., Shaw, J., Cole, T.J., and Fraser, D.R., 1990, Bone mineral content of Gambian and British children 0-36 months, *Bone and Mineral* **10:**211.

Prior, J.C., Vigna, Y.M., Schecter, M.T., Burgess, A.E., 1990, Spinal bone loss and ovulatory disturbances, *N. Engl. J. Med.* **323:**1221.

Pruzansky, A., Turano, M., Luckey, M., and Senie, R., 1990, Low body weight as a risk factor for hip fracture in both black and white women, *J. Orthop. Res.* **7:**192.

Reed, J.A., Anderson, J.J.B., Tylavsky, F.A., and Gallagher, P.N., Jr., 1994, Comparative changes in radial bone density of elderly female lactoovovegetarians and omnivores, *Am. J. Clin. Nutr.* **59**(suppl.):1197s

Reid, I.R., Plank, L.D., and Evans, M.C., 1992, Fat mass is an important determinant of whole body bone density in premenopausal women but not in men, *J. Clin. Endocrinol. Metab.* **75:**779.

Ribot, C., Tremollieres, F., Pouilles, J-M., Bonneau, M., Germain, F., and Louvet, J-P., 1988, Obesity and postmenopausal bone loss: The influence of obesity on vertebral density and bone turnover in postmenopausal women, *Bone* **8:**327.

Richards, R.J., Svec, F., Bao, W., Srinivasan, S.R., and Berenson, G.S., 1992, Steroid hormones during puberty: Racial (black-white) differences in androstenedione and estradiol—The Bogalusa Heart Study, *J. Clin. Endocrinol.* **75:**624.

Ross, P.D., Orimo, H., Wasnich, R.D., Vogel, J.M., MacLean, C.J., Davis, J.W., and Nomura, A., 1989, Methodological issues in comparing genetic and environmental influences on bone mass, *Bone and Mineral* *7:*67.

Russell-Aulet, M., Wang, J., Thornton, J.C., Colt, E.W.D., and Pierson, R.N., Jr., 1993, Bone mineral density and mass in a cross-sectional study of white and Asian women, *J. Bone Min. Res.* **8**:575.

Sandler, R.B., Slemenda, C.W., LaPorte, R.E., Cauley, J.A., Schramm, M.M., Barresi, M.L., and Kriska, A.M., 1985, Postmenopausal bone density and milk consumption in childhood and adolescence, *Am. J. Clin. Nutr.* **42**:270.

Seeman, E., Hopper, J.L., Bach, L.A., Cooper, M.E., Parkinson, E., Mckay, J., and Jerums, G.. 1989, Reduced bone mass in daughters of women with osteoporosis, *N. Engl. J. Med.* **320**:554.

Sentipal, J.M., Wardlaw, G.M., Mahan, J., and Matkovic, V., 1991, Influence of calcium intake and growth indexes on vertebral bone mineral density in young females, *Am. J. Clin. Nutr.* **54**:425.

Silverman, S.L., and Madison, R.E., 1988, Decreased incidence of hip fracture in Hispanics, Asians, and blacks: California hospital discharge data, *Am. J. Public Health* **78**:1482.

Slemenda, C.W., Hui, S. L., Williams, C.J., Christian, J.C., Meaney, F.J., and Johnston, C.C., Jr., 1990, Bone mass and anthropometric measurements in adult females, *Bone and Mineral* **11**:101.

Slemenda, C.W., Christian, J.C., Williams, C.J., Norton, J.A., and Johnston, C.C., Jr., 1991, Genetic determinants of bone mass in adult women, *J. Bone Min. Res.* **6**:561.

Slemenda, C.W., Christian, J.C., Reed, T., Reister, T.K., Williams, C.J., and Johnston, C.C., Jr., 1992, Long-term bone loss in men: Effects of genetic and environmental factors, *Ann. Int. Med.* **117**: 286.

Smith, D.M., Nance, W.E., Kang, K.W., Christian, J.C., and Johnston, C.C., Jr., 1973, Genetic factors in determining bone mass, *J. Clin. Invest.* **52**:2800.

Smith, D.M., Khairi, M.R.A., and Johnston, C.C., Jr., 1975, The loss of bone mineral with aging and its relationship to risk of fracture, *J. Clin. Invest.* **56**:311.

Sowers, M.F.R., Burns, T.L., and Wallace, R.B., 1986, Familial resemblance of bone mass in adult women, *Gen. Epidemiol.* **3**:85.

Steckel, R.D., 1986, A peculiar population: The nutrition, health, and mortality of American slaves from childhood to maturity, *J. Econ. Hist.* **46**:721.

Steckel, R.D., 1987, Growth depression and recovery: The remarkable case of American slaves, *Ann. Human Biol.* **14**:111.

Sugimoto, T., Tsutsumi, N., Fujii, Y., Kawakatsu, M., Negishi, H., Lee, M.C., Tsai, K-S., Fukase, M., and Fujita, T., 1992, Comparison of bone mineral content among Japanese, Koreans, and Taiwanese assessed by dual-photon absorptiometry, *J. Bone Min. Res.* **7**:153.

Trotter, M., Broman, G.E., and Peterson, R.R., 1960, Densities of bones of white and negro skeletons, *J. Bone Jt. Surg.* **42A**:50.

Trotter, M., and Peterson, R.R., 1970, Weight of the skeleton during postnatal development, *Am. J. Physical Anthropol.* **33**:313.

Tsunenari, T., Tsutsumi, M., Ohno, K., Yamamoto, Y., Kawakatsu, M., Shimogaki, K., Negishi, H., Sugimoto, T., Fukase, M., and Fujita, T., 1993, Age- and gender-related changes in body composition in Japanese subjects, *J. Bone Min. Res.* **8**:397.

Tylavsky, F.A., Bortz, A.D., Hancock, R.L., and Anderson, J.J.B., 1989, Familial resemblance of radial bone mass between premenopausal mothers and their college-age daughters, *Calcif. Tissue Int.* **45**:265.

Tylavsky, F.A., Anderson, J.J.B., Talmage, R.V., and Taft, T., 1992, Are calcium intakes and physical activity patterns during adolescence related to radial bone mass of white college-age females? *Osteoporosis Int.* **2**:232.

Venkataraman, P.S., and Duke, J.C., 1991, Bone mineral content of healthy, full-term neonates: Effect of race, gender, and maternal cigarette smoking, *Am. J. Dis. Child.* **145:**1310.

Wakefield, T., Disney, G.W., Mason, R.L., and Beauchene, R.E., 1980, Relationships among anthropometric indices of growth and creatinine and hydroxyproline excretion in pre-adolescent black and white girls, *Growth* **44:**192.

Wasnich, R.D., Ross, P.D., Heilbrun, L.K., and Vogel, J.M., 1985, Prediction of post-menopausal fracture risk with use of bone mineral measurements, *Am. J. Obstet. Gynecol.* **153:**745.

Weinstein, R.S., and Bell, N.H., 1988, Diminished rates of bone formation in normal black adults, *N. Engl. J. Med.* **319:**1698.

Yano, K., Heilbrun, L.K., Wasnich, R.D., Hankin, J.H., and Vogel, J.M., 1985, The relationship between diet and bone mineral content of multiple skeletal sites in elderly Japanese-American men and women living in Hawaii, *Am. J. Clin. Nutr.* **42:**877.

Chapter 9

# Suboptimal Vitamin D Status: a Risk Factor for Osteoporosis?

P. Lips

## 1. Introduction

Severe vitamin D deficiency of long duration leads to osteomalacia, characterized by bone pain, muscle weakness, fractures and pseudofractures or Looser zones (Frame and Parfitt, 1978). Osteomalacia is relatively rare and occurs mainly in certain risk groups, e.g., patients with malabsorption and immigrants to Northwestern Europe from Asian and African countries. About 25 years ago it was recognized by Chalmers *et al.* (1967) that elderly women also were at risk for osteomalacia. They described 37 patients, 34 women and 3 men, mean age 72 years, of whom 19 had previously undergone gastric surgery. Looser zones, indicating gross osteomalacia, were present in 30 patients. Chalmers *et al.* (1969) found an association between osteomalacia and hip fractures. Later on, the measurement of vitamin D metabolites became feasible and low serum concentrations of 25-hydroxyvitamin D (25-OHD) were reported in institutionalized elderly and geriatric patients (Preece *et al.*, 1975; Corless *et al.*, 1975).

Since then, it has become apparent that suboptimal vitamin D status occurs frequently in elderly people and is much more common than the clinical disease osteomalacia. Suboptimal vitamin D status leads to secondary hyperparathyroidism and increased bone resorption (Parfitt *et al.*, 1982). By this pathway, vitamin D deficiency may lead to hip fractures (Lips and Obrant, 1991). Other

P. Lips • Afdeling Endocrinologie, Academisch Ziekenhuis, Vrije Universiteit, 1007 MB Amsterdam

*Advances in Nutritional Research,* Vol. 9
Edited by Harold H. Draper
Plenum Press, New York, 1994

forms of osteoporosis, including vertebral osteoporosis, also may be more common in vitamin D deficient elderly.

The diagnosis of vitamin D deficiency is usually based on a low serum 25-OHD concentration even when symptoms are lacking. However, there is no agreement on the boundary between a vitamin D deficient and replete state (Lips, 1991).

Vitamin D supplementation can suppress parathyroid function by raising serum calcium and thereby may decrease the incidence of fractures. Fortification of certain foods with vitamin D probably reaches a larger part of the elderly population, but it is not known whether all elderly or only well-defined risk groups profit from an increased vitamin D intake.

## 2. Vitamin D Deficiency in the Elderly

When the measurement of serum concentrations of vitamin D metabolites became feasible, it was recognized that 25-OHD is the main circulating metabolite. Accordingly, serum 25-OHD is used as the main indicator of vitamin D status (DeLuca, 1990). Low levels ($\leq$20 nmol/L) suggest vitamin D deficiency, but are not synonymous with clinical disease. As serum 25-OHD depends on sunshine exposure, serum concentrations and reference values (mean $\pm$ 2 SD or 95% confidence limits in healthy adults) vary with latitude, and are usually higher in the United States than in northwestern European countries (McKenna *et al.*, 1985). In the Netherlands, the lower reference limit is considered to be 20 nmol/L in winter and 30 nmol/L in summer (Netelenbos *et al.*, 1985). It has been suggested that overt osteomalacia is observed only when serum 25-OHD is below 12.5 nmol/L (5 ng/ml). However, subclinical osteomalacia may occur at higher levels, and the appearance of osteomalacia also depends on calcium intake.

Most studies have yielded lower values for serum 25-OHD in elderly people than in young adults. The many studies on this subject recently have been reviewed (Bouillon *et al.*, 1991). A general downward trend is visible from healthy, independently living elderly to inhabitants of homes for the elderly, to residents of nursing homes and geriatric patients (Corless *et al.*, 1975; McKenna *et al.*, 1985; Vir *et al.*, 1978; Chapuy *et al.*, 1987; Lips *et al.*, 1987; Lips *et al.*, 1988). Data from several studies carried out in the Netherlands are shown in Figure 1. If a serum 25-OHD level of less than 20 nmol/L indicates vitamin D deficiency and a level between 20 and 30 nmol/L is considered borderline, then 60% of independently living elderly subjects, but less than 25% of institutionalized elderly subjects, are in adequate vitamin D status.

Patients with hip fractures form a heterogeneous group with a large range of serum 25-OHD concentrations from deficient to adequate (Lips *et al.*, 1982; Brown *et al.*, 1976; von Knorring *et al.*, 1982). Although hip fracture trauma may influence serum vitamin D binding protein and 25-OHD to some degree, this cannot explain these differences (Lips *et al.*, 1985). Elderly patients with hip fractures also differ from elderly subjects living independently and geriatric patients. This points to the question whether vitamin D deficiency is a cause of hip fracture or merely a coincidence.

## 3. Determinants of Vitamin D Status and Causes of Vitamin D Deficiency

Vitamin D status in the elderly depends upon sunshine exposure and nutrition. Elderly people do not go out into the sunshine as frequently as young people. Serum 25-OHD correlates with sunshine exposure in healthy elderly subjects and in patients with hip fracture (Lips *et al.*, 1987). Furthermore, the capacity of the skin to synthesize vitamin $D_3$ under the influence of ultraviolet radiation decreases with age (MacLaughlin and Holick, 1985). The synthesis of vitamin $D_3$ in response to a whole body dose of ultraviolet radiation in elderly volunteers was about 30% of that in young volunteers (Holick, 1990).

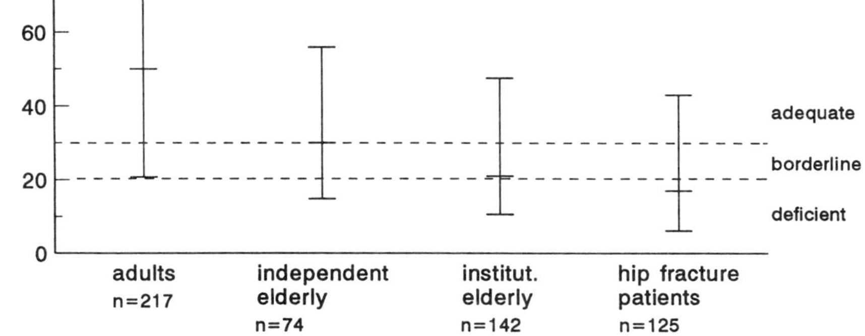


Figure 1. Vitamin D status, as indicated by serum 25-OHD, in healthy adults and various elderly subgroups. Values are expressed as median and 95% confidence intervals. The dashed lines indicate the lower reference limits in summer (30 nmol/L) and winter (20 nmol/L). Data from Netelenbos *et al.* (1985), Lips *et al.* (1988) and Lips *et al.* (1982).

When sunshine exposure is low or negligible, nutrition becomes the more important source of vitamin D. Foods containing vitamin D in reasonable amounts are fatty fish, eggs, margarine (when fortified) and milk (when fortified) (McLeod *et al.*, 1974). From several studies in the United Kingdom and other European countries it appears that the mean dietary intake of vitamin D is generally not more than 100 IU/day ( Lips, 1991; Lips *et al.*, 1987; McLeod *et al.*, 1974). The intake is higher in the United States (up to 200 IU/day) due to the fortification of milk with vitamin D (400 IU/quart) (Parfitt *et al.*, 1982; Omdahl *et al.*, 1982). The recommended dietary allowance for adults, including the elderly, is 100 IU/day in the United Kingdom and the Netherlands and 400 IU/day in the USA, reflecting average current intake (Lips *et al.*, 1987; Omdahl *et al.*, 1982).

In patients with hip fracture who were not exposed to sunshine, a positive correlation was observed between dietary vitamin D intake and serum 25-OHD (Lips *et al.*, 1987). From this relationship it appeared that vitamin D intake in the absence of sunshine should be 200 to 300 IU/day in order to achieve an acceptable vitamin D status (serum 25-OHD >30 nmol). This may be an underestimate, as not all foods containing vitamin D were recorded. Multivitamin preparations are another source of vitamin D. This may explain higher serum 25-OHD levels in Denmark and the United States (Omdahl *et al.*, 1982; Lund *et al.*, 1975).

Other causes of vitamin D deficiency, although rare, are malabsorption, increased loss of vitamin D due to nephrotic syndrome and increased catabolism, which may follow liver enzyme induction by antiepileptic drugs. Hyperparathyroidism can lead to an increased conversion of 25-OHD to inactive metabolites, thereby decreasing vitamin D stores and inducing a deficiency (Clements *et al.*, 1987).

## 4. Changes in Vitamin D Metabolism with Aging

The conversion of vitamin D to 25-OHD does not decrease appreciably with aging, but further hydroxylation to 1,25-dihydroxyvitamin D ($1,25(OH)_2D$) in the kidney may be impaired. Serum $1,25(OH)_2D$ decreases with age as a result of the decrease in renal function with aging (Gallagher *et al.*, 1979). The formation of $1,25(OH)_2D$ is influenced by parathyroid hormone (PTH). Administration of the synthetic fragment PTH(1-34) showed less response in elderly people and especially in patients with hip fracture (Tsai *et al.*, 1984). Recently, however, it was found that serum $1,25(OH)_2D$ was not decreased in a selected group of healthy elderly people, suggesting that the decrease with age is caused by disease (Sherman *et al.*, 1990).

The serum concentrations of 25-OHD and $1{,}25(OH)_2D$ do not correlate in healthy adults, as $1{,}25(OH)_2D$ is under negative feedback control, but a positive correlation between these metabolites is found in vitamin D deficient elderly and in patients with hip fracture (Lips *et al.*, 1987; Bouillon *et al.*, 1987) as shown in Figure 2. This relationship indicates a substrate-dependent synthesis of $1{,}25(OH)_2D$. Treatment with vitamin D or 25-OHD in such circumstances leads to an increase of $1{,}25(OH)_2D$ (Lips *et al.*, 1988; Bouillon *et al.*, 1987).

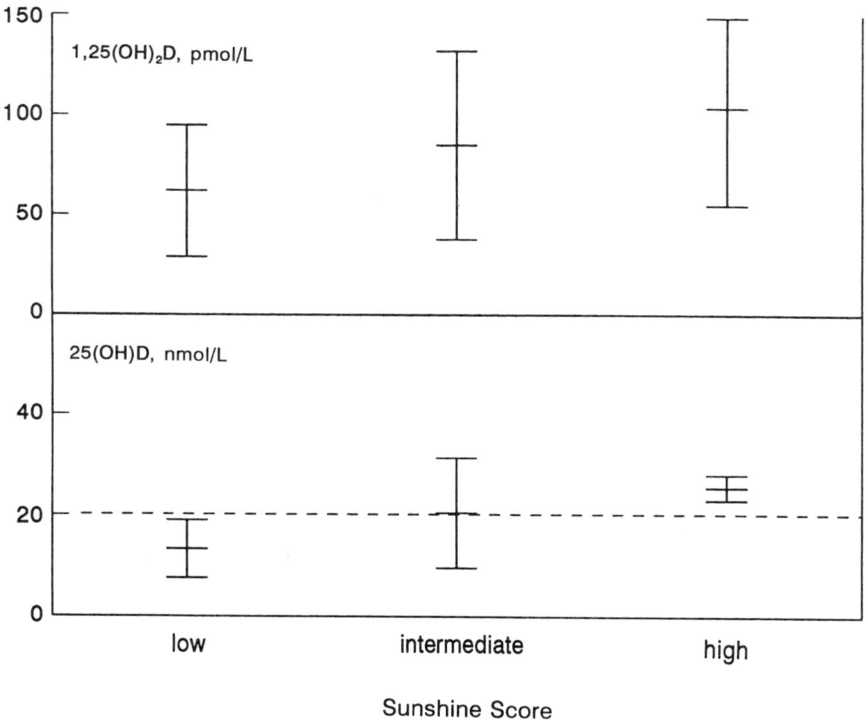


Figure 2. Serum concentrations (mean ± SD) of vitamin D metabolites in patients with hip fracture grouped according to sunshine exposure. The dashed line indicates the lower reference limit for serum 25-OHD in winter. Serum $1{,}25(OH)_2D$ follows serum 25-OHD, and both metabolites show a significant positive correlation ($r = 0.49$, $P<0.001$), confirming the substrate dependent synthesis of $1{,}25(OH)_2D$. Data from Lips *et al.* (1987).

Very low serum concentrations of $1{,}25(OH)_2D$ can be found in residents of nursing homes (Corless *et al.*, 1975; McKenna *et al.*, 1985; Lips *et al.*, 1988). An additional explanation for low serum $1{,}25(OH)_2D$ may be immobility, which showed a positive relationship with walking score in residents of nursing homes

(Lips *et al.*, 1990a). Increased bone resorption due to immobility suppresses the formation of $1{,}25(OH)_2D$. Thus, low serum $1{,}25(OH)_2D$ in elderly people may be multifactorial, i.e., it may be caused by decreased renal function, vitamin D deficiency, immobility or chronic disease.

## 5. Consequences of Vitamin D Deficiency

The negative effects of vitamin D deficiency on bone may be mediated by a decrease in serum $1{,}25(OH)_2D$ concentration and by an increase in parathyroid gland activity (Parfitt *et al.*, 1982; Lips and Obrant, 1991). When vitamin D status is suboptimal, the synthesis of $1{,}25(OH)_2D$ is limited to some degree, as it becomes substrate-dependent. A small decrease in serum $1{,}25(OH)_2D$ leads to a decrease in calcium absorption from the intestine. These changes are compensated by an increase in PTH secretion (Parfitt *et al.*, 1982). The increase in PTH stimulates the renal hydroxylation of 25-OHD to form $1{,}25(OH)_2D$. Thus an almost normal $1{,}25(OH)_2D$ level may be maintained at the expense of an increase in parathyroid gland activity. Several investigators have observed a negative correlation between serum concentrations of 25-OHD and PTH in elderly people and in patients with hip fracture (Lips, 1991; Chapuy *et al.*, 1987; Lips *et al.*, 1988; von Knorring *et al.*, 1982; Krall *et al.*, 1989). However, this correlation is rather weak, indicating that the response of PTH to a lowering of vitamin D status is variable.

Vitamin D levels are maximal in summer and lowest at the end of winter. An inverse seasonal variation in serum PTH has been observed in elderly people and in patients with hip fracture, with a maximum at the end of winter (Krall *et al.*, 1989; Lips *et al.*, 1983). Increased PTH secretion in incipient vitamin D deficiency leads to an increase in bone turnover, which results in bone loss (Parfitt *et al.*, 1982). This is most marked at the cortical-endosteal site where bone loss is visible mainly as cortical thinning (Lips *et al.*, 1982; Fonseca *et al.*, 1988). The mineral content of hyperparathyroid bone is lower than normal because it contains more young bone in which mineralization is not yet complete (Parfitt, 1980). In addition, decreased calcium absorption due to vitamin D deficiency may lead to delayed secondary mineralization. After synthesis of the bone matrix (the osteoid) by osteoblasts, the first phase of mineralization (the so-called primary mineralization) proceeds within a few days, leading to deposition of about half of the bone mineral. Secondary mineralization may take about 6 months and may remain incomplete when vitamin D status is suboptimal (Parfitt, 1980).

Decreased bone mineral content in vitamin D deficient patients has been found in several cross-sectional studies. In a group of postmenopausal women, vertebral bone density correlated positively with serum 25-OHD and negatively with serum PTH (Villareal *et al.*, 1991). In the study of Khaw *et al.* (1992) bone density of the lumbar spine and upper femur in middle-aged women showed a similar positive relationship to serum 25-OHD and negative relationship to serum PTH and these relationships were independent of possible confounding factors. The difference in bone density between the highest and lowest tertile of vitamin D status was 5–10%.

In the case of severe long-term vitamin D deficiency, primary mineralization of the newly formed bone matrix is severely depressed, leading to an accumulation of osteoid tissue. Increased thickness of osteoid seams is the most characteristic sign of osteomalacia. It may take many years for osteomalacia to develop (Frame and Parfitt, 1978).

In addition to stimulating calcium absorption, $1,25(OH)_2D$ may act directly on the osteoblasts to stimulate the formation of osteocalcin, as shown in in vitro studies (DeLuca, 1990). Low serum $1,25(OH)_2D$ may lead to decreased osteoblastic activity, but the clinical relevance is unknown (Duda *et al.*, 1987). Vitamin D metabolites are involved in muscular function (Boland, 1986). Vitamin D deficiency and osteomalacia are associated with myopathy. It has been suggested that this myopathy could lead to falls. However, testing of muscle strength in elderly people before and after vitamin D supplementation revealed no significant differences (Corless *et al.*, 1985). Immunologic incompetence has been mentioned as another consequence of vitamin D deficiency, but its practical significance for elderly people is unknown.

## 6. Is Vitamin D Deficiency a Risk Factor for Hip Fracture?

In 1969 Chalmers *et al.* observed histological osteomalacia in 20% of a large series of patients with hip fracture. In subsequent studies in the United Kingdom, osteomalacia was found in varying degrees, from zero to 37%, in patients with hip fracture (Hodkinson, 1971; Jenkins *et al.*, 1973; Aaron *et al.*, 1974; Faccini *et al.*, 1976; Wootton *et al.*, 1979; Wilton *et al.*, 1987; Hordon and Peacock, 1990). In studies carried out in various countries the percentage of histological osteomalacia range from zero to 24% (Lips *et al.*, 1982; Sokoloff, 1978; Hoikka *et al.*, 1982; Johnston *et al.*, 1985; Wicks *et al.*, 1982).

This large variance is explained by the use of different criteria for diagnosing osteomalacia (Lips and Obrant, 1991). The frequency of osteomalacia was higher when osteoid volume or surface area was used as the criterion and lower when osteoid seam thickness was the main criterion. In a series of bone biopsies on 89 hip fracture patients in Amsterdam, increased osteoid volume was observed in 10%, increased osteoid surface in 17% and increased osteoid seam thickness in none (Lips *et al.*, 1982). Whereas increased osteoid seam thickness is the best criterion of osteomalacia, increased osteoid surface and volume are also observed in hyperparathyroidism and other states of hyperosteoidosis (Meunier *et al.*, 1977). In fact, subclinical or incipient osteomalacia and hyperparathyroid bone disease cannot be distinguished on histological grounds.

According to Parfitt *et al.* (1981), the first stage of bone disease caused by vitamin D deficiency is characterized by signs of secondary hyperparathyroidism, increased bone turnover and cortical bone loss, whereas in later stages decreased mineralization predominates. In patients with hip fractures, increased bone turnover predicted reduced cortical bone mass as measured by photon absorptiometry (Lips *et al.*, 1990b). It can be concluded that vitamin D deficiency contributes to the pathogenesis of hip fractures either by increased bone loss or decreased mineralization (Lips and Obrant, 1991).

As has been mentioned before, another consequence of vitamin D deficiency that may lead to hip fractures is myopathy leading to falls. Various hypothetical pathways by which vitamin D deficiency may result in osteoporosis and hip fractures are summarized in Figure 3.

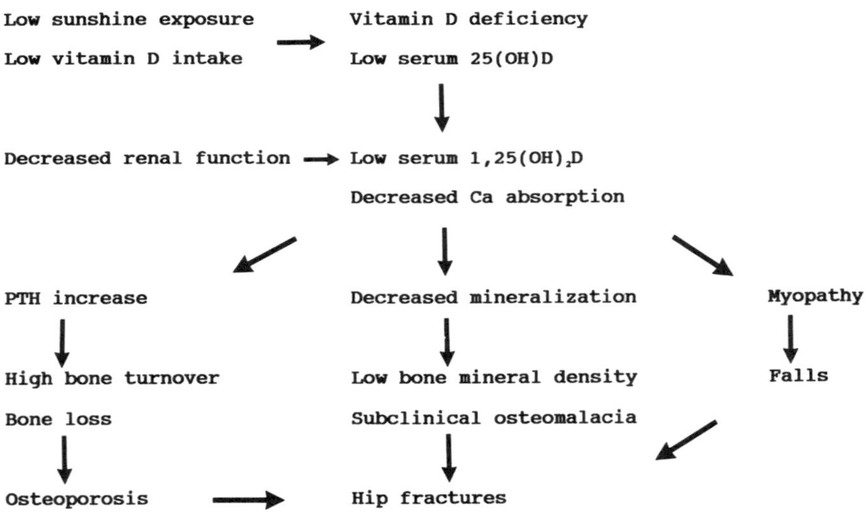


Figure 3. Schematic presentation of various pathways from vitamin D deficiency to osteoporosis and hip fractures.

## 7. The Effect of Vitamin D Supplementation

Because consensus is lacking on the consequences of suboptimal vitamin D status in elderly people, studies have been done on the ability of vitamin D and 25-OHD supplements to correct metabolic abnormalities and improve bone health. Most investigators have studied the effect of a low dose of vitamin $D_3$ (400 or 800 IU per day) (Chapuy *et al.*, 1987; Lips *et al.*, 1988; McLennan and Hamilton, 1977). In general, malabsorption was not observed, and an adequate rise in serum 25-OHD was seen after administration of the lower dose. Intermittent large doses of vitamin D (100,000 IU) led to adequate levels of serum 25-OHD for up to half a year and did not cause side effects (Davies *et al.*, 1985). However, in one study a daily dose of 2000 IU led to hypercalcemia in 2 of 63 elderly subjects (Johnson *et al.*, 1980).

The effect of vitamin D supplementation on serum $1{,}25(OH)_2D$ depends on preexisting vitamin D status. Supplementation of vitamin D deficient elderly subjects with 25-OHD led to an increase in serum $1{,}25(OH)_2D$ within one week (Bouillon et al., 1987). In a study on the effect of a low dose of vitamin $D_3$ (400–800 IU/day) on 142 institutionalized elderly subjects, a small but significant increase in serum $1{,}25(OH)_2D$ was observed, but only when the initial serum 25-OHD was lower than 30 nmol/L, indicating substrate dependent synthesis of $1{,}25(OH)_2D$ as seen in vitamin D deficiency (Lips *et al.*, 1988). Serum parathyroid hormone concentration decreased significantly after supplementation in this study as well as in other supplementation studies on vitamin D deficient elderly subjects (Chapuy *et al.*, 1987; Lips *et al.*, 1988). On the other hand, a recent study on American nursing home residents showed neither an increase in serum $1{,}25(OH)_2D$ nor a decrease in serum PTH after vitamin D supplementation; however, the initial mean serum 25-OHD level was more than 40 nmol/L, indicating that these elderly were vitamin D replete (Himmelstein *et al.*, 1990).

Recently, data have been published on the effect of vitamin D supplementation on bone mineral density and fracture incidence. A low dose of vitamin D (400 IU/day) led to a decrease in wintertime bone loss in healthy postmenopausal women (Dawson-Hughes *et al.*, 1991). A large double-blind study on more than 3000 institutionalized elderly subjects in France showed that vitamin D (800 IU/day) in combination with calcium (1 gram/day) led to a significant decrease in hip and other peripheral fractures in comparison with a placebo group (Chapuy *et al.*, 1991). Bone loss in the hip was significantly less in the supplemented than in the placebo group. It is not possible to determine from this study whether vitamin D, calcium or (more probably) both was responsible for the decrease in fracture rate and bone loss. In a Finnish study ((Heikinheino *et al.*, 1992), an annual intramuscular injection of vitamin $D_2$ (150,000-300,000 IU) was given to independently living and institutionalized elderly subjects. During follow-up the occurrence of fractures was lower in the vitamin D group than in the control group (16.4% vs 21.8%, $P<0.05$). When the data were analyzed according to fracture site, the difference was significant only for fractures of the upper limb.

## 8. What is Normal Vitamin D Status?

A sharp definition of normality and a strict boundary between a deficient and a replete state are important tools in developing preventive strategies. The hallmark for the diagnosis of osteomalacia is still an increased osteoid seam thickness in an undecalcified bone biopsy. A decrease in osteoid parameters following vitamin D therapy may confirm that vitamin D deficiency was the cause (Hosking *et al.*, 1983). Measurement of serum 25-OHD concentration is a much simpler diagnostic test, which has given much information. However, reference values for serum 25-OHD in healthy adults vary with sunshine exposure and diet. Mean serum 25-OHD is usually much higher in the USA than in European countries (Omdahl *et al.*, 1982; Himmelstein *et al.*, 1990; Dawson-Hughes *et al.*, 1991), due to a more southern latitude, use of vitamin D fortified milk and the consumption of vitamin tablets. Synthesis of the active metabolite $1,25(OH)_2D$ depends on renal function and on the availability of sufficient 25-OHD substrate. In healthy adults the synthesis of $1,25(OH)_2D$ is not substrate dependent. Studies on elderly people (Lips *et al.*, 1988) indicate that the serum 25-OHD level at which substrate dependency becomes visible is about 30 nmol/L.

An increase in serum PTH in vitamin D deficient subjects may be another way of defining inadequate vitamin D status (Fonseca *et al.*, 1988). Mean serum PTH is higher in vitamin D deficient than in vitamin D replete elderly subjects and serum PTH shows an inverse relationship with serum 25-OHD (Lips *et al.*, 1988). However, serum PTH usually remains within normal reference limits, even in suboptimal vitamin D status (Lips, 1991). Improvement of vitamin D status, e.g., by increasing dietary intake or by supplementation, suppresses parathyroid function and may blunt seasonal variation (Lips *et al.*, 1988). Krall *et al.* (1989) found that the increase in serum PTH in winter was prevented at a vitamin D intake of 220 IU per day and a serum 25-OHD level of 95 nmol/L. According to this criterion, the level distinguishing between a vitamin D deficient and replete state would be very high and most people in Europe would be classified as vitamin D deficient. In any event, the same investigators showed that bone loss from the spine during the winter could be decreased by increasing serum 25-OHD from 60 to 90 nmol/L and suppressing serum PTH by 10% (Dawson-Hughes et al., 1991). This discrepancy may be explained by a different methodology for the measurement of vitamin D metabolites, or by differences in calcium intake, a higher intake suppressing parathyroid hormone secretion.

It follows from the above that it is virtually impossible to define a clear boundary between a vitamin D deficient and replete state (Heaney, 1986). Even slight elevations in serum PTH caused by a less than optimal vitamin D state and low calcium intake may increase bone turnover and bone loss and may be responsible for a 5 to 10% decrease in bone mass (Khaw *et al.*, 1992).

## 9. Prevention

Supplementation of vitamin D deficient elderly subjects with vitamin $D_3$ leads to a small increase of serum $1,25(OH)_2D$ and suppresses parathyroid function. In addition, evidence is accumulating that supplementation leads to a small increase in bone mineral density by increasing mineralization and decreasing bone loss, and to a decrease in fracture incidence (Dawson-Hughes *et al.*, 1991; Meunier *et al.*, 1991; Heikinheino *et al.*, 1992). These positive changes may follow correction of a suboptimal vitamin D status. The serum 25-OHD level below which elderly people profit from vitamin D therapy is uncertain.

From dietary surveys and supplementation studies it appears that the vitamin D requirement is about 400 IU/day in the absence of sunshine exposure (Lips *et al.*, 1987). It is advisable to increase the recommended dietary allowance for vitamin D to 400 IU/day for elderly people exposed to little or no sunshine. Institutionalized elderly subjects constitute the most important risk group.

Vitamin D deficiency can be prevented by sunshine exposure, artificial ultraviolet irradiation, diet or supplementation. Exposure to sunlight (e.g., head and arms 15 minutes daily in spring and summer) is advisable, but sunshine is unpredictable and higher exposures are required in the elderly than in young adults (Holick, 1990). Artificial ultraviolet irradiation (e.g., by fluorescent lighting in geriatric wards) has been used in several studies. It is effective but time-consuming and the risks for the skin and eyes are not well known (Corless *et al.*, 1978; Toss *et al.*, 1982).

The vitamin D content of the diet can be increased by regular consumption of fatty fish; however, it is difficult to ensure a dietary vitamin D intake of 400 IU/day. Thus, fortification of certain foods with vitamin D or supplementation are better alternatives. In the Netherlands and the United Kingdom, margarine is fortified with vitamin $D_3$ (3 IU/gram), but this seldom provides more than 50 to 100 IU/day. Fortification of milk with vitamin $D_3$ (400 IU/quart) is practiced in the United States (Holick, 1990; Omdahl *et al.*, 1982). This type of fortification is more effective and can easily provide 200 IU/day.

Vitamin D supplementation is the most reliable way of providing for an adequate vitamin D status. It has been shown in several studies that 400 IU/day leads to an adequate level of serum 25-OHD and suppresses parathyroid function (Lips *et al.*, 1988). Higher doses are not necessary and a dose of 2000 IU/day may occasionally cause hypercalcemia (Johnson *et al.*, 1980). In some studies a high dose (e.g., 100,000 IU) given to institutionalized elderly subjects once every 6 months, led to adequate serum 25-OHD levels (about 60 nmol/L gradually decreasing to 25 nmol/L) without apparent hypercalcemia (Davies *et al.*, 1985; Toss *et al.*, 1983). Although this is a very practical solution, it may be less safe than low dose supplementation. A low dose of vitamin D does not influence renal function or serum cholesterol concentration and side effects have not been observed (Lips *et al.*, 1988). Vitamin D supplementation should be considered only for well defined risk groups, such as the elderly who are never exposed to sunshine, or elderly people living in institutions. The diagnosis of vitamin D deficiency can be confirmed by measuring serum 25-OHD before starting supplementation, but this is not recommended because of its high cost in comparison with the low cost of supplementation.

## 10. Conclusion

Vitamin D status in the elderly can be improved by encouraging exposure to sunshine and consumption of certain foods (e.g., fatty fish). Milk fortified with vitamin D (400 IU/L) should be more widely available. Supplementation with a low dose of vitamin $D_3$ (400 IU/day) should be considered for certain risk groups, especially institutionalized elderly subjects.

## References


Aaron, J.E., Gallagher, J.C., Anderson J., Stasiak, L., Longton, E.B., Nordin, B.E.C., and Nicholson M., 1974, Frequency of osteomalacia and osteoporosis in fractures of the proximal femur, *Lancet* **1**:229.

Boland, R., 1986, Role of vitamin D in skeletal muscle function, *Endocrinol. Rev.* **7**:434.

Bouillon, R.A., Auwerx, J.H., Lissens, W.D., and Pelemans, W.K., 1987, Vitamin D status in the elderly; seasonal substrate deficiency causes 1,25-dihydroxycholecalciferol deficiency, *Am. J. Clin. Nutr.* **45**:755.

Bouillon, R., Pelemans, W., and Quesada, J.M., 1991, "Vitamin D deficiency in the elderly," in: *Nutritional Aspects of Osteoporosis* (P. Buckhardt and R.P. Heaney, eds.), Serono Symposia **85**, pp. 245-256, Raven Press, New York.

Brown, I.R.F., Bakowska, A., and Millard, P.H., 1976, Vitamin D status of patients with femoral neck fracture, *Age Aging* **5**:127.

Chalmers, J., Conacher, W.D.H., Gardner, D.L., and Scott, P.J., 1967, Osteomalacia, a common disease in elderly women, *J. Bone Joint Surg.* **49B**:403.

Chalmers, J., Barclay, A., Davidson, A.M., Macleod, D.A.D., and Williams, D.A., 1969, Quantitative measurements of osteoid in health and disease. *Clin. Orthop.* **63:**196.

Chapuy, M.C., Chapuy, P., and Meunier, P.J., 1987, Calcium and vitamin D supplements: effects on calcium metabolism in elderly people, *Am. J. Clin. Nutr.* **46:**324.

Chapuy, M.C., Arlot, M.E., Duboeuf, F., Brun, J., Crouzet, B., Arnaud, S., Delmas, P.D., and Meunier, P.J., 1992, Vitamin $D_3$ and calcium to prevent hip fractures in elderly women, *N. Engl. J. Med.* **327:**1637.

Clements, M.R., Davies, M., Fraser, D.R., Lumb, G.A., Mawer, E.B., and Adams, P.H., 1987, Metabolic inactivation of vitamin D is enhanced in primary hyperparathyroidism, *Clin. Sci.* **73**:659.

Corless, D., Beer, M., Boucher, B.J., Gupta, S.P., and Cohen, R.D., 1975, Vitamin D status in long-stay geriatric patients, *Lancet* **1:**140.

Corless, D., Gupta, S.P., Switala, S., Boucher, B.J., Barragry, J.M., and Cohen, R.D., 1978, Response of plasma 25-hydroxyvitamin D to ultraviolet irradiation in long-stay geriatric patients, *Lancet* **2:**649.

Corless, D., Dawson, E., Fraser F., Ellis, M., Evans, S.J.W., Perry, J.D., Reisner, C., Silver, C.P., Beer, M., Boucher, B.J., and Cohen, R.D., 1985, Do vitamin D supplements improve the physical capabilities of elderly hospital patients? *Age Aging* **14:**76.

Davies, M., Mawer, E.B., Hann, J.T., Stephens, W.P., and Taylor, J.L., 1985, Vitamin D prophylaxis in the elderly: a simple effective method suitable for large populations, *Age Ageing* **14:**349.

Dawson-Hughes, B., Dallal, G.E., Krall, E.A., Harris, S., Sokoll, L.J., and Falconer, G., 1991, Effect of vitamin D supplementation on wintertime and overall bone loss in healthy postmenopausal women, *Ann. Intern. Med.* **115:**505.

DeLuca, H.F., 1990, Osteoporosis and the metabolites of vitamin D, *Metabolism* **39**:Suppl. 1:3.

Duda, R.J., Kumar, R., Nelson, K.I., Zinsmeister, A.R., Mann, K.G., and Riggs, B.L., 1987, 1,25-Dihydroxyvitamin D stimulation test for osteoblast function in normal and osteoporotic postmenopausal women, *J. Clin. Invest.* **79:**1249.

Faccini, J.M., Exton-Smith, A.N., and Boyde, A., 1976, Disorders of bone and fracture of the femoral neck, *Lancet* **I:**1089.

Fonseca, V., Agnew, J.E., Nag, D., and Dandona, P., 1988, Bone density and cortical thickness in nutritional vitamin D deficiency; effect of secondary hyperparathyroidism, *Ann. Clin. Biochem.* **25:**271.

Frame, B., and Parfitt, A.M., 1978, Osteomalacia: current concepts, *Ann. Intern. Med.* **89**:966.

Gallagher, J.C., Riggs, B.L., Eisman, J., Hamstra, A., Arnaud, S.B., and DeLuca, H.F., 1979, Intestinal calcium absorption and serum vitamin D metabolites in normal subjects and osteoporotic patients, *J. Clin. Invest.* **64**:729.

Heaney, R.P., 1986, "Calcium, bone health and osteoporosis," in: *Bone and Mineral Research 4* (W. Peck, ed.), pp. 255-301, Elsevier, Amsterdam and New York.

Heikinheimo, R.J., Inkovaara, J.A., Harju, E.J., Haavisto, M.V., Kaarela, R.H., Kataja, J.M., Kokko, A.M.L., Kolho, L.A., and Rajala, S.A., 1992, Annual injection of vitamin D and fractures of aged bones, *Calcif. Tissue Int.* **51:**105.

Himmelstein, S., Clemens, T.L., Rubin, A., and Lindsay R., 1990, Vitamin D supplementation in elderly nursing home residents increases 25-OHD but not $1,25(OH)_2D$, *Am. J. Clin. Nutr.* **52:**701.

Hodkinson, H.M., 1971, Fracture of the femoral neck in the elderly: Assessment of the role of osteomalacia, *Gerontol. Clin.* **13:**153.

Hoikka, V., Althava, E.M., Savolaimen, K., and Parviamen, M., 1982, Osteomalacia in fractures of the proximal femur, *Acta. Orthop. Scand.* **53**:255.

Holick, M.F., 1990, "Vitamin D and the skin: photobiology, physiology and therapeutic efficacy for psoriasis," in: *Bone and Mineral Research 7,* (J.N.M. Heersche and J.A. Kanis, eds.), pp. 313-366, Elsevier, Amsterdam.

Hordon, L.D., and Peacock, M., 1990, Osteomalacia and osteoporosis in femoral neck fracture, *Bone Mineral* **11**:147.

Hosking, D.J., Kemm, J.R., Knight M.E., Campbell, G.A., Cotton, R.E., Berryman, R., and Boyd, R.V., 1983, Screening for subclinical osteomalacia in the elderly: normal ranges or pragmatism? *Lancet* **2**:1290.

Jenkins, D.H.R., Roberts, J.G., Webster, D., and Williams, E.O., 1973, Osteomalacia in elderly patients with fracture of the femoral neck, *J. Bone Jt. Surg. (Br.)* **55**:575.

Johnson, K.R., Jobber, J., and Stonawski, B.J., 1980, Prophylactic vitamin D in the elderly, *Age Ageing* **9**:121.

Johnston, C.C., Norton, J., Khairi, M.R., Kernek, C., Edouard, C., Arlot, M., and Meunier, P.J., 1985, Heterogeneity of fracture syndromes in postmenopausal women, *J. Clin. Endocrinol. Metab.* **61**:551.

Khaw, K.T., Sneyd, M.J., and Compston, J., 1992, Bone density, parathyroid hormone and 25-hydroxyvitamin D concentrations in middle aged women, *Br. Med. J.* **305**:273.

Knorring, J. von, Slätis, P., Weber, T.H., and Helenius, T., 1982, Serum level of 25-hydroxyvitamin D, 24,25-dihydroxyvitamin D and parathyroid hormone in patients with femoral neck fracture in Southern Finland, *Clin. Endocrinol.* **17**:189.

Krall, E.A., Sahyoun, N., Tannenbaum, S., Dallal, G.E., and Dawson-Hughes, B., 1989, Effect of vitamin D intake on seasonal variations in parathyroid hormone secretion in postmenopausal women, *N. Engl. J. Med.* **321**:1777.

Lips, P., 1991, "Vitamin D nutrition in the elderly; problems and recommendations," in: *Vitamin D: Gene Regulation, Structure-Function Analysis and Clinical Application,* (A.W. Norman, R. Bouillon, M. Thomasset, eds.), pp. 757-764, Walter de Gruyter, Berlin, New York.

Lips, P., and Obrant, K., 1991, The pathogenesis and treatment of hip fractures, *Osteoporosis Int.* **I**:218.

Lips, P., Netelenbos, J.C., Jongen, M.J.M., van Ginkel, F.C., Althuis, A.L., van Schaik, C.L., Van der Vijgh, W.J.F., Vermeiden, J.P.W., and Van der Meer, C., 1982, Histomorphometric profile and vitamin D status in patients with femoral neck fracture, *Metab. Bone Dis. Rel. Res.* **4**:85.

Lips, P., Hackeng, W.H.L., Jongen, M.J.M., van Ginkel, F.C., and Netelenbos, J.C., 1983, Seasonal variation in serum concentrations of parathyroid hormone in elderly people, *J. Clin. Endocrinol. Metab.* **57**:204.

Lips, P., Bouillon, R., Jongen, M.J.M., van Ginkel, F.C., van der Vijgh, W.J.F., and Netelenbos J.C., 1985, The effect of trauma on serum concentrations of vitamin D metabolites in patients with hip fracture, *Bone* **6**:63.

Lips, P., van Ginkel, F.C., Jongen, M.J.M., Rubertus, A., van der Vijgh, W.J.F., and Netelenbos, J.C., 1987, Determinants of vitamin D status in patients with hip fracture and elderly control subjects, *Am. J. Clin. Nutr.* **46**:1005.

Lips, P., Wiersinga, A., van Ginkel, F.C., Jongen, M.J.M., Netelenbos, J.C., Hackeng, W.H.L., Delmas, P.D., and Van der Vijgh, W.J.F., 1988, The effect of vitamin D supplementation on vitamin D status and parathyroid function in elderly subjects, *J. Clin. Endocrinol. Metab.* **67**:644.

Lips, P., van Ginkel, F.C., Netelenbos, J.C., Wiersinga, A., and van der Vijgh, W.J.F., 1990a, Lower mobility and markers of bone resorption in the elderly, *Bone and Mineral* **9**:49.

Lips, P., Hesp, R., Reeve, J., Wootton, R., Green, J.R., and Klenerman, L., 1990b, High indices of remodelling in iliac trabecular bone predict reduced forearm cortical bone mass indices in patients with proximal femoral fractures, *Bone and Mineral* **11**:93.

Lund, B., Sorensen, O.H., and Christensen, A.B., 1975, 25-hydroxy-cholecalciferol and fractures of the proximal femur, *Lancet* **2**:300.

MacLaughlin, J., and Holick, M.F., 1985, Aging decreases the capacity of human skin to produce vitamin $D_3$, *J. Clin. Invest.* **76**: 1536.

McKenna, M.J., Freaney, R., Meade, A., and Muldowney, F.P., 1985, Hypovitaminosis D and elevated serum alkaline phosphatase in elderly Irish people, *Am. J. Clin. Nutr.* **41**:101.

McLennan, W.J., and Hamilton, J.C., 1977, Vitamin D supplements and 25-hydroxyvitamin D concentrations in the elderly, *Br. Med. J.* **2**:859.

McLeod, C.C., Judge, T.G., and Caird, F.I., 1974, Nutrition of the elderly at home; II: Intake of vitamins, *Age Aging* **3**:209.

Meunier, P.J., Edouard, C., Richard, D., and Laurent, J., 1977, "Histomorphometry of osteoid tissue: the hyperosteoidoses," in: *Second International Workshop on Bone Histomorphometry*, (P.J. Meunier, ed.), pp. 249-262, Armour Montagu, Paris.

Meunier, P.J., Chapuy, M.C., Arlot, M.E., Delmas, P.D., Duboeuf, F., Brun, J., Crouzet, B., and Arnaud, S., 1991, Effects of a calcium and vitamin $D_3$ supplement on non-vertebral fracture rate, femoral bone density, and parathyroid function in elderly women, *J. Bone Min. Res.* **S:135** (abstract).

Netelenbos, J.C., Jongen, M.J.M., van der Vijgh, W.J.F., Lips, P., and van Ginkel, F.C., 1985, Vitamin D status in urinary calcium stone formation, *Arch. Intern. Med.* **145**:681.

Omdahl, J.L., Garry, P.J., Hunsaker, L.A., Hunt, W.C., and Goodwin, J.S., 1982, Nutritional status in a healthy elderly population, *Am. J. Clin. Nutr.* **36**:1225.

Parfitt, A.M., 1980, Morphological basis of bone mineral measurements: transient and steady state effects of treatment in osteoporosis, *Min. Electrolyte Metab.* **4**:273.

Parfitt, A.M., Matthews, C., Rao, D., Frame, B., Kleerekoper, M., and Villanueva, A.R., 1981, "Impaired osteoblast function in metabolic bone disease," in: *Osteoporosis, Recent Advances in Pathogenesis and Treatment*, (H.F. DeLuca, H.M. Frost, W.S.S. Jee, C.C. Johnston and A.M. Parfitt, eds.), pp. 321-30, University Park Press, Baltimore.

Parfitt, A.M., Gallagher, J.C., Heaney, R.P., Johnston, C.C., Neer, R., and Whedon, G.D., 1982, Vitamin D and bone health in the elderly, *Am. J. Clin. Nutr.* **36**:1014.

Preece, M.A., Tomlinson,S., Ribot, C.A., Pietrek, J., Korn, H.T., Davies, D.M., Ford, J.A., Dunnigan, M.G., and O'Riordan, J.L.H., 1975, Studies of vitamin D deficiency in men, *Quart. J. Med.* **44**:575.

Sherman, S.S., Hollis, B.W., and Tobin, J., 1990, Vitamin D status and related parameters in a healthy population: the effects of age, sex and season, *J. Clin. Endocrinol. Metab.* **71**:405.

Sokoloff, L., 1978, Occult osteomalacia in American (USA) patients with fracture of the hip, *Am. J. Surg. Pathol.* **2**:21.

Toss, G., Andersson, R., Diffey, B.L., Fall, P.A., Larkö, O., and Larsson, L., 1982, Oral vitamin D and ultraviolet radiation for the prevention of vitamin D deficiency in the elderly, *Acta Med. Scand.* **212**:157.

Toss, G., Larsson, L., and Lindgren, S., 1983, Serum levels of 25-hydroxyvitamin D in adults and elderly humans after a prophylactic dose of vitamin $D_2$, *Scand. J. Clin. Lab. Invest.* **43**:329.

Tsai, K.S., Heath, H., Kumar, R., and Riggs, B.L., 1984, Impaired vitamin D metabolism with aging in women; possible role in pathogenesis of senile osteoporosis, *J. Clin. Invest.* **73**:1668.

Villareal, D.T., Civitelli, R., Chines, A., and Avioli, L.V., 1991, Subclinical vitamin D deficiency in postmenopausal women with low vertebral bone mass, *J. Clin. Endocrinol. Metab.* **72**:628.

Vir, S.C., and Love, A.H.G., 1978, Vitamin D status of elderly at home and institutionalized in hospital, *Int. J. Vit. Nutr. Res.* **48**: 123.

Wicks, M., Garrett, R., Vernon-Roberts, B., and Fazzalari, N., 1982, Absence of metabolic bone disease in the proximal femur in patients with fracture of the femoral neck, *J. Bone Joint Surg. (Br.)* **64**:319.

Wilton, T.J., Hosking, D.J., Pawley, F., Stevens, A., and Harvey, I., 1987, Osteomalacia and femoral neck fractures in the elderly, *J. Bone Joint Surg. (Br.)* **69**:388.

Wootton, R., Brereton, P.J., Clark M.B., Hesp. R., Hodkinson, H.M., Klenerman L., Reeve, J., Slavin, G., and Tellez-Judilevich, M., 1979, Fractured neck of femur in the elderly: an attempt to identify patients at risk, *Clin. Sci.* **57**:93.

Chapter 10

# Protein Intake and Calcium Homeostasis

Jane E. Kerstetter and Lindsay H. Allen

## 1. Introduction

The earliest study documenting the relationship between dietary protein and urinary calcium was published 70 years ago. Sherman (1920) reported that an all-meat diet fed to humans increased urinary calcium. Twenty years later McCance *et al.* (1942) confirmed this observation by showing that peptones, gluten, gelatin or egg white added to the diet increased urinary loss of calcium. Another 20 years elapsed until Engstrom and DeLuca (1963) reported that a doubling of dietary protein induced excess urinary calcium excretion, negative calcium balance, and reduced bone ash in rats. Up to this time there was little concern over protein-induced hypercalciuria, because it was thought that dietary protein enhanced the intestinal absorption of calcium and this would offset the urinary calcium loss (Yuen *et al.*, 1984).

However, when it was recognized that protein did not influence intestinal calcium absorption under normal conditions (Shenolikar, 1974), this stimulated interest in obtaining a better understanding of the relationship between dietary protein and urinary calcium. Recent concern over osteoporosis has further fueled interest in nutritional regulators of calcium balance. The purpose of this chapter is to review the effect of dietary protein on urinary calcium excretion, calcium balance, bone density, and osteoporosis-related bone fractures.

---

Jane E. Kerstetter • School of Allied Health Professions, University of Connecticut, Storrs, CT 06269. Lindsay H. Allen • Department of Nutrition, University of California at Davis, Davis, CA 95616.

*Advances in Nutritional Research,* Vol. 9
Edited by Harold H. Draper
Plenum Press, New York, 1994

## 2. Magnitude of Protein-induced Urinary Calcium Loss

### 2.1. Longer Term Human Studies

Figure 1 summarizes 19 separate human studies and illustrates the relationship between dietary protein intake and urine calcium excretion in over 150 adult subjects (Johnson *et al.*, 1970; Walker and Linkswiler, 1972; Anand and Linkswiler, 1974; Allen *et al.*, 1979a; Schuette *et al.*, 1980; Hegsted *et al.*, 1981; Hegsted and Linkswiler, 1981; Spencer *et al.*, 1978; Spencer *et al.*, 1983; Margen *et al.*, 1974; Chu *et al.*, 1975; Kim and Linkswiler, 1979; Lutz and Linkswiler, 1981; Zemel *et al.*, 1981; Schuette *et al.*, 1981; Schuette and Linkswiler, 1982; Draper *et al.*, 1991; Lutz, 1984; Trilok and Draper, 1989). This figure summarizes data from all reports in which urinary calcium and dietary protein relationships were measured in adults. Each point on the graph represents the mean of a group of subjects from one of the above studies. All the studies were included in order to avoid bias, even though each had a unique protocol and sometimes differing conclusions concerning the hypercalciuretic effect of protein. Figure 1 is similar to one previously presented (Kerstetter and Allen, 1979) except for three differences. Study groups consuming more than 200 g protein daily are eliminated, three additional studies are included (Draper *et al.,* 1991; Lutz, 1984; Trilok and Draper, 1989) and phosphorus intakes are delineated.

In the 19 studies included in Figure 1, protein sources consisted of wheat gluten, beef, milk, egg, soy, lactalbumin or casein in the form of purified proteins or mixed foods, and study periods ranged from 12 to 60 days. Despite a wide variety of experimental protocols, a relationship between protein intake and urinary calcium becomes obvious. With protein intakes below 175 g/day, the relationship between dietary protein and urine calcium is linear (r = 0.70); for each 50 g increment of protein consumed, an extra 60 mg of urinary calcium is lost.

Some related studies were excluded from Figure 1, either because the actual values for urinary calcium were not reported or the subjects were not adults. Robertson *et al.* (1979) reported that an increase of 34 g of animal protein/day increased urinary calcium by 23%. Schwartz *et al.* (1973) demonstrated consistent hypercalciuria in adolescent boys consuming a high protein diet, although there was no change in calcium balance attributable to the level of protein intake. In a small group of renal stone-formers, increasing dietary protein from 57 to 142 g was associated with a 35% rise in 24-hour urinary calcium concurrent with elevations in urinary cAMP and hydroxyproline (Fellström *et al.,* 1984). The latter changes indicate that the higher protein diet stimulates PTH secretion and bone resorption. Both Brockis *et al.* (1982) and Heaney and Recker (1982) found a significant positive correlation between dietary protein intake and urinary calcium in subjects consuming their typical diet.

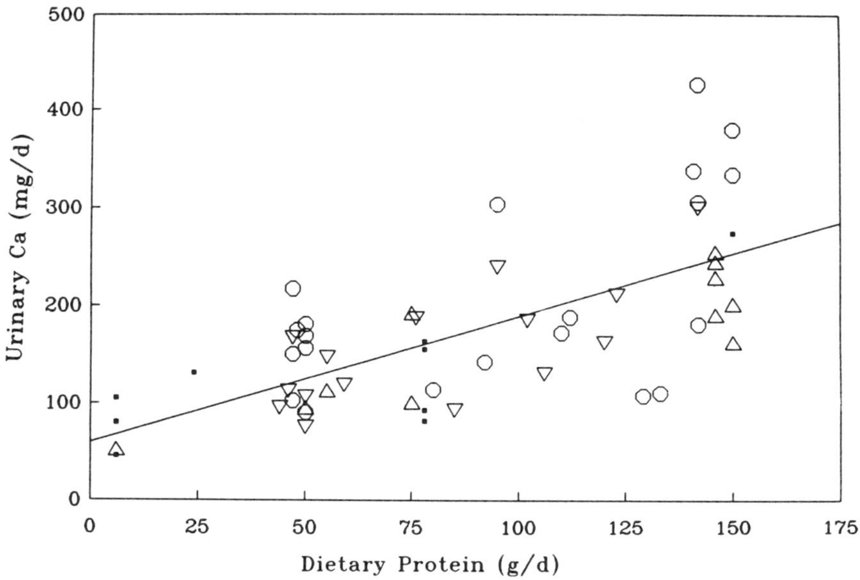


Figure 1. Relationship of dietary protein and phosphorus to urinary calcium. Each point represents the mean of a group of subjects from one of the 17 studies cited in the text. Phosphorus intakes are: ▿ < 1000 mg/day; ○ 1000–1500 mg/day; ▵ > 1500 mg/day; ■ unreported.

## 2.2. Short Term Human Studies

The renal response to dietary protein is rapid. Hypercalciuria is observed within 30 minutes after protein ingestion (Block *et al.,* 1980). Therefore, measurement of acute changes in an individual's urinary calcium after consuming protein has been a useful model for studying mechanisms of the calciuretic response. For example, this approach was used to examine the specific role of sulfur amino acids in protein on calciuria (Block *et al.,* 1980), to compare the calciuretic effects of meals high in protein vs. sucrose or starch (Holl and Allen, 1988), and to relate concomitant hormonal changes to the magnitude of urinary calcium response (Allen *et al.,* 1981). Administration of amino acids by total parenteral nutrition also increases urinary calcium. Bengoa *et al.* (1983) found that infusing amino acids at 2 g/kg vs. 1 g/kg increased calcium excretion in the urine from 287 mg/day to 455 mg/day.

## 2.3. Animal Studies

The rat has been used to study protein-induced hypercalciuria, although it is less than a perfect model. The amount of calcium excreted via the urine is very small, less than 1% of dietary calcium compared to about 20% in humans. Even though a calciuretic response to protein is evident, it is less likely to affect calcium homeostasis or bone mass. Bell *et al.* (1975) "deep labeled" bones with $^{45}Ca$ and fed diets varying in protein level. Radioactive calcium loss from bone was not dependent on dietary protein. The source of urinary calcium was increased intestinal absorption and a shift in route of excretion from feces to urine. Bone mineralization was not affected by a high level of dietary protein fed to young rats for 32 days (Allen and Hall, 1978) or for 10 months (Whiting and Draper, 1981). Yet Engstrom and DeLuca (1963) reported a decreased ash content in bone when dietary egg white protein was increased from 18% to 36% in rats. In comparison to humans, rats are better able to increase acid excretion as ammonium ions, which may contribute to their ability to withstand an acid load (Trilok and Draper, 1989).

The acute response of rats to a high protein diet appears to be similar to that of humans, although the hypercalciuretic response disappears with time. In a short term study in rats by Wood *et al.* (1991), a 38% protein diet increased urinary calcium by 72% in comparison to the 18% protein diet. Allen and Hall (1978) found that in 200 g rats doubling dietary casein from 18% to 36% increased urinary calcium from 0.7 to 1.7 mg/day at 3 days but the increase was only 0.4 mg/day at 14 days. At 29 days, urinary calcium had returned to the level of control diets. Calvo *et al.* (1982) reported a marked but transient calciuretic response in 7-month old rats fed a 24% vs. a 6% protein diet. After 56 days, urinary calcium returned to control levels indicating adaptation. In an 8-week experiment conducted by Whiting and Draper (1980), the peak response in calcium excretion occurred at 2 days when adult male rats were fed a 36% protein diet (lactalbumin, egg white, casein or gelatin) compared to an 18% protein diet. Subsequently urinary calcium declined, but persisted to a moderate degree until the end of the long experiment.

There are several rodent studies in which feeding high protein caused persistent hypercalciuria, even though the age of the animals and protein levels were similar to those in the experiments where adaptation occurred. Bell *et al.* (1975) reported a hypercalciuretic response in rats through 9–10 weeks. At 10 weeks, animals receiving 40% dietary protein excreted approximately 10 times more calcium than those fed 10% protein. Whiting and Draper (1981) found that urine calcium peaked at 2 days and moderate hypercalciuria persisted for 8 weeks when 35% vs. 15% protein diets were fed. And finally, chronic hypercalciuria lasted for 52 weeks in mice consuming a 30% vs. a 15% protein diet (Yuen and Draper, 1983) but, again, bone calcium content was unaffected.

## 3. Mechanism Of Protein-Induced Renal Calcium Loss

Urinary calcium excretion is a function of the calcium entering the ultrafiltrate vs. the amount that is subsequently reabsorbed by the nephron. The amount of calcium entering the ultrafiltrate is dependent primarily upon serum ionized calcium concentration and glomerular filtration rate (GFR). Dietary protein has no measurable influence on serum ultrafiltrable calcium (Schuette *et al.*, 1980; Hegsted and Linkswiler, 1981; Allen *et al.*, 1979b). However, dietary protein does significantly increase GFR up to 20% in humans (Schuette *et al.*, 1980; Kim and Linkswiler, 1979; Allen *et al.,* 1979a; Hegsted and Linkswiler, 1981; Hegsted *et al.,* 1981; Zemel *et al.,* 1981) within several hours following a meat meal (Hostetter, 1986). When GFR is increased by 20% above an average of 100 ml/min (assuming constant serum ultrafiltrable calcium), an extra 1.7 g of calcium enters the ultrafiltrate per day. While GFR is contributory, it cannot totally explain the calciuretic response, particularly in short-term studies. Allen *et al.* (1979b) showed that a high protein meal can cause an immediate hypercalciuria without a change in GFR or serum calcium.

Protein consumption results in a reduction of renal calcium reabsorption (Allen *et al.* 1979b; Schuette *et al.* 1980). Frequently, the decrease in renal fractional reabsorption appears to be minuscule (approximately 1%), but considering that the total amount of calcium filtered by the human kidney is 10 g/day, a 1% loss is substantial. There are several possible mechanisms by which dietary protein may decrease renal calcium reabsorption.

The most substantiated explanation is related to the acid generated from protein metabolism. Oxidation of the sulfur amino acids, methionine and cysteine, yields inorganic sulfate ($SO_4^{-2}$) and two hydrogen ions ($H^+$). Oxidation of excessive sulfur amino acids accounts for the increase in acid production when a high protein diet is fed (Trilok and Draper, 1989). Sulfate is excreted in the urine and the hydrogen ions combine with ammonia and are excreted in the form of ammonium. Minor amounts of hydrogen ions are excreted as dihydrogen phosphate. Metabolic acidosis, regardless of the cause, is associated with elevated urinary calcium (Lemann *et al.*, 1967). Likewise, the addition of a base reduces urinary calcium losses (Lutz, 1984). Evidence suggests that between half and all of the urinary calcium excreted following a high protein diet can be attributed to the acid load (Trilok and Draper, 1989; Schuette *et al.*, 1981).

Sulfate may inhibit calcium reabsorption by forming a weak complex with calcium in the renal filtrate (Walser and Browder, 1959). Approximately 12% of urinary calcium is normally bound to sulfate (Robertson, 1969). Urinary sulfate and calcium excretion are closely related under a variety of experimental conditions (Hunt, 1956; Lemann and Relman, 1959; Breslau *et al.*, 1988; Tschöpe and Ritz, 1985). In rats, the calciuretic response to dietary protein is proportional to the sulfur amino acid content of the diet: lactalbumin > egg white

> casein > gelatin (Whiting and Draper, 1980). Similarly, soy protein foods that are low in sulfur-containing amino acids help to preserve calcium equilibrium despite high protein and low calcium intakes (Zemel, 1988).

It is not clear that the sulfur and acid loads from protein ingestion fully explain the calciuretic response. Margen *et al.* (1974) were unable to detect any difference in calcium excretion from various amino acid mixtures attributable to their various levels of sulfur amino acids. In an experiment by Block *et al.* (1980), meals contained either 15 g protein, 45 g protein (high protein), or 15 g protein plus sulfur amino acids equivalent to those in the high protein diet. The sulfur amino acid supplement had no effect on calcium excretion or reabsorption. Schuette *et al.* (1981) studied calciuresis in adults using three dietary treatments: 50 g protein diet; 150 g protein diet; and a 50 g protein diet supplemented with sulfur amino acids to equal the 150 g diet. Urinary calcium doubled between the first two treatments. The third diet elevated urinary calcium, but by only 43% of the increase caused by feeding the high protein diet alone. Zemel *et al.* (1981) reported that sulfur amino acid supplements added to a low protein diet increased urinary calcium. However, the change was only half of the hypercalciuria caused by feeding the high protein diet.

Circulating levels of 1,25-dihydroxyvitamin D do not change in response to dietary protein level, nor does parathyroid hormone (Allen *et al.*, 1979a; Kim and Linkswiler, 1979; Schuette *et al.*, 1980; Schuette *et al.*, 1981). In fact, a calciuretic response to arginine infusion was observed even in parathyroidectomized rats (Gollaher *et al.*, 1984). Dietary protein and amino acids are well known insulin secretagogues (Floyd *et al.*, 1966). From animal and human experimentation there is some evidence that serum insulin may be involved in the mechanism of protein-induced hypercalciuria (Allen *et al.*, 1981). The strongest evidence for this is that the hypercalciuretic effect of arginine infusion is blocked by streptozotocin inhibition of insulin secretion in rats (Wood and Allen, 1983).

## 4. Protein Intake and Calcium Balance

The calciuretic response of humans to dietary protein does not diminish with time. Johnson *et al.* (1970), Allen *et al.* (1979a) and Hegsted and Linkswiler (1981) showed sustained hypercalciuria through experimental periods lasting 45, 48 and 60 days, respectively. It is therefore not surprising that protein-induced hypercalciuria results in negative calcium balance, especially given the lack of evidence for an accompanying response in intestinal absorption (Wood *et al.*, 1991). Figure 2 illustrates calcium balance data from 15 separate studies in which dietary protein was manipulated in more than 100 human adults (Johnson *et al.*, 1970; Walker and Linkswiler, 1972; Anand and Linkswiler, 1974; Allen *et al.*, 1979a; Schuette *et al.*, 1980; Hegsted *et al.*, 1981; Hegsted and Linkswiler, 1981; Spencer *et al.*, 1983; Spencer *et al.,* 1978; Chu *et al.*, 1975; Kim and Linkswiler, 1979; Lutz and Linkswiler, 1981; Schuette and Linkswiler, 1982; Draper *et al.*, 1991; Lutz, 1984). At low protein intakes (25–74 g/day), calcium balance is close to equilibrium when calcium intake is between 500 and 1400 mg/day. For healthy individuals weighing less than 90 kg, the recommended dietary allowance (RDA) for protein is included in this "low protein" range. Protein intakes between 75 and 124 g/day, typical for most adult Americans, resulted in negative calcium retention in 10 of the 12 group averages reported.

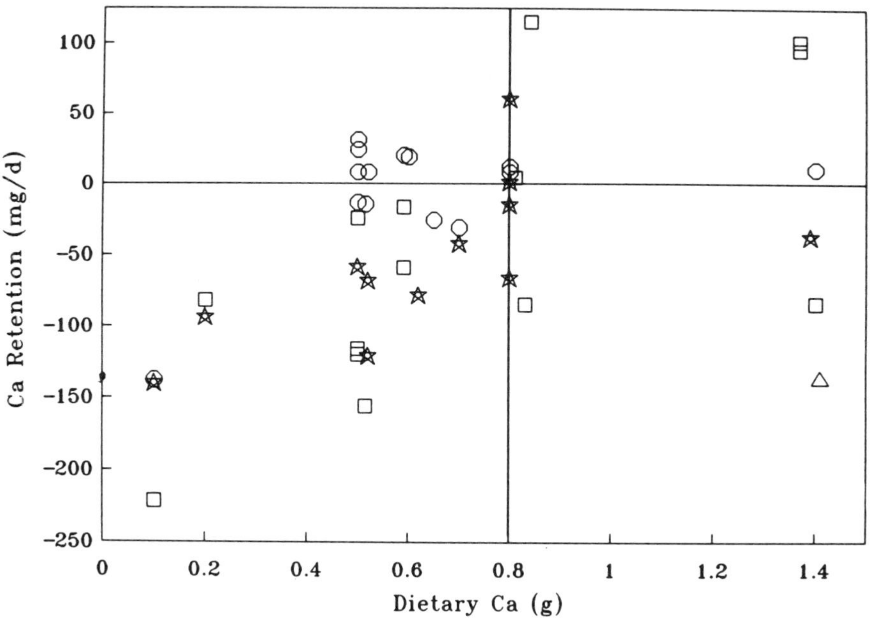


Figure 2. Relationship of dietary protein and calcium intakes to calcium retention. Each point represents the mean of a group of subjects from one of the studies in the text. Protein intakes are ○ 25–74 g/day; ☆ 75-124 g/day; □ 125-174 g/day; △ >175 g/day.

Surprisingly, additional dietary calcium has a minimal effect on calcium balance when protein intake is high. Of the 27 group means when protein intake was above 75 g/day (stars, boxes, and triangles, Figure 2), only 6 were in positive calcium balance or equilibrium. Clearly, when protein intake was higher than 75 g/day and calcium intake less than 600 mg/day, negative balance was inevitable. In half the experiments in Figure 2 intakes as high as 1400 mg calcium failed to restore calcium balance when dietary protein was high. Margen *et al.* (1974) concluded that varying protein intake between 0 and 560 g/day resulted in an 8-fold increase in urinary calcium that was not reversible by calcium supplementation up to 2300 mg/day.

Is the negative calcium balance seen in these carefully controlled experimental conditions similar to that of average adults consuming a mixed diet? Heaney and Recker (1982) studied calcium balance in 170 healthy middle-aged women every 5 years. Notably, diets during the 8-day balance studies were constructed to match the subjects' usual intake of protein, phosphorus and calcium. For each 50% increase in dietary nitrogen, there was a shift in calcium balance of -32 mg/day. Brockis *et al.* (1982) also reported a significant correlation between usual protein intake and urinary calcium excretion.

Phosphorus clearly decreases urinary calcium losses (Hegsted *et al.*, 1981; Schuette *et al.*, 1981). Meats and dairy foods are high in both protein and phosphorus, so the concomitant intake of phosphorus negates or blunts the calciuretic response to protein. The degree to which this occurs, however, is unclear. Spencer and colleagues have shown repeatedly that the addition of dietary red meat or dairy products fails to affect urinary calcium because of their high phosphorus content (Spencer *et al.*, 1978; Spencer *et al.,* 1983; Spencer *et al.*, 1988). For example, a high protein intake (2 g/kg·day) from meat consumed over several weeks had no effect on the urinary calcium of 14 adult males (Spencer *et al.*, 1978). In a later study, Spencer *et al.* (1983) measured calcium balance in seven adults fed two levels of dietary protein, primarily as meat. Urinary and fecal excretion of calcium did not differ between low and high protein groups over a period of 132 days, nor was negative calcium balance associated with high protein intake. From Figure 1, it appears that phosphorus intakes above 1500 mg/day reduce calciuria in comparison to phosphorus intakes between 1000 and 1500 mg/day, when dietary protein is near 150 g/day. However, this relationship is not as prominent when protein intake is less than 150 g/day.

Although increasing dietary phosphorus blunts protein-induced hypercalciuria, resulting in improved calcium retention, an ability to restore calcium balance is not evident from the data shown in Figure 3. A more positive calcium retention would be expected at the higher phosphorus intakes. Hegsted *et al.* (1981) measured calcium balance in adults fed two levels of dietary protein (50 and 150 g) and phosphorus (1010 and 2525 mg). At both levels of protein intake, phosphorus reduced urinary calcium losses by approximately 40% and shifted calcium balance in a positive direction. Yet, calcium balance remained negative (-25 mg/day) on the high protein, high phosphorus diet. There are other examples of a protein and phosphorus load given as animal flesh being associated

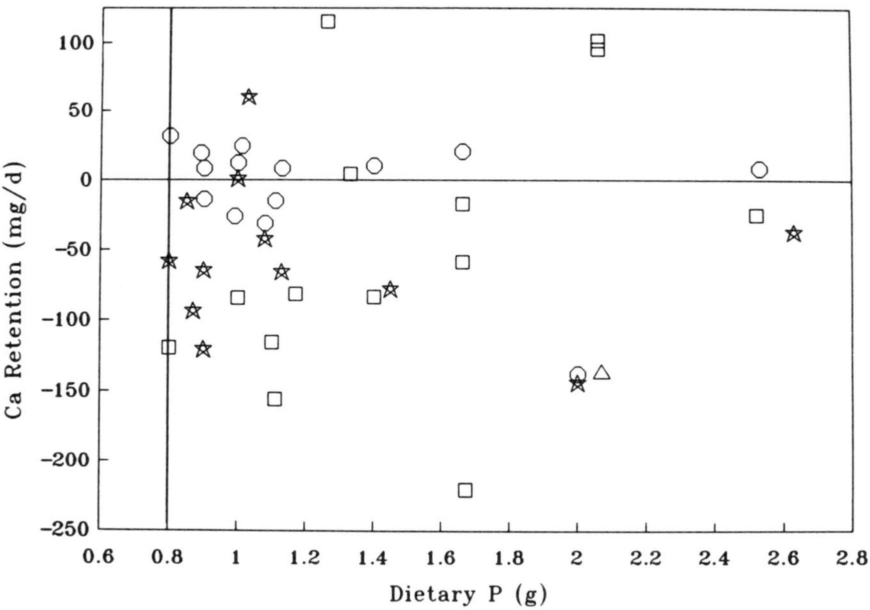


Figure 3. Relationship of dietary protein and phosphorus intakes to calcium retention. Each point represents the mean of a group of subjects from one of the 15 studies in Figure 2. Protein intakes are ○ 25–74 g/day; ☆ 75–124 g/day; □ 125-174 g/day; △ >175 g/day.

with small but significant increases in urine calcium losses (Robertson *et al.*, 1979; Brockis *et al.*, 1982; Cummings *et al.*, 1979; Breslau *et al.*, 1988; Fellström *et al.*, 1984).

The average adult body contains approximately 1,000 to 1,200 g calcium, 99% of which resides in the skeleton. The remaining 1% regulates vital functions (blood clotting, nerve excitation, intracellular signalling, secretory processes, membrane permeability and muscle contractions) and it is meticulously regulated to maintain stability. Therefore, the source of calcium lost during periods of negative calcium balance is ultimately bone. A daily negative balance of 25–30 mg represents a loss of 10 g of calcium per year, equivalent to 1% of body calcium annually or 10% over a decade. This daily loss of calcium may manifest itself as accelerated bone loss. Because protein intake from the typical Western diet is almost twice the RDA, this diet may constitute a risk factor for osteoporotic bone disease (Yuen *et al.*, 1984; Heaney and Recker, 1982; Allen *et al.*, 1979a; Abelow *et al.*, 1992; Hegsted, 1986).

## 5. Dietary Protein and Bone Density

If this proposal is correct, populations that consistently consume high protein diets should have a higher prevalence of bone disease and osteoporotic fractures. In some studies, but not all (Ellis *et al.* 1974; Sanchez *et al.* 1980), bone loss in older vegetarians was reported to be less than that of omnivores. It is often assumed that one explanation for this difference is a lower consumption of protein by vegetarians; however, in most studies this was not the case (Marsh *et al.*, 1988). In others, protein intake was not measured (Ellis *et al.* 1972; Marsh *et al.,* 1980). In addition, a difference in meat and calcium content is only one difference between lacto-ovo-vegetarians (frequently Seventh Day Adventists) and omnivores. These groups may also have distinctly different physical activity patterns, genetic backgrounds, and intakes of other dietary constituents that affect bone density.

From an international perspective, most studies do indicate a relationship between bone health and protein intake. An exception is an early study by Garn *et al.* (1969), in which no difference was found in the rate of metacarpal bone loss between populations in Europe, Asia, and Central and South America, despite presumed differences in protein intake. In contrast, several more recent cross-cultural studies have provided evidence of a relationship between usual protein intake and hip fracture rate. Abelow *et al.* (1992) have published the most convincing epidemiological evidence for this association. They performed a meta-analysis of 34 publications on the incidence of hip fractures in women over 50 years of age in 16 countries. Estimates of animal protein intake were calculated from Food Balance Sheets published by the Food and Agriculture Organization of the United Nations. There was a strong positive association between animal protein intake and age-adjusted female hip fracture rates (Figure 4). The more industrialized countries (United States, northern and western Europe) had higher protein intakes and hip fracture rates than less industrialized countries in Asia and Africa. This observation agrees with the earlier analysis of Hegsted (1986) in which, again, epidemiological data were cited indicating a positive relationship between hip fractures and protein intake.

## 6. Conclusions

There is little doubt that increasing dietary protein in humans across a range of zero up to several hundred grams per day stimulates urinary calcium excretion. The calciuretic response is extremely rapid and persists for at least several months. The hypocalciuretic effect of dietary phosphorus from milk and meat significantly blunts, but does not eliminate, the hypercalciuretic effect of protein. Increasing dietary calcium intake alone does little to compensate for the higher urinary calcium losses. There is no single explanation for protein-induced hypercalciuria and multiple factors are probably involved: changes in GFR, renal calcium reabsorption, acid-base balance, sulfate excretion, and serum insulin level. Long term calcium balance studies indicate that high protein intakes increase urinary calcium losses to the point that negative calcium balance ensues. Calcium absorption from the gastrointestinal tract does not improve or offset these losses.

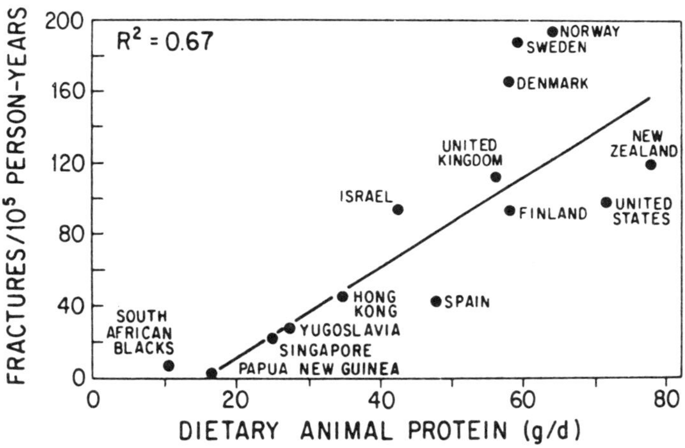


Figure 4. Plot of age-adjusted hip fracture incidence in women over 50 years of age vs. estimated per capita dietary protein intake. Reproduced with permission from Abelow *et al.* (1992).

Epidemiological evidence for a relationship between bone fractures and usual level of dietary protein is consistent with the adverse effect of protein on calcium balance. However, meta-analysis of cross-cultural data indicates that there are other differences between populations that may explain, at least in part, the relationship between protein intake and bone loss.

Based on this evidence, from the perspective of bone health, increasing protein intake for perceived health benefits should be discouraged. The average adult protein intake in industrialized countries is already approximately twice the level recommended for prevention of protein deficiency. Amino acid or protein supplements promoted to young athletes with promises of enhanced performance have the potential to reduce calcium accretion. Because this age group should still be laying down large quantities of bone calcium, a high protein diet may compromise peak bone mass and increase risk for bone fractures later in life.

It is concluded that there is a *mean* increase in urinary calcium by adults of 60 mg/day for each 50 g increment in dietary protein. High variability in the calciuretic response to dietary protein indicates that some individuals incur a much greater negative calcium balance and bone loss. Whether individuals may benefit from lowering their usual protein intake needs to be investigated in a prospective clinical trial. However, there is already sufficient evidence to consider high dietary protein among the risk factors for osteoporosis.

## References

Abelow, B.J., Holford, T.R., and Insogna, K.L., 1992, Cross-cultural association between dietary animal protein and hip fracture: a hypothesis, *Calcif. Tissue Int.* **50:**14.

Allen, L.H., and Hall, T.E., 1978, Calcium metabolism, intestinal calcium-binding protein, and bone growth of rats fed high protein diets, *J. Nutr.* **108:**967.

Allen, L.H., Oddoye, E.A., and Margen, S., 1979a, Protein-induced hypercalciuria: a longer term study, *Am. J. Clin. Nutr.* **32:**741.

Allen, L.H., Bartlett, R.S., and Block, G.D., 1979b, Reduction of renal calcium reabsorption in man by consumption of dietary protein, *J. Nutr.* **109:**1345.

Allen, L.H., Block, G.D., Wood, R.J., and Bryce, G.F., 1981, The role of insulin and parathyroid hormone in the protein-induced calciuria of man, *Nutr. Res.* **1:**3.

Anand, C.R., and Linkswiler, H.M., 1974, Effect of protein intake on calcium balance of young men given 500 mg calcium daily, *J. Nutr.* **104:**695.

Bell, R.R., Engelmann, D.T., Sie, T.-L., and Draper, H.H., 1975, Effect of a high protein intake on calcium metabolism in the rat, *J. Nutr.* **105:**475.

Bengoa, J.M., Sitrin, M.D., Wood, R.J., and Rosenberg, I.H., 1983, Amino acid-induced hypercalciuria in patients on total parenteral nutrition, *Am. J. Clin. Nutr.* **38:**264.

Block, G.D., Wood, R.J., and Allen, L.H., 1980, A comparison of the effects of feeding sulfur amino acids and protein on urine calcium in man, *Am. J. Clin. Nutr.* **33:**2128.

Breslau, N.A., Brinkley, L., Hill, K.D., and Pak, C.Y.C., 1988, Relationship of protein-rich diet to kidney stone formation and calcium metabolism, *J. Clin. Endocrinol. Metab.* **66:**140.

Brockis, J.G., Levitt, A.J., and Cruthers, S.M., 1982, The effects of vegetable and animal protein diets on calcium, urate and oxalate excretion, *Br. J. Urol.* **54:** 590.

Calvo, M.S., Bell, R.R., and Forbes, R.M., 1982, Effect of protein-induced calciuria on calcium metabolism and bone status in adult rats, *J. Nutr.* **112:**1401.

Chu, J.-Y., Margen, S., and Costa, F.M., 1975, Studies in calcium metabolism. II. Effects of low calcium and variable protein intake on human calcium metabolism, *Am. J. Clin. Nutr.* **28:**1028.

Cummings, J.H., Hill, M.J., Jivraj, T., Houston, H., Branch, W.J., and Jenkins, D.J.A., 1979, The effect of meat protein and dietary fiber on colonic function and metabolism. I. Changes in bowel habit, bile acid excretion, and calcium absorption, *Am. J. Clin. Nutr.* **32:**2086.

Draper, H.H., Piché, L.A., and Gibson R.S., 1991, Effects of a high protein intake from common foods on calcium metabolism in a cohort of postmenopausal women, *Nutr. Res.* **11**:273.

Ellis, F.R., Holesh, S., and Ellis, J.W., 1972, Incidence of osteoporosis in vegetarians and omnivores, *Am. J. Clin. Nutr.* **25**:555.

Ellis, F.R., Holesh, S., and Sanders, T.A.B., 1974, Osteoporosis in British vegetarians and omnivores, *Am. J. Clin. Nutr.* **27**:769.

Engstrom, G.W., and DeLuca, H.F., 1963, Effect of egg white diets on calcium metabolism in the rat, *J. Nutr.* **81**:218.

Fellström, B., Danielson, B.G., Karlström, B., Lithell, H., Ljunghall, S., Vessby, B., and Wide, L., 1984, Effects of high intake of dietary animal protein on mineral metabolism and urinary supersaturation of calcium oxalate in renal stone formers, *Br. J. Urol.* **56**:263.

Floyd, J.C., Fajans, S.S., Conn, J.W., Knopf, R.F., and Rull, J., 1966, Insulin secretion in response to protein ingestion, *J. Clin. Invest.* **45**:1479.

Garn, S.M., Robmann, C.G., Wagner, B., Dairla, H.G., and Ascoli, W., 1969, Population similarities in the onset and rate of adult endosteal bone loss, *Clin. Orth. Rel. Res.* **65**:51.

Gollaher, C.J., Wood, R.J., Holl, M.H., and Allen, L.H., 1984, A comparison of amino acid-induced hypercalciuria in sham-operated and parathyroidectomized rats, *J. Nutr.* **114**:622.

Heaney, R.P., and Recker, R.R., 1982, Effects of nitrogen, phosphorus, and caffeine on calcium balance in women, *J. Lab. Clin. Med.* **99**:46.

Hegsted, D.M., 1986, Calcium and osteoporosis, *J. Nutr.* **116**:2316.

Hegsted, M., and Linkswiler, H.M., 1981, Long-term effects of level of protein intake on calcium metabolism in young adult women, *J. Nutr.* **111**:244.

Hegsted, M., Schuette, S.A., Zemel, M.B., and Linkswiler, H.M., 1981, Urinary calcium and calcium balance in young men as affected by level of protein and phosphorus intake, *J. Nutr.* **111**:553.

Holl, M.G., and Allen, L.H., 1988, Comparative effects of meals high in protein, sucrose, or starch on human mineral metabolism and insulin secretion, *Am. J. Clin. Nutr.* **48**:1219.

Hostetter, T.H., 1986, Human renal response to a meat meal, *Am. J. Physiol.* **250**:F613.

Hunt, J.N., 1956, The influence of dietary sulfur on the urinary output of acid in man, *Clin. Sci.* **5**:119.

Johnson, N.E., Alcantara, E.N., and Linkswiler, H.M., 1970, Effect of level of protein intake on urinary and fecal calcium and calcium retention in young adult males, *J. Nutr.* **100**:1425.

Kerstetter, J.E., and Allen, L.H., 1988, Dietary protein increases urinary calcium, *J. Nutr.* **120**:134.

Kim, Y., and Linkswiler, H.M., 1979, Effect of level of protein intake on calcium metabolism and on parathyroid and renal function in the adult human male, *J. Nutr.* **109**:1399.

Lemann, J., and Relman, A.S., 1959, The relationship of sulfur metabolism to acid-base balance and of electrolytes excretion: the effects of DL-methionine in normal man, *J. Clin. Invest.* **38**:2215.

Lemann, J., Litzow, J.R., and Lennon, E.J., 1967, Studies of the mechanisms by which chronic metabolic acidosis augments urinary calcium excretion in man, *J. Clin. Invest.* **46**:1318.

Lutz, J., 1984, Calcium balance and acid-base status of women as affected by increased protein intake and by sodium bicarbonate ingestion, *Am. J. Clin. Nutr.* **39**:281.

Lutz, J., and Linkswiler, H.M., 1981, Calcium metabolism in postmenopausal and osteoporotic women consuming two levels of dietary protein, *Am. J. Clin. Nutr.* **34**:2178.

Margen, S., Chu, J.-Y., Kaufmann, N.A., and Calloway, D.H. 1974, Studies in calcium metabolism. I. The calciuretic effect of dietary protein, *Am. J. Clin. Nutr.* **27**:584.

Marsh, A.G., Sanchez, T.V., Mickelson, O., Keiser, J., and Mayor, S., 1980, Cortical bone density of adult lacto-ovo-vegetarian and omnivorous women, *J. Am. Diet. Assoc.* **76**:148.

Marsh, A.G., Sanchez, T.V., Mickelson, O., Chaffee, F.L., and Fagal, S.M., 1988, Vegetarian lifestyle and bone mineral density, *Am. J. Clin. Nutr.* **48**:837.

McCance, R.A., Widdowson, E.M., and Lehmann, H., 1942, The effect of protein intake on the absorption of calcium and magnesium, *Biochem. J.* **36**:686.

Robertson, W.G., 1969, Measurement of ionized calcium in biological fluids, *Clin. Chim. Acta.* **24**:149.

Robertson, W.G., Heyburn, P.J., Peacock, M., Hanes, F.A., and Swaminathan, R., 1979, The effect of high animal protein intake on the risk of calcium stone-formation in the urinary tract, *Clin. Sci.* **57**:285.

Sanchez, T.V., Mickelsen, O., Marsh, A.G., Garn, S.M., and Mayor, G.H., 1980, "Bone mineral mass in elderly vegetarian females and omnivorous females," in: *Proceedings of the Fourth International Conference on Bone Measurement* (R.B. Mazess, ed.), Bethesda MD: NIAMDD (NIH Publication no. 801938).

Schuette, S.A., and Linkswiler, H.M., 1982, Effects on Ca and P metabolism in humans by adding meat, meat plus milk, or purified proteins plus Ca and P to a low protein diet, *J. Nutr.* **112**:338.

Schuette, S.A., Zemel, M.B., and Linkswiler, H.M., 1980, Studies on the mechanism of protein-induced hypercalciuria in older men and women, *J. Nutr.* **110**:305.

Schuette, S.A., Hegsted, M., Zemel, M.B., and Linkswiler, H.M., 1981, Renal acid, urinary cyclic AMP, and hydroxyproline excretion as affected by level of protein, sulfur amino acid, and phosphorus intake, *J. Nutr.* **111**:2106.

Schwartz, R., Woodcock, N.A., Blakely, J.D., and MacKellar, I., 1973, Metabolic responses of adolescent boys to two levels of dietary magnesium and protein. II. Effect of magnesium and protein level on calcium balance, *Am. J. Clin. Nutr.* **26**:519.

Shenolikar, I.S., 1974, Protein nutrition and calcium absorption, *Nutr. Metab.* **16**:10.

Sherman, H.C., 1920, Calcium requirement of maintenance in men, *J. Biol. Chem.* **44**:21.

Spencer, H., Kramer, L., Osis, D., and Norris, C., 1978, Effect of a high protein (meat) intake on calcium metabolism in man, *Am. J. Clin. Nutr.* **31**:2167.

Spencer, H., Kramer, L., DeBartolo. M., Norris, C., and Osis, D., 1983, Further studies of the effect of a high protein diet as meat on calcium metabolism, *Am. J. Clin. Nutr.* **37**:924.

Spencer, H., Kramer, L., and Osis, D., 1988, Do protein and phosphorus cause calcium loss? *J. Nutr.* **118**:657.

Trilok, G., and Draper, H.H., 1989, Effect of high protein on acid-base balance in adult rats, *Calcif. Tissue Int.* **44**:339.

Trilok, G., and Draper, H.H., 1989, Sources of protein-induced endogenous acid production and excretion by human adults, *Calcif. Tissue Int.* **44**:335.

Tschöpe, W., and Ritz, E., 1985, Sulfur-containing amino acids are a major determinant of urinary calcium, *Min. Electrolyte Metab.* **11**:137.

Walker, R.M., and Linkswiler, H.M., 1972, Calcium retention in the adult human male as affected by protein intake, *J. Nutr.* **102**:1297.

Walser, M., and Browder, A.A., 1959, Ion association. II. The effect of sulfate infusion on calcium excretion, *J. Clin. Invest.* **38**:1404.

Whiting, S.J., and Draper, H.H., 1980, The role of sulfate in the calciuria of high protein diets in adult rats, *J. Nutr.* **110**:212.

Whiting, S.J., and Draper, H.H., 1981, Effect of chronic high protein feeding on bone composition in the adult rat, *J. Nutr.* **111**:178.

Wood, R.J., and Allen, L.H., 1983, Evidence for insulin involvement in arginine- and glucose-induced hypercalciuria in the rat, *J. Nutr.* **113**:1561.

Wood, R.J., Contois, J., and Wilkening, C., 1991. Intestinal adaptation to dietary calcium restriction in rats fed high protein diets, *Nutr. Res.* **11**:831.

Yuen, D.E., and Draper, H.H., 1983, Long-term effects of excess protein and phosphorus on bone homeostasis in adult mice, *J. Nutr.* **113**:1374.

Yuen, D.E., Draper, H.H., and Trilok G., 1984, Effect of dietary protein on calcium metabolism in man, *Nutr. Abst. Rev.* **54**:447.

Zemel, M.B., 1988, Calcium utilization: effect of varying level and source of dietary protein, *Am. J. Clin. Nutr.* **48**:880.

Zemel, M.B., Schuette, S.A., Hegsted, M., and Linkswiler, H.M., 1981, Role of the sulfur-containing amino acids in protein-induced hypercalciuria in men, *J. Nutr.* **111**:545.

Chapter 11

# The Effects of High Phosphorus Intake on Calcium Homeostasis

Mona S. Calvo

## 1. Introduction

Osteoporosis and related bone fractures are recognized as a serious cause of morbidity and mortality, largely among elderly women. The most preventable cause of fractures is low bone mass (Riggs and Melton, 1992), which is thought to be dependent on both the current rate of bone loss and peak bone mass (i.e., the amount of bone present at skeletal maturity) (Riggs and Melton, 1986). The most cost-effective approach to reducing the risk of osteoporosis is to maximize peak bone mass by optimizing bone accretion during teen and early adult life and later to slow the rate of bone loss with increasing age (Riggs and Melton, 1986, 1992; Ott, 1990; Matkovic *et al.,* 1990).

To implement this approach, we need a better understanding of the environmental factors that influence peak bone mass in order to enable young women to achieve their full genetic potential for optimal bone mass. While gender, race, genetics and other factors such as body habitus and hormonal influences are clearly important forces governing peak bone mass, certain dietary factors are also important (Heaney, 1988; Matkovic and Dekanic, 1989). There is growing

Mona S. Calvo • Department of Health and Human Services, Public Health Service, Food and Drug Administration, Washington DC 20204.

*Advances in Nutritional Research,* Vol. 9
Edited by Harold H. Draper
Plenum Press, New York, 1994

evidence that adequate dietary calcium influences peak bone mass during adolescence and early adulthood (Matkovic *et al.*, 1990; Matkovic, 1991). Others have shown that supplemental calcium can reduce the rate of bone loss in premenopausal women (Baran *et al.,* 1990) and in peri- and postmenopausal women (Elders *et al.,* 1991; Dawson-Hughes *et al.,* 1990; Reid *et al.,* 1993). However, before safe, effective, and inexpensive strategies can be developed to maximize peak bone mass and minimize subsequent bone loss, we need to understand how other nutrients in the existing dietary pattern influence bone mass. Some nutrients, either alone or in combination with low calcium intake, may impair peak bone mass or increase the rate of bone loss later in life. This chapter focuses on the theoretical and experimental evidence that the high phosphorus content of the diet of of industrialized countries may impede the achievement of maximal bone mass and accelerate the rate of bone loss when calcium intake is low.

## 2. Current Dietary Patterns

The dietary pattern of many American women is typically high in phosphorus relative to calcium. Concern for this dietary pattern stems from studies showing that high phosphorus, low calcium intakes cause secondary hyperparathyroidism and increased bone resorption, ultimately leading to osteoporosis in a number of animal models. National nutrition surveys show a large discrepancy between actual dietary intake of calcium and phosphorus and the recommended levels of intake for these minerals (Human Nutrition Information Service, 1983; Life Sciences Research Office, 1989). This point is illustrated in Figure 1, where data from the National Health and Nutrition Examination Survey II, 1976-80 (NHANES II) are compared to the 1989 recommended dietary allowance (RDA) for calcium and phosphorus at all ages (National Research Council, 1989; Calvo, 1993). For men, mean calcium intake remains near the RDA over age, while phosphorus intake exceeds the RDA for all ages. In contrast, mean calcium intake for women falls below the RDA as early as 12 years of age and continues to decrease with age, while phosphorus intake exceeds the RDA for most age groups. Another important point is that women 16 to 25 years of age who are still actively accreting bone have a considerable imbalance in their calcium and phosphorus intakes.

This imbalance between calcium and phosphorus intake raises an important question. Should primary concern focus on the absolute level of dietary calcium or on the dietary ratio of calcium to phosphorus (Ca:P)? Most investigators believe that the absolute level of calcium is the most critical factor, largely because less phosphorus is required to exacerbate bone loss when dietary calcium is limited. Thus, concern about the effects of a high phosphorus diet is particularly justified when the calcium intake is low. According to the 1989 RDA for children and adults, the desired Ca:P ratio of the American diet is 1:1, but the ratio of the actual food consumed is quite different (Draper and Scythes, 1981; Albanese *et al.,* 1986).

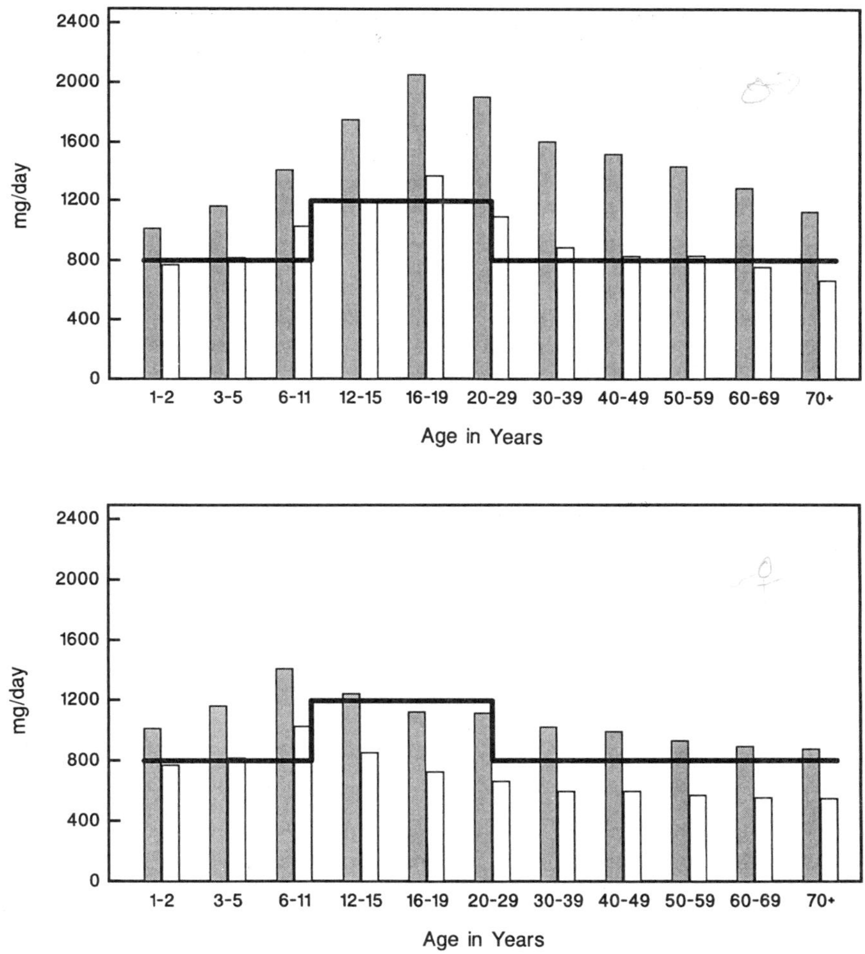


Figure 1. The mean phosphorus intake (shaded bars) and mean calcium intake (open bars) of men (upper panel) and women (lower panel) determined in the National Health and Nutrition Examination Survey II, 1976–80, plotted as a function of age. In both panels, the recommended dietary allowance (RDA) for calcium and phosphorus for various age groups is represented by the horizontal line. Reprinted from Calvo (1993) with permission.

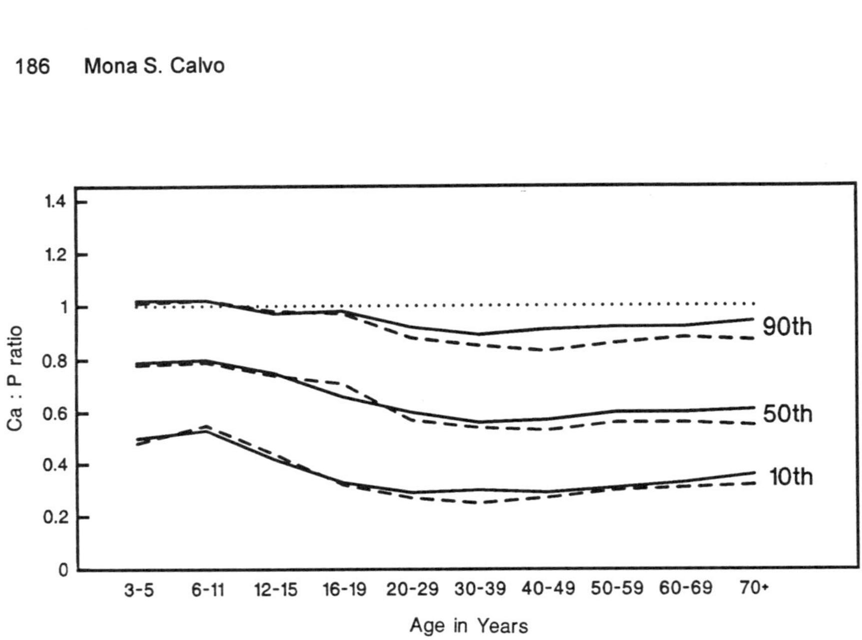


Figure 2. The 10th, 50th, and 90th percentiles of Ca:P ratios determined from daily intake as a function of age for men (---) and women (—). The Ca:P ratios were determined from 24–hr dietary recall data from the National Health and Nutrition Examination Survey II, 1976–80, by Dr. Anne Looker, National Center for Health Statistics, Hyattsville, MD. The dotted line represents the 1989 RDA. Reprinted from Calvo (1993) with permission.

As shown in Figure 2, data from the NHANES II survey indicate Ca:P ratios at the 50th percentile intake of 1:1.61 (0.62) for men and 1:1.56 (0.64) for women (Calvo, 1993). Men show slightly lower ratios than women and both sexes show an overall trend toward a lower ratio or greater imbalance between calcium and phosphorus intake with increasing age. The most important points to be gleaned from the survey data are that the Ca:P ratio in the daily diet of most Americans is well below the ideal 1:1 ratio and that calcium intake is low for a large subpopulation of women. Since this high phosphorus, low calcium intake is a prevailing dietary pattern, attention should focus on determining whether or not it has any adverse effect on bone. Defining the interaction between these two nutrients in the current diet is critical to the development of effective intervention. Optimizing peak bone mass or reducing the rate of bone loss may require both calcium supplementation and phosphorus restriction. Sorting out the specifics of the effects of low calcium intake from those due to a relatively high phosphorus intake is another important goal, since a relative excess of phosphorus may exacerbate the adverse effect of inadequate calcium intake on bone.

Several factors contribute to the observed disparity between calcium and phosphorus intake. Phosphorus is ubiquitously distributed in the food supply, while calcium is limited primarily to dairy foods. The estimated contribution by various food sources to the mean daily intake of Ca and P for women 20 to 29 years old, calculated from intake estimates made during the Continuing Survey of Food Intakes by Individuals conducted in 1985-86, is shown in Figure 3 (Life Science Research Office, 1989; Calvo, 1993). Ingested phosphorus comes from a variety of foods, while an estimated 80% of the calcium comes exclusively from dairy foods. Milk consumption has decreased over the last decade while soft drink consumption has increased dramatically, particularly for the cola beverages that are processed with phosphoric acid (Behlen, 1986; Guenther, 1986). Among teenage and young adult women cola beverages have supplanted milk, a primary source of calcium, with a product that supplies only phosphorus. The increased use of phosphorus-containing food additives in the processing of other foods (Dziezak, 1990) also contributes to the high intakes of phosphorus.

Greger and Krystofiak (1982) estimated that phosphate-containing food additive use accounted for 20 to 30% of the adult phosphorus intake in 1979. The International Food Additive Council estimated per capita phosphorus consumption that could be attributed to phosphorus-containing food additives used in food processing to be 400 mg/d in 1980 and 470 mg/d in 1990. However, despite the conservative estimate of a 17% increase in use over the last decade, the nutrient composition data bases used in calculating mineral intakes do not reflect the increased use of phosphorus-containing food additives (Oenning *et al.*, 1988).

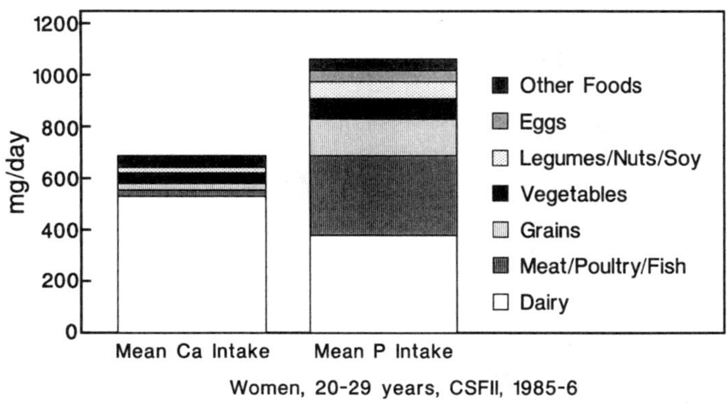


Figure 3. The graph demonstrates the relative contributions made by various food sources to the mean calcium and phosphorus intakes estimated for women 20–29 years of age in the Continuing Survey of Food Intakes by Individuals, 1985–86 (CSFII, 1985–86). (Reprinted with permission from Calvo, 1993).

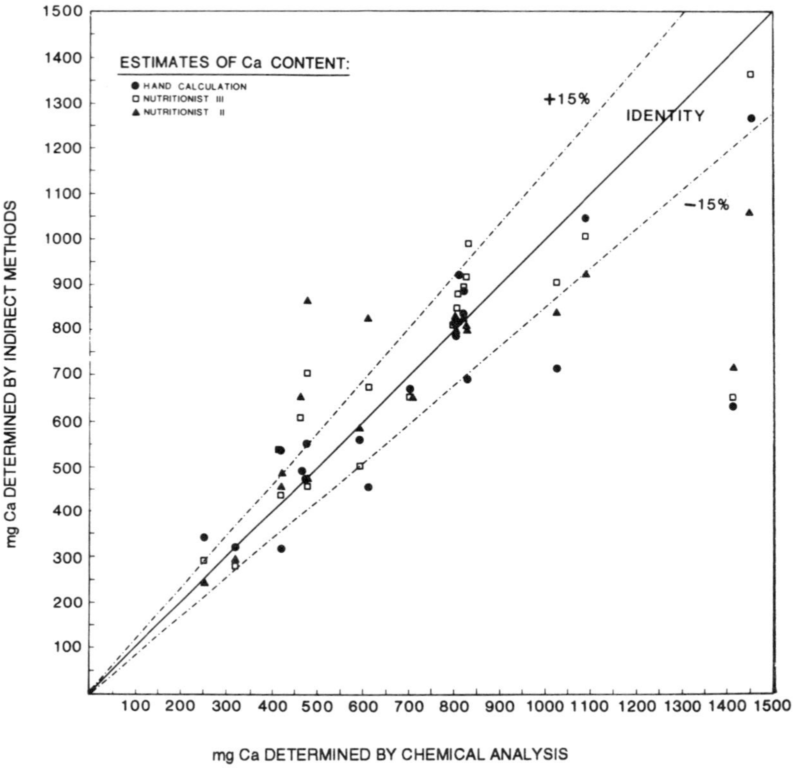


A. Calcium content

Figure 4 Calcium content (Figure 4A) and phosphorus content (Figure 4B) estimated by hand calculation (●), Nutritionist III (□), or Nutritionist II (▲) plotted as a function of the actual or chemically analyzed value for each of 20 daily menus. The line of identity is shown by the solid line and the broken lines indicate ± 15% variation from identity. The terms Nutritionist II and Nutritionist III describe the method of estimating calcium and phosphorus by computer calculation using the Nutritionist II and Nutritionist III computer software programs. Reprinted from Oenning *et al.* (1988) with permission.

Survey data that rely on these out-dated nutrient composition data bases underestimate actual phosphorus intake. When Oenning *et al.* (1988) compared estimations of calcium and phosphorus intake determined from nutrient data bases with those from direct chemical analyses of the same diets, they found that estimated calcium intakes correlated well with analytical values (Figure 4A). In contrast, the estimated phosphorus intakes determined from nutrient composition tables underestimated the analyzed diet phosphorus level by 15 to 25% (Figure 4B).

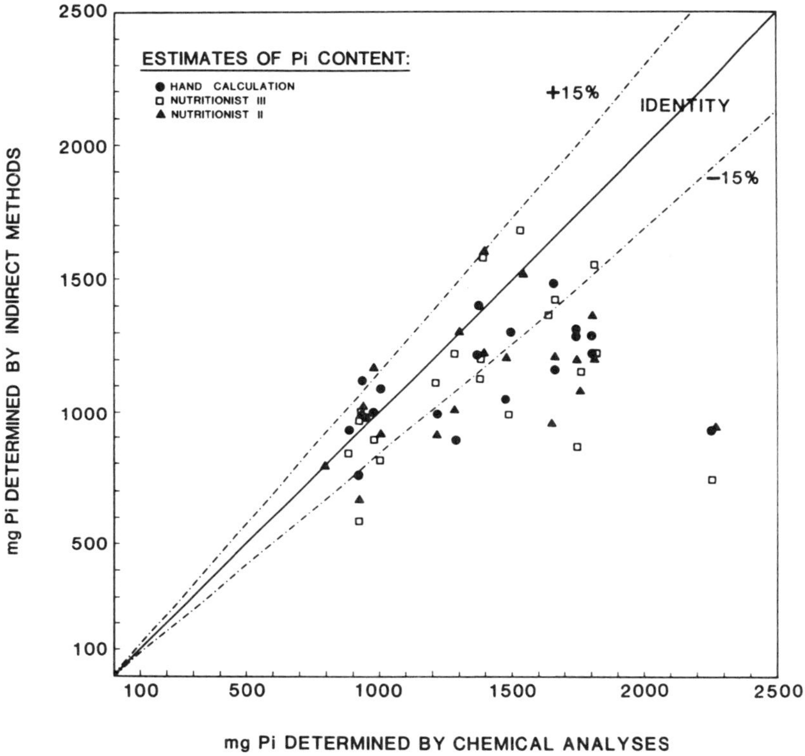


B. Phosphorus content.

# 3. Physiologic Effects of High Phosphorus, Low Calcium Intake on Bone

## 3.1. Findings from Human Populations

Although there is ample evidence for an imbalance in the Ca:P ratio of the American diet (Chinn, 1981; Albanese *et al.,* 1986), there is limited evidence from population studies that relates high phosphorus, low calcium intake to low bone mineral content or bone density. This question has largely been overshadowed by the singular interest in calcium, but a few investigators have also examined the influence of phosphorus intake. Tylavsky and Anderson (1988) reported that phosphorus intake contributed negatively to bone mineral content and bone density in postmenopausal women. Lukert and co-workers (1987) found a positive correlation between the dietary Ca:P ratio and bone density in a study of perimenopausal women. In a study of Japanese-Americans, dietary Ca:P ratios correlated positively with bone mineral content in older men, but not older women (Yano *et al.,* 1985). Although the role of the high phosphorus, low calcium Eskimo diet is confounded by the effects of their high protein intake, Mazess and Mather (1974, 1975) observed 10 to 15% less bone in adult Eskimos over 40 years old compared to age-matched Caucasians. It is important to keep in mind that epidemiologic evidence supporting a role for high phosphorus, low calcium intake in low bone mass is subject to the accuracy of estimates of phosphorus intake.

## 3.2. Findings from Animal Studies

Concern about the relatively high levels of phosphorus in the American diet and its role in osteoporosis relates to findings in animal studies (Chinn, 1981). High phosphorus intake, even with adequate calcium intake, has been shown repeatedly to lead to severe osteopenia in a variety of animal models. High phosphorus intake induced progressive bone loss in mice (Krishnarao and Draper, 1972; Bell *et al.,* 1980), rats (Clark, 1969; Anderson and Draper, 1972; Draper *et al.,* 1972; Sie *et al.,* 1974; Schaafsma and Visser, 1980), rabbits (Jowsey and Balasubramaniam, 1972), horses (Schryver *et al.,* 1971), cats ( Krook *et al.,* 1975), pigs (DeLuca *et al.,* 1976), and dogs (Saville and Krook, 1969; Krook *et al.,* 1971; LaFlamme and Jowsey, 1972; Cook *et al.,* 1983).

The bone loss reported in these studies is thought to be due to the limited control of phosphorus absorption compared to the tightly regulated absorption of calcium. Unlike excess dietary calcium which is predominantly dependent on active absorption, about 70% of dietary phosphorus is readily absorbed even at extremes of intake, largely by means of passive transport (Anderson, 1991). In these studies, the rapid absorption of excess phosphorus raised plasma phosphorus levels, causing a decrease in plasma calcium. The resultant hypocalcemia purportedly stimulates parathyroid hormone (PTH) secretion, which in turn accelerates bone resorption and, in experimental animals, has been shown to lead to osteoporosis (Draper and Bell, 1979; Jowsey *et al.,* 1974). The action of PTH on the renal tubules stimulates renal tubular resorption of calcium, decreasing urinary calcium loss. This action of PTH may relate to the high incidence of nephrocalcinosis reported in some animal models fed high levels of phosphorus. Large excesses of dietary phosphorus are associated with both osteoporosis and soft tissue calcification in rodents (Ritskes-Hoitinga *et al.,* 1989) and dogs (LaFlamme and Jowsey, 1972; Cook *et al.,* 1983).

The relevance of these findings in animals to humans has been criticized on several grounds, including the choice and sensitivity of the animal model, exaggerated levels of dietary phosphorus and/or calcium used, inability to distinguish between calcium and phosphorus effects when calcium levels are inadequate, use of diets deficient in other essential nutrients, and failure to determine the confounding effects of renal impairment on the results. One study in which growing dogs were supplemented with phosphorus revealed increased bone mass after 12 weeks of supplementation (Harris *et al.,* 1976). These findings contrast with those of studies in which high phosphorus diets were fed for a longer period (LaFlamme and Jowsey, 1972). Twelve weeks is too short a duration to accurately determine the lifetime effects of this diet on bone mass. This brings to mind findings from long-term clinical trials with sodium fluoride or intermittent low doses of PTH. Both of these agents stimulate bone formation and increase bone density in cancellous bone of the spine, but with time it is apparent that cancellous bone growth occurs at the expense of cortical bone mass, which has been shown to decrease progressively with time (Riggs and Melton, 1992).

Two studies on lower primates suggest that humans may not respond as dramatically as other animal models to imbalances in the Ca:P ratio of the diet. Anderson *et al.* (1977) studied the long-term (3–88 months) effects of high phosphorus diets on young and adult ringtail monkeys originating in the wild. Whether fed diets deficient (Ca:P ratio 1:4) or adequate (Ca:P ratio 1:2.2) in calcium, monkeys fed the high phosphorus diet did not develop significant clinical or radiographic evidence of bone loss and showed only minor microscopic changes suggestive of osteoporosis. While this study is often cited to disavow the role of excess phosphorus as a factor in the etiology of human osteoporosis (Marcus, 1982), it is subject to criticism. Bone remodeling in mature primates is a slow process and histological changes in the bones of animals of various ages fed test diets for various periods of time are not easily detected in the small samples (n = 2-4) used by these authors.

Pettifor and co-workers (1984) studied the effects of various dietary calcium and phosphorus intakes on calcium metabolism and bone histomorphometry in young vitamin D-replete baboons. The phosphorus content of three of their experimental diets was adequate, while the calcium content was high, medium or low, resulting in Ca:P ratios of 1:0.78, 1:2.2, and 1:7.7. In a fourth group, in which the calcium and phosphorus levels were both low, the ratio was 1:2.3. By approximately 8 months, the baboons fed the low calcium, normal diet were hyperphosphatemic and hypocalcemic. Both low calcium groups showed histologic evidence of hyperosteoidosis but there were no group differences in osteoclast numbers or in osteoclastic resorptive surface. The low calcium, low phosphorus group did not show hyperphosphatemia or hypocalcemia at 8 months. By 16 months, baboons fed the low calcium, normal phosphorus diet showed histologic evidence of hyperparathyroidism (increased osteoclast number and surface), whereas those fed the low calcium, low phosphorus diet showed only histological features of osteomalacia. These investigators pointed out that the early hyperosteoidosis at 8 months and later secondary hyperparathyroidism at 16 months are similar to the histologic changes observed in children with calcium deficiency rickets, who develop osteomalacia with secondary hyperparathyroidism. The diet of these children is low in calcium, but contains adequate levels of phosphorous and vitamin D (Marie *et al.,* 1982).

These primate studies showed a subtle response to perturbations in dietary Ca:P ratio and a histomorphometric response similar to that of calcium deficient children (Pettifor *et al.,* 1984). Osteomalacia with secondary hyperparathyroidism in primates is in contrast to the severe osteitis fibrosa cystica characteristic of phosphate-induced nutritional secondary hyperparathyroidism in other animal models fed exaggerated levels of dietary phosphorus (Cook *et al.,* 1983). With the exception of the study in young baboons, most of the studies in animals were carried out on mature animals. In humans, diet may have its greatest impact on peak bone mass during the period of consolidation, or the teens to early twenties, so it may be more appropriate to study the effects of high phosphorus, low calcium intake in a skeletally immature animal model.

Calvo *et al.* (1987) studied the response to a low dietary Ca:P ratio in skeletally immature female beagles. They fed a moderately low calcium (0.5%), high phosphorus (1.4%) diet to 6-month-old female beagles to determine if chronic intake of a high phosphorus, low calcium diet sustains secondary hyperparathyroidism and diminishes bone mass in a skeletally immature animal model. Unlike earlier studies in dogs, in which extreme Ca:P ratios, often as high as 1:10 were used (Saville *et al.,* 1969; Krook *et al.,* 1971), the Ca:P ratio of the test diet was approximately 1:3, which is consistent with the human dietary pattern observed in some teenagers and young adults. The control diet contained the recommended levels of calcium (1.0%) and phosphorus (0.8%) for the growing dog. Three groups of dogs were fed for 14 months. One group was fed the control diet, a second group the high phosphorus, low calcium or test diet, and a third group was fed the test diet for 7 months and then switched to a calcium-fortified control diet for 6 months (1.2% Ca, 0.8% P). The third group was designed to determine whether nutrition intervention could correct any adverse effects of the test diet on bone status.

As shown in Figure 5, a progressive increase in pre- and post-prandial levels of serum immunoreactive parathyroid hormone (iPTH) was observed in dogs fed the high phosphorus, low calcium diet relative to the dogs fed the control diet. Baseline measurements showed no significant differences in iPTH levels between the dietary groups. At one month of feeding the high phosphorus, low calcium diet, iPTH levels were elevated only postprandially, but they increased progressively with time to a 4-fold increase postprandially at 6 months. After 7 months, half of the test-fed dogs were switched to the calcium-fortified control diet for the remaining 5 months of study. At 12 months, this group showed a significant suppression of iPTH relative to the group fed the test diet. Dogs fed the test diet had significantly larger parathyroid glands on histological examination, evidence that the chronic mild nutritional insult produced frank secondary hyperparathyroidism and glandular hyperplasia. While calcium supplementation resulted in a significant decrease in plasma levels of 1,25-$(OH)_2D$ (calcitriol), there were no significant differences in plasma levels of the active metabolite of vitamin D among all dietary groups at 12 or 14 months. This finding suggests that with time, chronic phosphorus excess impairs homeostatic mechanisms needed to compensate for a low calcium intake. The dogs fed the test diet had significantly higher kidney calcium content and histologic evidence of metastatic calcification, although normal serum creatinine levels indicated that their renal function was not impaired. The extent to which metastatic calcification may have impaired the renal synthesis of calcitriol is unclear.

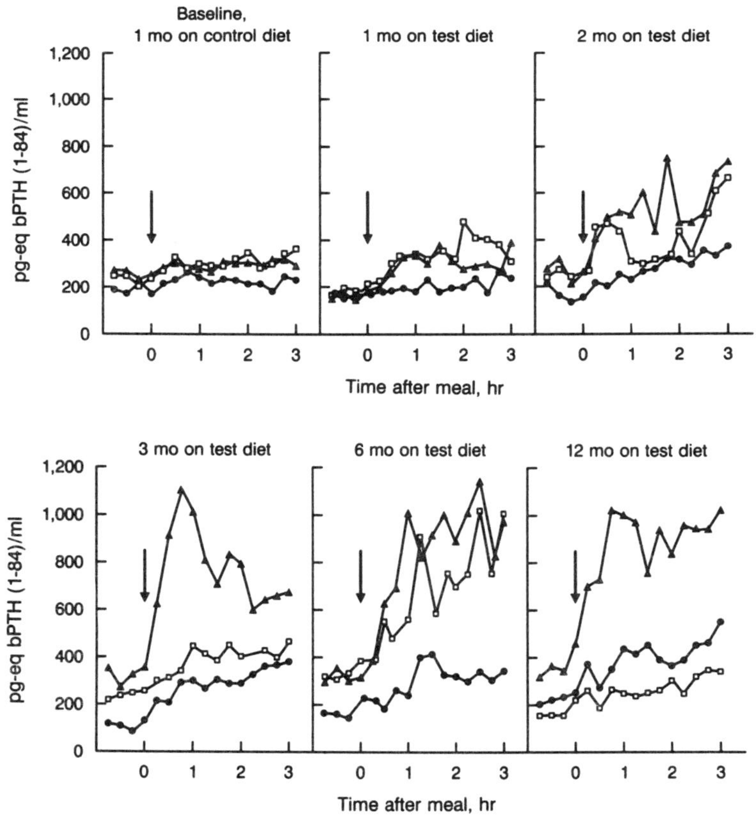


Figure 5. Group mean serum iPTH levels measured before (time 0) and at hourly intervals after consumption of the daily meal in dogs fed the control (●—●), high phosphorus, low calcium (▲—▲), and the high phosphorus low calcium /calcium supplemented control (□—□) diets. Hormonal responses are shown for baseline (all dogs on the control diet), 1, 2, 3, 6 and 12 months on the dietary regimens. Arrows indicate when meals were fed. Reprinted from Calvo *et al.* (1987) with permission.

Histomorphometric analyses of transiliac crest biopsies taken at 14 months were consistent with the findings of Pettifor *et al.* (1984) for dogs fed the high phosphorus, moderately low calcium diet. Calvo *et al.* (1987) also obtained evidence of increased percent osteoid and osteoid volume, and a greater number of osteoclasts and a higher percent of bone resorbing surface, although the differences in the latter two measures did not reach statistical significance. In the dogs fed the test diet, there was a small but significant reduction in bone mass of the second lumbar vertebra (2.5%), but there was no difference in the vertebral bone mass of the calcium-supplemented dogs and that of the control group. It was concluded that ingestion of a moderately low calcium, high phosphorus diet designed to mimic the diets of some teens and young adults caused progressive secondary hyperparathyroidism and a modest reduction in trabecular bone mass in young female beagles that could be restored with calcium supplementation.

## 4. Physiologic Effects of High Phosphorus Intake on Calcium Homeostasis

At this time there is no clear evidence that high phosphorus consumption, with or without adequate calcium intake, causes secondary hyperparathyroidism and bone loss in humans. There is much confusion about the physiologic effects of dietary phosphorus. Some investigators have dismissed the idea that high phosphorus intake may be deleterious to bone, based on findings that high concentrations of phosphorus in cell culture media inhibit PTH-induced bone resorptive activity of isolated osteoclasts (Raisz and Lorenzo, 1980; Heaney *et al.,* 1982; Heaney, 1988). In a more recent study, Yates *et al.* (1991) showed that increasing the phosphorus concentration of cell culture media over the normal physiologic range significantly inhibits osteoclast activity. The *in vivo* situation is, however, very different. The natural variation in extracellular phosphorus concentration may explain why the effects of high levels of phosphorus are different *in vivo.*

Acute exposure to high levels of phosphorus results in a secondary stimulation of parathyroid hormone secretion. Phosphate infusion (Herbert *et al.,* 1966), high phosphorus diets (Smith and Nordin, 1964), and oral loads of phosphate salts administered to adults (Reiss *et al.,* 1970), but not to young adults (Calvo and Heath, 1988) have been found to depress serum calcium and stimulate iPTH release. Reiss *et al.* (1970) reported a 60 to 125% increase in iPTH within 1 hr of an oral phosphate load in adults. The effects of chronic exposure to high phosphorus intake were first studied in relation to calcium balance.

## 4.1. Findings from Calcium Balance Studies

Earlier human studies (Malm, 1953) indicated that there was no net effect of high phosphorus intakes on calcium balance, while in later studies (Heaney and Recker, 1982) a notable decrease in urinary calcium excretion was found with a higher intestinal secretion of calcium, producing a negligible effect on calcium balance (Spencer *et al.,* 1978; Hegsted *et al.,* 1981). Spencer *et al.* (1965, 1978) found that varying the intake of dietary phosphate, mostly as orthophosphate, had little effect on calcium absorption when calcium intake was low or adequate. The various phosphate-containing food additives have since been shown to differ in their effects on calcium balance. Calcium absorption was significantly reduced when polyphosphate supplements were fed to young male volunteers; orthophosphate supplements improved calcium balance, but equilibrium was not achieved (Zemel and Linkswiler, 1981). More recently, studies in postmenopausal women on self-selected intakes revealed no net influence of a high phosphorus intake on calcium balance (Hasling *et al.,* 1992). Nevertheless, the finding of Matkovic and co-workers (1986) raises speculation that the effects of phosphorus intake on calcium balance may vary with age. They reported a negative correlation between phosphorus intake and calcium absorption during calcium balance in teenage girls. The girls with higher phosphorus intakes had less positive calcium balances.

The disadvantage in using calcium balance methodology to study nutrient interaction is that harmful effects can remain undetected, because the technique determines only the quantity of mineral retained in the body, while the fate of the mineral (deposition in bone or soft tissue) is unknown. The fate of the mineral is strongly influenced by the age of the subject, since growing individuals have stronger demands for bone deposition. For example, Goldsmith *et al.* (1976) found a large discrepancy between bone biopsy and calcium balance data obtained from postmenopausal osteoporotic women fed phosphate supplements for 12 to 14 months. Phosphate supplementation improved calcium balance but decreased bone-forming surface area while increasing the bone-resorbing surface area. In this case there was no evidence of soft tissue calcification, but relatively insensitive, noninvasive techniques were used. In these women, phosphate supplementation may have promoted greater calcium retention, but the histologic evidence suggests that it was not deposited in bone.

## 4.2. Findings from Clinical Studies

Chronic phosphorus supplementation has been found to influence the parathyroid-vitamin D axis, but not always in the direction anticipated. Chronic oral phosphate therapy was first used in disease states such as osteoporosis (Goldsmith *et al.,* 1976), idiopathic hypercalciuria (Van den Berg *et al.,* 1980), primary hyperparathyroidism (Broadus *et al.,* 1983) and to conserve calcium during prolonged bed rest (Hulley *et al.,* 1971). Administration of 2 g of phosphorus per day in patients with idiopathic hypercalciuria surprisingly decreased plasma calcitriol levels, despite slight elevations in serum iPTH and, as anticipated, decreased urinary calcium excretion (Van den Berg *et al.*, 1980). One year of oral phosphate therapy (1.5 g P per day) in patients with primary hyperparathyroidism significantly reduced circulating calcitriol levels and increased circulating levels of iPTH (Broadus *et al.,* 1983).

Silverberg *et al.* (1986) examined the hormonal response to five days of phosphate supplementation (2 g P/d) in young adult men and women and later in postmenopausal normal and osteoporotic women (1989). The young adults responded with a significant reduction in total serum calcium and a 50% increase in iPTH after five days of supplementation. In older women with osteoporosis, iPTH levels increased by only 43% in response to 2 g P given for five consecutive days, whereas iPTH levels rose 250% in the age-matched control group without osteoporosis (Silverberg *et al.,* 1989). In these acute studies, Silverberg and co-workers detected no change in circulating calcitriol levels in healthy old or young subjects, but calcitriol levels decreased 50% in the osteoporotic subjects after acute oral phosphate loading. The authors concluded that postmenopausal women with osteoporosis are not able to mount a parathyroid hormone response appropriate for their age. Other investigators also have observed poor iPTH response to peroral phosphate loading in postmenopausal osteoporotic women (Ittner *et al.,* 1986; Mallette *et al.,* 1989). The lower level of parathyroid hormone was inadequate to override the suppressive effects of phosphate on renal 1-alpha hydroxylase enzyme, resulting in lower circulating levels of the active metabolite of vitamin D (Silverberg *et al.,* 1989).

Parathyroid hormone is generally held to be the most important regulator of calcitriol levels. Recent studies have demonstrated that parathyroid hormone regulation of calcitriol is subject to modification by serum phosphorus levels. Portale *et al.* (1986) reported that high levels of dietary phosphate (3 g P/d for 10 days) caused a 30% reduction in plasma calcitriol levels of normal men from levels observed during their usual dietary intake of 1.5 g P/d and 850 mg Ca/d. These investigators also examined the role of phosphorus intake within the observed dietary range as a determinant of plasma calcitriol levels (Portale *et al.,* 1989). As shown in Figure 6, decreasing phosphorus intake from 2.3 g P/d to 0.625 g/d resulted in a 58% increase in plasma levels of calcitriol. Portale and co-workers concluded from this and their previous studies (1984, 1986, 1987, 1989) that dietary phosphorus throughout the normal range and beyond can finely regulate the renal production and plasma concentration of calcitriol. A recent study in women suggests that plasma phosphorus levels in the upper normal range, even in the presence of elevated iPTH levels, can influence plasma calcitriol levels. Dawson-Hughes *et al.* (1991) examined the association between

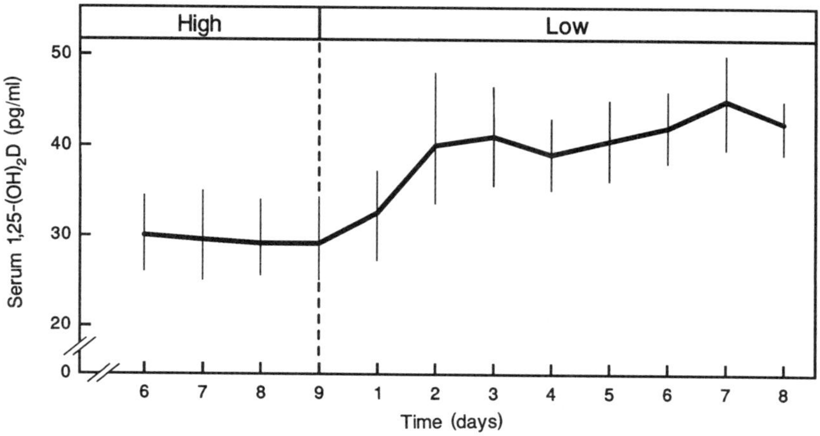


Figure 6. Effect of decreasing dietary phosphorus within its normal range on the daily serum concentration of 1,25(OH)$_2$D in 7 normal men. Dietary phosphorus was first maintained at 2,300 mg/d for 9 d *(High)*, then decreased to 625 mg/d *(Low)* for 8 d. Mean ± SEM. Redrawn from the data of Portale *et al.* (1989).

serum ionized calcium, iPTH, phosphorus and calcitriol levels in a cross-sectional study of 275 healthy postmenopausal women. Using partial correlation multiple regression analysis, they found that the relationshp between PTH and calcitriol is attenuated at both low and high-normal concentrations of phosphorus.

Changing dietary phosphorus within the normal range also alters the normal circadian rhythm of serum phosphorus. The circadian rhythm of serum phosphorus in subjects consuming self-selected diets (Portale *et al.,* 1987; Calvo *et al.,* 1991) follows a biphasic pattern with peaks in the afternoon and the late evening. After decreasing phosphorus intake, Portale and co-workers (1989) found a 35% decrease in the afternoon peak, a 12% decrease in the morning serum level, but no decrease in the morning fasting level of phosphorus. Serum iPTH has been shown to follow a circadian rhythm that closely parallels that of serum phosphorus (Figure 7) (Calvo *et al.,* 1991). For this reason, the time of blood sampling appears to be a crucial factor in determining the effect of phosphate loading on parathyroid function. The need for multiple blood sampling was first suggested by Smith and Nordin (1964), who were unable to demonstrate by single daily blood sampling that high phosphorus diets (1.5 g P) depressed serum calcium levels. However, when they drew 4 daily blood samples from each subject, they were able to demonstrate a significant depression in serum calcium.

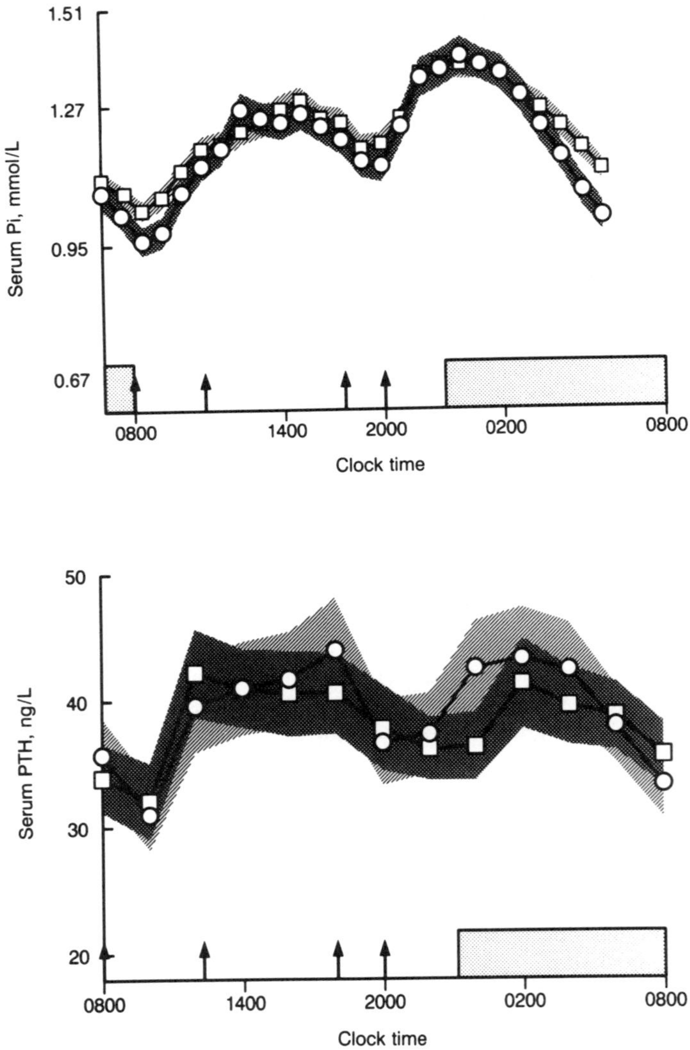


Figure 7. Top: Circadian variation in serum phosphorus (Pi) measured hourly in women (□—□; n = 25) and men (○—○; n = 24). Bottom: Circadian variation in serum intact PTH measured every other hour in women (□—□; n = 25) and men (○—○; n = 24). Means ± SEM (shaded area) are shown. Meals and snacks are indicated by the arrows, and the stippled bar indicates recumbent posture. Reprinted from Calvo *et al.*, 1991, with permission.

Changes in food consumption also may alter the circadian rhythm of calcium, phosphorus, and iPTH, particularly the diurnal secretory pattern of PTH. The chronic effect of consuming foods containing phosphate additives was examined in young adult males and females by Bell *et al.* (1977). Volunteers ate a basal diet containing no phosphorus additives (900 mg P; 700 mg Ca) for 4 weeks followed by a 4-week intake of foods containing phosphate additives (2,100 mg P; 700 mg Ca) that were substituted for similar items in the basal diet. Serum calcium concentrations measured 2 hr after the start of the evening meal were significantly depressed during the high phosphorus period, but there was no difference in serum iPTH levels. Using a similar dietary design, Calvo *et al.* (1988) examined the effects of high phosphorus (1.6 g/d), low calcium (400 mg/d) diets made from commercially available foods on PTH secretion and action. Young adult men and women were fed a high phosphorus, low calcium diet or control diet (900 mg P/d; 800 mg Ca/d) in random order, each for 8 days. Blood samples were taken at 2-hr intervals over 24 hr on the eighth and sixteenth days of study. This method of multiple blood sampling revealed a sustained increase in serum iPTH over most of the day in both men and women during the consumption of the test diet. The 24-hr mean serum iPTH level increased 11% in men and 22% in women during the test diet, and serum phosphorus, plasma calcitriol, and urinary hydroxyproline and cyclic AMP excretion were significantly increased in both sexes. Thus, in young adults, short-term ingestion of a diet typical of current U.S. levels of calcium and phosphorus intake resulted in elevated levels of serum iPTH and indices of PTH action as well as a significant increase in circulating calcitriol levels.

The effect of a longer period of consumption of the high phosphorus diet was also determined. Calvo *et al.* (1990) studied the 24-hr mineral and hormonal response to a high phosphorus, low calcium intake (1,700 mg P/d, 400 mg Ca/d) and a control diet (900 mg P/d, 800 mg Ca/d). All subjects consumed the control diet for an initial 4 weeks and then 10 subjects were switched to the high phosphorus, low calcium diet for 4 weeks, while 5 subjects continued to consume the control diet. Serum iPTH levels increased significantly after 4 weeks of consuming the test diet, but did not change in the control group (Figure 8). Urinary cyclic AMP, a marker of parathyroid action in the kidney, rose significantly after test diet consumption. There was a small increase in urinary hydroxyproline, a marker for parathyroid action in bone, but it was not statistically significant. In contrast to their earlier 8-day study (Calvo *et al.*, 1988), plasma calcitriol levels did not change in either group. There was a modest 6% mean increase in plasma calcitriol levels after chronic high phosphorus, low calcium intake. Thus, an important adaptive change to low calcium intake that occurred in the short-term study was not observed with longer intake. The authors concluded that prolonged high phosphorus intake may impair the usual homeostatic mechanisms involved in adaptation to a limited dietary calcium intake.

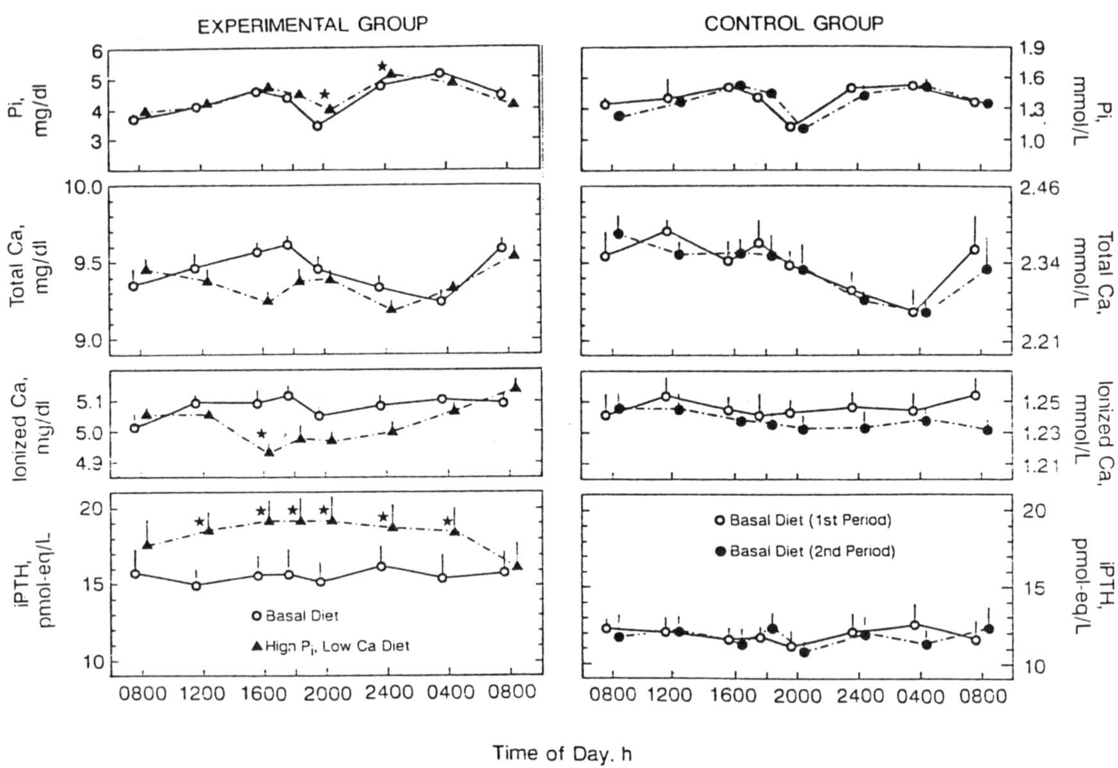


Figure 8. Mean (± SE) serum phosphorus (Pi), total calcium (Ca), ionized Ca, and immunoreactive parathyroid hormone (iPTH) concentrations during ingestion of the basal diet (○—○) and the high phosphorus, low calcium diet (▲---▲) after the experimental protocol (left panel; n =10) or during the first 4 weeks (○—○) and second 4 weeks (●---●) of basal diet, ingestion after the control protocol (right panel; n = 5). $P \leq 0.006$, within-group comparisons, vs. test or 8-week basal diet. (Reprinted with permission from Calvo *et al.*, 1990.)

To date, there have been no studies on the effects of this dietary pattern on calcium homeostasis and rate of bone resorption in postmenopausal women. Relevant to the question of how changes in phosphorus consumption may influence calcium homeostasis in postmenopausal women, Luz Villa *et al.* (1991), found that circulating levels of calcitriol are exquisitely sensitive to reductions in serum phosphorus levels in normal postmenopausal women. In earlier studies, these researchers observed that estrogen replacement therapy in postmenopausal women lowered renal phosphorus reabsorption and serum phosphorus levels, changes that were correlated with a rise in circulating calcitriol (Cheema *et al.*, 1989; Packer *et al.*, 1990). In this later study, they tested the hypothesis that estrogen increases circulating calcitriol levels through a reduction in plasma phosphorus by feeding aluminum hydroxide, a phosphorus-binding antacid, to postmenopausal women (Luz Villa *et al.*, 1991). This effectively restricted dietary phosphorus absorption, lowered plasma phosphorus, and elevated plasma calcitriol without altering other regulators of calcitriol production including serum ionized calcium, total calcium, and iPTH. A 17% reduction in plasma phosphorus was associated with a 58% rise in circulating calcitriol level, which is consistent with the findings of Portale *et al.* (1989).

## 5. Summary

Survey data confirm that the dietary pattern of many American women who are at high risk of developing osteoporosis is typically high in phosphorus and low in calcium. The imbalance between calcium and phosphorus intake may become more pronounced with continued changes in food preferences and the growing use of phosphorus-containing food additives. Recent studies in young women have shown that a high phosphorus diet moderately low in calcium results in a mild secondary hyperparathyroidism that persists over 4 weeks. Plasma levels of calcitriol did not change despite changes in PTH and serum ionized calcium. Studies on men have shown that dietary phosphorus at levels within the normal range of intakes can affect the renal production and serum concentration of calcitriol. High phosphorus intakes for ten days reduced their plasma calcitriol levels; a 70% reduction in phosphate intake significantly increased their plasma calcitriol.

Thus, several lines of evidence indicate that prolonged high phosphorus intake may impair the usual homeostatic mechanisms that come into play when dietary calcium is limited. This, in turn, could impair achievement of maximal bone mass or accelerate bone loss. Although no clinical studies have linked high phosphorus intake with lower bone mass or higher rates of bone loss in humans, this relationship has been demonstrated in animal models. For example, young beagles fed high phosphorus, moderately low calcium diets showed a significant reduction in vertebral bone mass.

Current dietary patterns of high phosphorus, low calcium consumption result in persistent changes in calcium regulating hormones that are not conducive to maximizing peak bone mass during growth or slowing the rate of aging bone loss. The net effect of the present dietary pattern on bone status, particularly in teenage and young adult women, needs to be determined. Optimal nutrition early in life, which may include higher calcium and lower phosphorus intakes, together with adequate exercise, may be the most cost-effective approach to the prevention of osteoporotic fractures.

## References

Albanese, A.A., Lorenze, E.J., Edelson, A.H., Tarlow, A., Wein, E.H., and Carroll, L., 1986, Effects of dietary calcium:phosphorus ratios on utilization of dietary calcium for bone synthesis in women 20-75 years of age, *Nutr. Rep. Int.* **33**:879.

Anderson, J.J.B., 1991, Nutritional biochemistry of calcium and phosphorus, *J. Nutr. Biochem.* **2**:300.

Anderson, G.H., and Draper, H.H., 1972, Effect of dietary phosphorus on calcium metabolism in intact and parathyroidectomized adult rats, *J. Nutr.* **102**:1123.

Anderson, M.P., Hunt, R.D., Griffiths, H.F., McIntyre, K.W., and Zimmerman, R.E., 1977, Long-term effect of low dietary calcium:phosphate ratio on the skeleton of cebus albifrons monkeys, *J. Nutr.* **107**:834.

Baran, D., Sorensen, A., Grimes, J., Lew, R., Karellas, A., Johnson, B., and Roche, J., 1990, Dietary modification with dairy products for preventing vertebral bone loss in premenopausal women: a three-year prospective study, *J. Clin. Endocrinol. Metab.* **70**:264.

Behlen, P., 1986, Calcium in women's diets, *Nat. Food Rev.* **34**:16.

Bell, R.R., Draper, H.H., Tzeng, D.Y.M., Shin, H.K., and Schmidt, G.R., 1977, Physiological responses of human adults to foods containing phosphate additives, *J. Nutr.* **107**:42.

Bell, R.R., Tzeng, D.Y., and Draper, H.H., 1980, Long-term effects of calcium phosphorus and forced exercise on the bones of mature mice, *J. Nutr.* **110**:1161.

Broadus, A.E., Magee, J.S., Mallette, L.E., Horst, R.L., Lang, R., Jensen, P.S., Gertner, J.M., and Baron, R., 1983, A detailed evaluation of oral phosphate therapy in selected patients with primary hyperparathyroidism, *J. Clin. Endocrinol. Metab.* **56**:953.

Calvo, M.S., 1993, Dietary phosphorus, calcium metabolism and bone, *J. Nutr.* **123**:1627.

Calvo, M.S., and Heath, H. III., 1988, Acute effects of oral phosphate-salt ingestion on serum phosphorus, serum ionized calcium and parathyroid hormone in young adults, *Am. J. Clin. Nutr.* **47**:1025.

Calvo, M.S., Harstad, L., Laakso, K.J., and Heath, H. III, 1987, Chronic low calcium (Ca), high phosphorus (P) intake during adolescence causes secondary hyperparathyroidism (2HPT) and reduces bone mass in female beagles, *J. Bone Mineral Res.* **2**:s464 (abstract).

Calvo, M.S., Kumar, R., and Heath, H. III., 1988, Elevated secretion and action of serum parathyroid hormone in young adults consuming high phosphorus, low calcium diets assembled from common foods, *J. Clin. Endocrinol. Metab.* **66**:823.

Calvo, M.S., Kumar, R., and Heath, H. III., 1990, Persistently elevated parathyroid hormone secretion and action in young woman after four weeks of ingesting high phosphorus low calcium diets, *J. Clin. Endocrinol. Metab.* **70**:1334.

Calvo, M.S., Eastell, R., Offord, K.P., Bergstrahl, E.J., and Burritt, M.F., 1991, Circadian variations in ionized calcium and intact parathyroid hormone: evidence for sex differences in calcium homeostasis, *J. Clin. Endocrinol. Metab.* **72**:69.

Cheema, C., Grant, B.F., and Marcus, R., 1989, Effects of estrogen on circulating "free" and total 1,25-dihydroxyvitamin D and on the parathyroid vitamin D axis in postmenopausal women, *J. Clin. Invest.* **83**:537.

Chinn, H. I., 1981, *Effects of dietary factors on skeletal integrity in adults: calcium, phosphorus, vitamin D and protein*, Life Sciences Research Office, Federation of American Societies for Experimental Biology, Bethesda.

Clark, I., 1969, Importance of dietary Ca:$PO_4$ ratios on skeletal Ca, Mg and $PO_4$ metabolism, *Am. J. Physiol.* **217**:865-70.

Cook, S.D., Skinner, H.B., and Haddad, R.J. Jr., 1983, A quantitative histologic study of osteoporosis produced by nutritional secondary hyperparathyroidism in dogs, *Clin. Orthop. Relat. Res.* **175**:105.

Dawson-Hughes, B., Dallal, G.E., Krall, E.A., Sadowski, L., Sayhoun, N., and Tannenbaum, S., 1990, A controlled trial of the effect of calcium supplementation on bone density in menopausal women, *N. Engl. J. Med.* **323**:878.

Dawson-Hughes, B., Harris, S., and Dallal, G.E., 1991, Serum ionized calcium, as well as phosphorus and parathyroid hormone, is associated with the plasma 1,25-dihydroxyvitamin $D_3$ concentration in normal postmenopausal women, *J. Bone Mineral Res.* **6**:461.

DeLuca, H.F., Castillo, L., Jee, W.S., Oldham, S., and Grummer, R.H. 1976, Studies on high phosphate diets. *Food Research Institute Annual Report, University of Wisconsin-Madison,* **1976**:394.

Draper, H.H., and Bell, R.R., 1979, "Nutrition and osteoporosis," in *Advances in Nutritional Research* (H. Draper, ed.), pp.79-106, Plenum Press, New York.

Draper, H.H., and Scythes, C.A., 1981, Calcium, phosphorus and osteoporosis, *Fed. Proc.* **40**:2434.

Draper, H.H., Sie, T-L., and Bergan, J.G., 1972, Osteoporosis in aging rats induced by high phosphorus diets, *J. Nutr.* **102**:1133.

Dziezak, J.D., 1990, Phosphates improve many foods, *Food Technol.* **90**:80.

Elders, P.J.M., Netelenbos, J.C., Lips, P., van Ginkel, F.C., Khoe, E., Leeuwenkamp, O.R., Hackeng, W.H.L., and van der Stelt, P.F., 1991, Calcium supplementation reduces vertebral bone loss in perimenopausal women: A controlled trial in 248 women between 46 and 55 years of age, *J. Clin. Endocrinol. Metab.* **73**:533.

Goldsmith, R.S, Jowsey, J., Dube, W.J., Riggs, B.L., Arnaud, C.D., and Kelly, P.J., 1976, Effects of phosphorus supplementation on serum parathyroid hormone and bone morphology in osteoporosis, *J. Clin. Endocrinol. Metab.* **43**:523.

Greger, J.L., and Krystofiak, M., 1982, Phosphorus intake of Americans, *Food Technol.* **36**:78.

Guenther, P.M., 1986, Beverages in the diets of American teenagers, *J. Am. Diet. Assoc.* **86**:493.

Harris, W.H., Heaney, R.P, Davis, L.A, Weinberg, E.H., Coutts, R.D., and Schiller, A.L., 1976, Stimulation of bone formation in vivo by phosphate supplementation, *Calcif. Tissue Res.* **22**:85.

Hasling, C., Sondegaard, K., Charles, P., and Mosekilde, L., 1992, Calcium metabolism in postmenopausal osteoporotic women is determined by dietary calcium and coffee intake, *J. Nutr.* **122**:1119.

Heaney, R., 1988, "Nutritional Factors in Bone Health," in: *Osteoporosis: Etiology, Diagnosis, and Management* (B. L. Riggs and L.J. Melton III, eds.), pp. 359-372, Raven Press, New York.

Heaney, R.P., and Recker, R., 1982, Effects of nitrogen, phosphorus, and caffeine on calcium balance in women, *J. Lab. Clin. Med.* **99**:46.

Heaney, R., Gallagher, C., Johnston, C., Neer, R., Parffit, M., and Whedon, G., 1982, Calcium nutrition and bone health in the elderly, *Am. J. Clin. Nutr.* **36**:986.

Hegsted, M., Schuette, S.A., Zemel, M.B., and Linkswiler, H.M., 1981, Urinary calcium and calcium balance in young men as affected by level of protein and phosphorus intake, *J. Nutr.* **111**:553.

Herbert, L.A., Lemann, J., Petersen, J.R., and Lennon, E.J., 1966, Studies of the mechanism by which phosphate infusion lowers serum calcium concentration, *J. Clin. Invest.* **48**:1886.

Hulley, S.B., Vogel, J.M., Donaldson, C.J., Bayers, J.H., Friedman, R.J., and Rosen, S.N., 1971, The effect of supplemental oral phosphate on bone mineral changes during prolonged bed rest, *J. Clin. Invest.* **50**:2506.

Human Nutrition Information Service, U.S. Department of Agriculture, 1983, *Food intakes: individuals in 48 states, year 1977-78*. U.S. Department of Agriculture, Washington, DC, (Nationwide food consumption survey 1977-78, report #I-1.)

Ittner, J., Dambacher, M.A., Muff, R., Ruegsegger, P., Trechsel, U., and Fisher, J.A., 1986, Reduced parathyroid hormone response to peroral phosphate in osteoporotic patients, *Miner. Electrolyte Metab.* **12**:199.

Jowsey, J., and Balasubramaniam, P., 1972, Effect of phosphate supplements on soft tissue calcification and bone turnover, *Clin. Sci.* **42**:289.

Jowsey, J., Reiss, E., and Canterbury, J.M., 1974, Long-term effects of high phosphate intake on parathyroid hormone levels and bone metabolism, *Acta Orthop. Scand.* **45**:801.

Krishnarao, G.V.G., and Draper, H.H., 1972, Influence of dietary phosphate on bone resorption in senescent mice, *J. Nutr.* **102**:1143.

Krook, L., Lutwak, L., Henrikson, P-A., Kallfelz, F., Hirsch, C., Romanus, B., Belanger, L.F., Marier, J.R., and Sheffy, B.E., 1971, Reversibility of nutritional osteoporosis: physicochemical data on bones from an experimental study in dogs, *J. Nutr.* **101**:233.

Krook, L., Whalen, J.P., Lesser, G.V., and Berens, D.L., 1975, Experimental studies on osteoporosis, *Methods Achiev. Exp. Pathol.* **7**:72.

LaFlamme, G.H, and Jowsey, J., 1972, Bone and soft tissue changes with oral phosphate supplements, *J. Clin. Invest.* **51**:2834.

Life Sciences Research Office, Federation of American Societies for Experimental Biology, 1989, *Nutrition Monitoring in the United States—An Update Report on Nutrition Monitoring*, Prepared for the U.S. Department of Agriculture and the U.S. Department of Health and Human Services. DHHS Publication No. (PHS) 89-1255. Public Health Service. U.S. Government Printing Office, Washington, DC.

Lukert, B.P., Carey, M., McCarthy, B., Tiemann. S., Goodnight, L., Helm, M., Hassanein, R., Stevenson, C., Stoskopf, M., and Doolan, L., 1987, Influence of nutritional factors on calcium-regulating hormones and bone loss, *Calcif. Tissue Int.* **40**:119.

Luz Villa, M., Packer, E., Cheema, M., Holloway, L., and Marcus, R., 1991, Effects of aluminium hydroxide on the parathyroid-vitamin D axis of postmenopausal women, *J. Clin. Endocrinol. Metab.* **73**:1256.

Mallette, L.E., LeBlanc, A.D., Pool, J.L., and Mechanick, J.I., 1989, Cyclic therapy of osteoporosis with neutral phosphate and brief, high-dose pulses of etidronate, *J. Bone Mineral Res.* **4**:143.

Malm, O.J., 1953, On phosphate and phosphoric acid as dietary factors in the calcium balance of man, *Scand. J. Clin. Lab. Invest.* **5**:75.

Marcus, R., 1982, The relationship of dietary calcium to the maintenance of skeletal integrity in man. An interface of endocrinology and nutrition, *Metabolism* **31**:93.

Marie, P.J., Pettifor, J.M., Ross, F.P., and Glorieux, F.N., 1982, Histological osteomalacia due to dietary calcium deficiency in children, *N. Engl. J. Med.* **307**:584.

Matkovic, V., 1991, Calcium metabolism and calcium requirements during skeletal modeling and consolidation of bone mass, *Am. J. Clin. Nutr.* **54**:245s.

Matkovic, V., and Dekanic, D., 1989, "Developing strong bones: the teenage female," in: *Clinical Disorders of Bone and Mineral Metabolism* (M. Kleerekoper and S.M. Krane, eds.), pp. 165-70, Mary Liebert, Inc, New York.

Matkovic, V., Fontana, D., Tominac, C., Lehmann, J., and Chestnut, C., 1986, Influence of calcium on peak bone mass: a pilot study, *J. Bone Mineral Res.* **1**:s168 (abstract).

Matkovic, V., Fontanna, D., Tominac, C., Goel, P., and Chestnut, C.H., 1990, Factors that influence peak bone mass formation: a study of calcium balance and the inheritance of bone mass in adolescent females, *Am. J. Clin. Nutr.* **52**:878.

Mazess, R.B., and Mather, W., 1974, Bone mineral content of North Alaskan Eskimos, *Am. J. Clin. Nutr.* **27**:916.

Mazess, R.B., and Mather, W., 1975, Bone mineral content in Canadian Eskimos, *Human Biol.* **47**:45.

National Research Council, 1989, *Recommended Dietary Allowances.* 10th ed. National Academy Press, Washington, DC.

Oenning, L.J., Vogel, J., and Calvo, M.S., 1988, Accuracy of methods estimating calcium and phosphorus intake in daily diets, *J. Am. Diet. Assoc.* **88**:1076.

Ott, S.M., 1990, Editorial: attainment of peak bone mass, *J. Clin. Endocrinol. Metab.* **71**:1082A.

Packer, E., Holloway, L., Newhall, K., Kanwar, G., Butterfield, G., and Marcus, R., 1990, Effects of estrogen on daylong circulating calcium, phosphorus, 1,25-dihydroxyvitamin D, and parathyroid hormone in postmenopausal women, *J. Bone Mineral Res.* **5**:877.

Pettifor, J.M., Marie, P.J., Sly, M.R., du Bruyn, D.B., Ross, F., Isdale, J.M., De Klerk, W., and van der Walt, W.H., 1984, The effects of differing dietary calcium and phosphorus contents on mineral metabolism and bone histomorphometry in young vitamin D-replete baboons, *Calcif. Tissue Int.* **36**:668.

Portale, A.A., Booth, B.E., Halloran, B.P., and Morris, R.C., Jr., 1984, Effect of dietary phosphorus on circulating concentrations of 1,25-dihydroxyvitamin D and immunoreactive parathyroid hormone in children with moderate renal insufficiency, *J. Clin. Invest.* **73**:1580.

Portale, A.A., Halloran, B.P., Murphy, M.M., and Morris, R.C., Jr., 1986, Oral intake of phosphorus can determine the serum concentration of 1,25-dihydroxyvitamin D by determining its production rate in humans, *J. Clin. Invest.* **77**:7.

Portale, A.A., Halloran, B.H., and Morris, R.C., Jr., 1987, Dietary intake of phosphorus modulates the circadian rhythm in serum concentration of phosphorus, *J. Clin. Invest.* **80**:1147.

Portale, A.A., Halloran, B.P., and Morris, R.C., Jr., 1989, Physiologic regulation of the serum concentration of 1,25-dihydroxyvitamin D by phosphorus in normal men, *J. Clin. Invest.* **83**:1494.

Raisz, L.G., and Lorenzo, J.A. 1980, "Interactions of hormones, ions, and drugs in the regulation of osteoclastic bone resorption," in: *Phosphate and Minerals in Health and Disease, Advances in Experimental Medicine and Biology* (S. G. Massry, E. Ritz, and H. Jahns, eds.), pp. 579-96, Plenum Press, New York.

Reid, I.R., Ames, R.W., Evans, M.C., Gamble, G.D., and Sharpe, S.J., 1993, The effect of calcium supplementation on bone loss in postmenopausal women, *N. Engl. J. Med.* **328**:460.

Reiss, E., Canterbury, J.M., Bercovitz, M.A., and Kaplan, E.L., 1970, The role of phosphate in the secretion of parathyroid hormone in man, *J. Clin. Invest.* **49**:146.

Riggs, B.L., and Melton, L.J., 1986, Involutional osteoporosis, *N. Engl. J. Med.* **314**:1676.

Riggs, B.L., and Melton, L.J., 1992, Prevention and treatment of osteoporosis, *N. Engl. J. Med.* **327**:620.

Ritskes-Hoitinga, J., Lemmens, A.G., Danse, L.H.J.C., and Beynen, A.C., 1989, Phosphorus-induced nephrocalcinosis and kidney function in female rats, *J. Nutr.* **119**:1423.

Saville, P. D., and Krook, L., 1969, Gravimetric and isotopic studies in nutritional hyperparathyroidism in beagles, *Clin. Orthop.* **62**:15.

Schaafsma, G., and Visser, R., 1980, Nutritional interrelationships between calcium, phosphorus, and lactose in rats, *J. Nutr.* **110**:1101.

Schryver, H.F., Hintz, H.F., and Craig, P.H., 1971, Calcium metabolism in ponies fed a high phosphorus diet, *J. Nutr.* **101**:259.

Sie, T-L., Draper, H.H., and Bell, R.R., 1974, Hypocalcemia, hypoparathyroidism and bone resorption in rats induced by dietary phosphate, *J. Nutr.* **104**:1195.

Silverberg, S.J., Shane, E., Clemens, T.L., Dempster, D.W., Segre, G.V., Lindsay, R., and Bilezikian, J.P., 1986, The effect of oral phosphate administration on major indices of skeletal metabolism in normal subjects, *J. Bone Mineral Res.* **1**:383.

Silverberg, S. J., Shane, E., Luz del la Cruz, R.N., Segre, G.V., Clemens, T.L., and Bilezikian, J.P., 1989, Abnormalities in parathyroid hormone secretion and 1,25-dihydroxyvitamin $D_3$ formation in women with osteoporosis, *N. Engl. J. Med.* **320**:277.

Smith, D.A, and Nordin, B.E.C., 1964, The effect of high phosphorus intake on total and ultra-filtrable plasma calcium and on phosphate clearance, *Clin. Sci.* **26**:479.

Spencer, H., Menczel, J., Lewin, I., and Samachson, J., 1965, Effect of high phosphorus intake on calcium and phosphorus metabolism in man, *J. Nutr.* **86**:125.

Spencer, H., Kramer, L., Osis, D., and Norris, C., 1978, Effect of phosphorus on the absorption of calcium and on the calcium balance in man, *J. Nutr.* **108**:447.

Tylavsky, F.A., and Anderson, J.J.B., 1988, Dietary factors in bone health of elderly lacto-ovovegetarians and omnivorous women, *Am. J. Clin. Nutr.* **48**:842.

Van den Berg, C.J., Kumar, R., Wilson, D.M., Heath, H. III, and Smith, L,H., 1980, Orthophosphate therapy decreases urinary calcium excretion and serum 1,25-dihydroxyvitamin D concentrations in idiopathic hypercalciuria, *J. Clin. Endocrinol. Metab.* **51**:998.

Yano, K., Heilbrun L, Wasnich R., Hankins, J., and Voegel, J., 1985, The relation between diet and bone mineral content of multiple skeletal sites in elderly Japanese-American men and women living in Hawaii, *Am. J. Clin. Nutr.* **42**:877.

Yates, J.A., Oreffo, R.O.C., Mayor, K., and Mundy, G.R., 1991, Inhibition of bone resorption by inorganic phosphate is mediated by both reduced osteoclast formation and decreased activity of mature osteoclasts, *J. Bone Mineral Res.* **6**:473.

Zemel, M.B., and Linkswiler, H.M., 1981, Calcium metabolism in the young adult male as affected by level and form of phosphorus intake and level of calcium intake, *J. Nutr.* **111**:315.

Chapter 12

# The Effect of Sodium on Calcium Requirement

B. E. Christopher Nordin and Allan G. Need

## 1. Introduction

Ever since Walser (1961) demonstrated the powerful effect of sodium infusion on calcium excretion in dogs, there has been a continuing interest in the relationship between sodium and calcium excretion in human subjects and in various animal species. It has become increasingly clear that sodium intake is an important determinant of obligatory calcium loss, and this in turn has aroused interest in the role of sodium in the pathogenesis of osteoporosis and the role of salt restriction in its management. At the same time, renal physiologists have sought to describe and explain (with only partial success) the nature of the mechanisms which link sodium and calcium transport in the kidneys. The present position is that more is known about the empirical effects of sodium intake on urine calcium than on the nature of the mechanism which governs this relationship. This review is therefore necessarily more concerned with outcomes than with mechanisms, but the latter will be reviewed in brief.

---

B.E. Christopher Nordin • Division of Clinical Biochemistry, Institute of Medical and Veterinary Science, Frome Road, Adelaide, South Australia 5000, and Department of Pathology, The University of Adelaide. Allan G. Need • Division of Clinical Biochemistry, Institute of Medical and Veterinary Science, Adelaide, South Australia.

*Advances in Nutritional Research,* Vol. 9
Edited by Harold H. Draper
Plenum Press, New York, 1994

## 2. Renal Handling of Sodium and Calcium

There are strong similarities between the renal clearance of sodium and calcium from plasma, but also certain important differences. Whereas the plasma sodium is totally ultrafiltrable, only about 60% of the plasma calcium passes through the glomerular membrane; the rest is bound to protein. A typical plasma calcium of 2.4 mmol/L (9.6 mg/dL) comprises 0.8 mmol/L (3.2 mg/dL) bound to protein (mainly albumen) and 1.6 mmol/L (6.4 mg/dL) in the ultrafiltrate. The latter fraction in turn contains about 1.20 mmol/L (4.8 mg/dL) of ionized calcium and 0.4 mmol/L (1.6 mg/dL) of "complexed" calcium bound to bicarbonate and other anions. These fractions can be measured but more simply (and probably more precisely) calculated (Nordin *et al.,* 1989).

After allowing for these differences in the plasma distribution of sodium and calcium, the renal handling of these two electrolytes is similar but not identical. About 99% of the filtered calcium and sodium, and a similar proportion of plasma water are reabsorbed in the renal tubules with the result that sodium and calcium clearances are generally similar and their concentrations in the urine of the same order as those in the plasma water (Table 1). However, although the calcium and sodium clearances are each generally about 1 ml/min, the ratio between them does vary with sodium intake because urine sodium can fall virtually to zero in states of sodium deprivation, whereas there is always obligatory loss of calcium in the urine regardless of calcium or sodium intake. This is discussed below.

Table 1. Sodium and Calcium Concentrations in Plasma Water and Urine

| Electrolyte | Plasma water concentration | Urine concentration | Average clearance |
|---|---|---|---|
| Sodium | 140 mmol/L | 50-150 mmol/L | 1 ml/min |
| Calcium | 1.6 mmol/L | 1-5 mmol/L | 1 ml/min |

The similarities do not end there. About 60% of the filtered sodium is reabsorbed in the proximal tubules, 30% in the thick ascending limb of the Henle's loop, 7% in the distal convoluted tubule, and 2–3% in the collecting system (Berry and Rector, 1991). The corresponding figures for calcium are estimated as 70% in the proximal tubules, 20% in the thick ascending limb, 5–10% in the distal nephron and less than 5% in the collecting tubules (Agus and Goldfarb, 1985). Clearly these values are essentially the same. Reabsorption of both ions involves both active and passive transport mechanisms but the general pattern appears to be that calcium follows salt and water (with a perceptible time lag) in the proximal tubules but is subject to independent regulation (notably by parathyroid hormone (Bourdeau and Burg, 1980; Costanzo and Windhager, 1980) in the distal tubules. Sodium reabsorption is subject to independent regulation by aldosterone in the distal tubules (Berry and Rector, 1991).

Despite the large amount of information available from micropuncture, tubule perfusion and other studies, it is still not clear where or how the positive effect of sodium load on urine calcium is mediated. There is substantial evidence that loop diuretics (e.g., frusemide) inhibit both calcium and sodium reabsorption in the loop of Henle, whereas thiazides and amiloride act more distally to inhibit sodium reabsorption without reducing (probably even enhancing) reabsorption of calcium (Costanzo and Weiner, 1976; Edwards *et al.,* 1973; Costanzo, 1984; Hanson *et al.,* 1982). But these studies do not explain where or how a sodium load reduces calcium reabsorption. Volume expansion *per se* is probably not an adequate explanation (Wong and Quamme, 1990), since the sodium effect is still present in volume-depleted dogs (Massry *et al.,* 1967). A possible explanation may be found in the work of Brunette *et al.* (1992) who reported that the uptake of calcium by the luminal membranes of distal tubules was inhibited by sodium in the bathing fluid; this would suggest that the competition occurs in the distal tubules.

## 3. The Relation Between Urine Sodium and Calcium

It is an empirical fact that urine sodium and calcium are significantly related in normal and hypercalciuric subjects on free diets, both in 24-hr collections and in fasting samples (Nordin and Polley, 1987; Goulding *et al.,* 1986; Sabto *et al.,* 1984; Kleeman *et al.,* 1964; Epstein, 1968). This is illustrated in Figure 1 and Table 2.

Table 2. The Relation Between Calcium/Creatinine and Sodium/Creatinine Ratios in 24-hr and Fasting Urine Samples in Young Women

| Sample | n | Slope of Ca/Cr on Na/Cr | Intercept | r | P |
|---|---|---|---|---|---|
| 24-hr Urine | 224 | 0.0081 | 0.26 | 0.25 | $<0.001$ |
| Fasting Urine | 77[a] | 0.0068 | 0.11 | 0.42 | $<0.001$ |
| 't' | | 0.4 | 4.2 | | |
| P | | ns | $<0.001$ | | |

[a] FSH $\leq$45 U/L

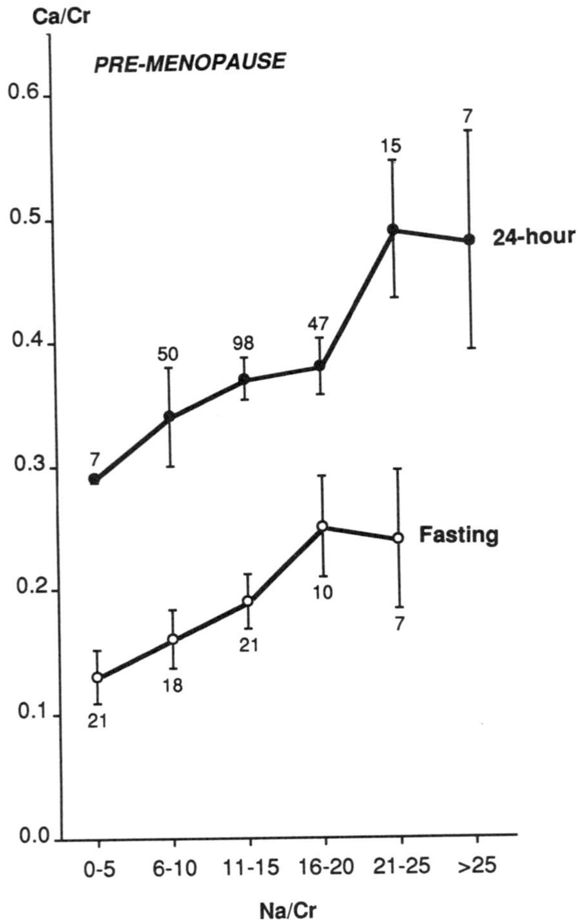


Figure 1. The relation between urine sodium and calcium in 24-hr collections and fasting samples from premenopausal women. The numbers in each group are indicated

The slope of urine calcium on sodium is similar in the fasting and 24-hr collections but the intercept on the calcium axis (i.e., urine calcium at zero sodium excretion) is much higher in the 24-hr than in the fasting samples due to the contribution of absorbed dietary calcium. The coefficient of correlation is much higher in the fasting specimens than in the 24-hr urines, though both are significant. This simply means that sodium is a more important determinant of obligatory loss of calcium, i.e., calcium that is being "pulled out" of the body, than of excess urine calcium that is being "pushed out." In respect of both

Table 3. Mean 24-hr and Fasting Ca/Cr and Na/Cr ± SE in Young Women

| Collection | n | Ca/Cr | Na/Cr |
|---|---|---|---|
| 24-hr | 224 | 0.37 ± 0.011 | 14.1 ± 0.38 |
| Fasting | 77[a] | 0.18 ± 0.013 | 10.8 ± 0.80 |
| 't' | | 11.2 | 3.7 |
| P | | <0.001 | 0.001 |

[a] FSH ≤45 U/L

sodium and calcium, the fasting excretion rates are lower than the 24-hr rates because of the overnight deprivation of sodium and calcium. This is shown in Table 3 which suggests that the overnight fall in urine calcium (Ca/Cr) may be proportionately greater than the overnight fall in Na/Cr, at least in young women.

This is also apparent from the data of Simpson *et al* (1978) and Goulding (1981). In her premenopausal women, the mean fasting Na/Cr (11.2) is 8% below the 24-hr Na/Cr (12.1) but the mean fasting Ca/Cr (0.24) is 17% below the 24-hr mean (0.29). Statistical analysis is not possible from the published data.

Because of the positive intercepts of the calcium/sodium regression slopes on the calcium axis (Figure 1), the calcium and sodium clearances diverge as the urine sodium approaches zero. This is shown in Figure 2. The assumed identity of the two clearances is only operative when the urine sodium/creatinine ratio is over about 10; below that, the calcium clearance frequently exceeds the sodium clearance.

That the calcium/sodium relationship is not simply the result of an association between calcium and sodium intakes is demonstrated by studies in which dietary sodium has been raised or lowered and urine calcium has followed suit. In 17 young normal men, urine calcium fell from 5.3 ± 1.5 (SE) mmol/day to 4.4 ± 0.8 after 12 days on a sodium intake of 56 mmol/day (King *et al.,* 1964). In 14 men with idiopathic hypercalciuria and renal stones and 10 young controls, urine calcium rose and fell with sodium intake both on low and normal calcium intakes (Phillips and Cooke, 1967). The slope of urine calcium on sodium was steeper in the stone-formers than in the controls but the intercepts appeared to be similar. In 6 normal men on a calcium intake of 10 mmol/day, urine calcium rose from 1.5 ± 0.5 (SE) to 6.5 ± 1.3 mmol/day as sodium intake was raised from 10 to 1500 mmol/day (McCarron *et al.,* 1981). When urine sodium rose from 18 ± 5 to 249 ± 2 mmol, urine calcium rose from 1.5 ± 0.50 to 3.1 ± 0.88 mmol. This yields a slope of 0.007 and a calcium intercept of 1.4 mmol/day. At higher sodium intakes, the effect on urine calcium became progressively less marked.

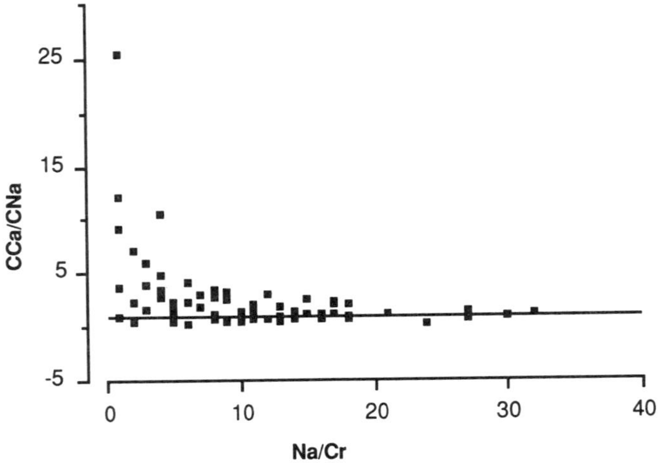


Figure 2. The calcium/sodium clearance ratio as a function of urine sodium in premenopausal women. The horizontal line represents a clearance ratio of 1.

In 21 normal subjects on a 10 mmol calcium diet (Breslau *et al.*, 1982), urine calcium rose from 2.8 ± 0.35 (SE) to 4.2 ± 0.40 mmol/day when diet sodium intake was increased from 10 to 250 mmol/day. The slope of urine calcium on sodium was 0.0064 and the calcium intercept at zero sodium was 1.4 mmol/day. (In 2 cases of hypoparathyroidism, the slope of urine calcium on sodium was only 0.0068 but the calcium intercept was very high at 6.6 mmol/day). In another study, 14 male and 4 female hypercalciuric stone-formers on unrestricted calcium intakes were observed on sodium intakes of 80 and 120 mmol/day (Muldowney *et al.*, 1982). Their urine calcium rose from 7.0 ± 0.46 (SE) to 9.6 ± 0.56 mmol/day ($P<0.001$). This yields a slope of calcium on sodium of 0.025 with an intercept on the calcium axis of 1.8 mmol/day. In another study of 4 hypercalciuric stone-formers (Silver *et al.*, 1983), the slope of urine calcium on sodium was about 0.03 with a calcium intercept of about 2.0 mmol/day.

One can conclude from these studies that 24-hour urine calcium is significantly related to urine sodium and causally related to dietary sodium, particularly when dietary calcium is restricted. The slopes of changes in urine calcium on changes in sodium have a gradient of about 1% in normal subjects. The slope is steeper in hypercalciuric stone-formers but the intercept on the calcium axis is unchanged; in hypoparathyroidism, the slope is normal but the calcium intercept is high.

Although most of the work on this subject has dealt with the effect of sodium on calcium, there are situations in which calcium intake, and/or absorption may influence sodium excretion, again presumably by the same kind of competitition between sodium and calcium for tubular reabsorption somewhere in the nephron. The classical example of this is the dehydration and salt depletion associated with hypercalcemia (Harinck *et al.,* 1987). Sodium excretion due to a calcium load is not, however, the simple reciprocal of calcium excretion secondary to a sodium load. Sodium-dependent calcium excretion can continue indefinitely, is clearly manifest in the steady state, is met by increased calcium absorption and/or bone resorption, and has little if any effect on the plasma calcium. Calcium-driven sodium loss, on the other hand, cannot be offset by sodium absorption (which is virtually 100% of intake anyway) or by bone resorption; there is no feedback loop to transmit the signal and very little sodium in bone or other tissue. The effect is therefore to cause volume contraction and/or hyponatremia and a return of sodium excretion to the intake level. Absolute calcium-induced natriuria is therefore seen only as an acute transient effect although relative natriuria continues. The acute effect was noted by Reid *et al.* (1986) when they administered a calcium load to normal subjects and recorded an acute rise in urine sodium. The molar ratio of the increase in urine sodium relative to calcium was 20:1 compared with the 1:100 Ca/Na ratio that applies when a sodium load takes out calcium. In an unpublished study, administration of a 1 g calcium supplement, of which about 5% appeared in the urine, raised urine sodium acutely in a sodium:calcium ratio of about 10:1 (Table 4).

Table 4. Effect of a 1 g Calcium Load (as Carbonate) on Mean (±SE) 24-hr Urinary Calcium and Sodium (mmol/day) in 30 Normal Premenopausal Women [a]

| Day | Urine calcium | Urine sodium |
|---|---|---|
| Control | 3.23 ± 0.34 | 87.7 ± 7.0 |
| Calcium | 5.48 ± 0.50 | 109 ± 9.6 |
| Change | 2.25 ± 0.36 | 21.0 ± 9.2 |
| P | <0.001 | <0.05 |

[a] Need & Nordin, unpublished

## 4. Effect of Sodium Intake on Calcium Requirement

The calcium requirement of young adults has traditionally been defined as the calcium intake at which calcium intake and output are equal. Another way of expressing the same concept is to define requirement as the value at which absorbed and excreted calcium are equal. Based on balance studies, absorbed calcium (y) is related to calcium intake (x) by the term:

$$y = \frac{12.3\ x}{7.2 + x} + 0.06x - 5.2 \text{ mmol}$$

Urine calcium (z) is related to intake by the term:

$$z = .051x + 3.2 \text{ mmol}$$

(Nordin and Marshall, 1988). From these relationships, the mean calcium requirement of young adults can be estimated to be 14.5 mmol (580 mg). However, this does not allow for skin losses of calcium estimated at 1 mmol (40 mg) (Charles *et al.,* 1983; Harden, 1964) which if added to the urinary component as an obligatory loss, increase the mean calcium requirement from 14.5 to 21 mmol (840 mg) (Figure 3).

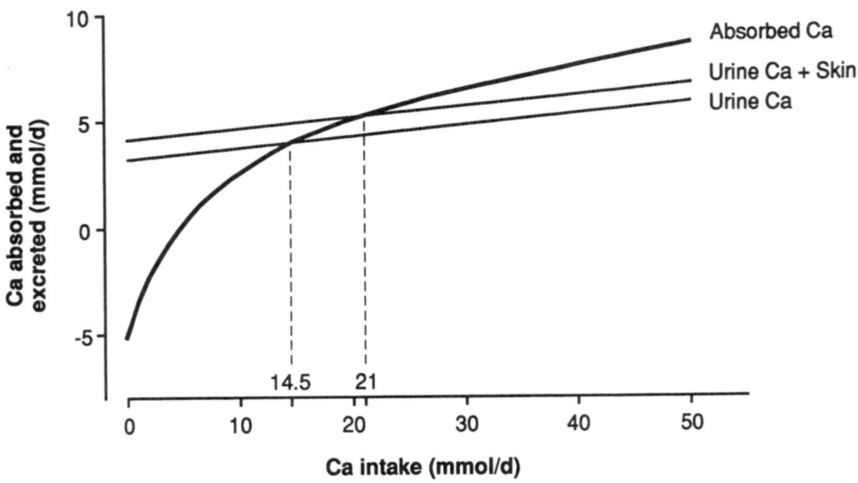


Figure 3. Calcium intake, absorption and excretion calculated from calcium balances in normal young adults. The estimated mean calcium requirement is the intake at which absorbed and excreted calcium are equal, i.e., 21 mmol. Allowing for a 1 mmol loss through the skin increases the requirement to 26 mmol.

The sodium intake on which the relevant balance studies were performed is not known, but the effect of changing the sodium intake upwards or downwards can nonetheless be computed, assuming a slope of urine calcium on sodium of about 1%. This has been done in Figure 4, which shows that reducing the sodium intake by 50 mmol would reduce the calcium requirement by 3.5 mmol (140 mg), and increasing the sodium intake by 50 mmol would increase the theoretical calcium requirement by 5 mmol (200 mg). Thus sodium intake may be a significant determinant of calcium requirement and therefore of calcium balance.

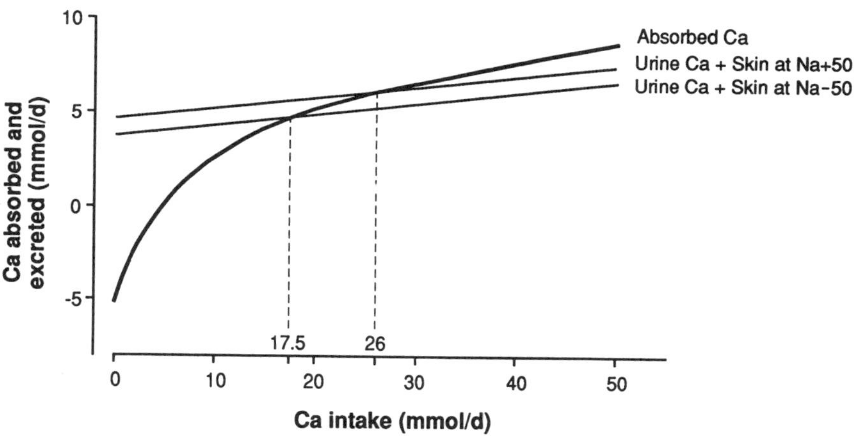


Figure 4. The effect of varying sodium intakes on calculated calcium requirement in normal young adults. Increasing sodium intake by 50 mmol/day increases urine calcium and raises the calcium requirement from 21 to 26 mmol. Reducing sodium intake by the same amount lowers the requirement to 17.5 mmol.

## 5. Effect of Menopause

It is well known that urine calcium rises at the menopause (Nordin and Polley, 1987; Goulding, 1981; Stepan *et al.*, 1987; Kelly *et al.*, 1989). This is apparent both in 24-hr collections, which rise at least 0.5 mmol (20 mg), and more particularly in the fasting Ca/Cr, which rises by about 50% (Table 5) (Nordin *et al.*, 1990). (These figures, from Adelaide, are virtually identical with the corresponding figures from New Zealand (Goulding, 1981; Simpson *et al.*, 1978). This extra obligatory loss of 0.5 mmol in the urine increases the mean calcium requirement from 21 mmol (840 mg) to 26 mmol (1040 mg) (Figure 5).

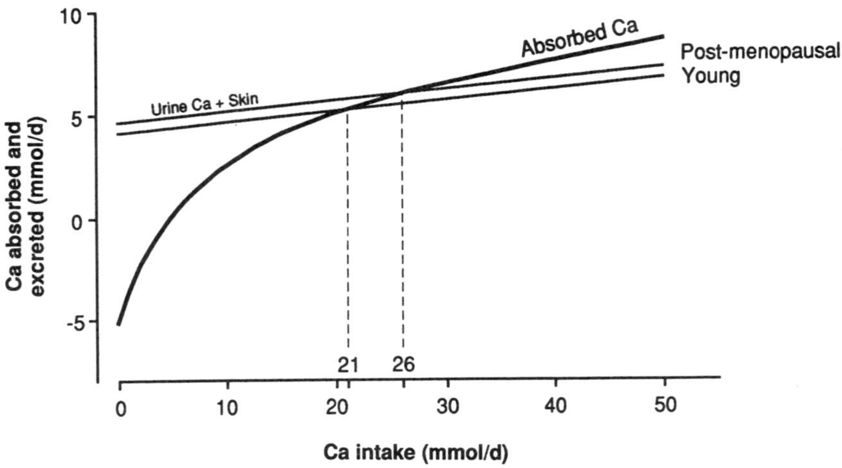


Figure 5. The effect of a 0.5 mmol increase in urine calcium at the menopause on estimated calcium requirement assuming no change in calcium absorption. Note that the effect of menopause is equivalent to an increase of 50 mmol in sodium intake in young adults (cf. Figure 4).

Table 5. Mean 24- hr and Fasting Urine Calcium (± SE) in Normal Premenopausal Women and Postmenopausal Women Within 10 Years of Menopause

| Variable | n | Premenopause | n | Postmenopause | P |
|---|---|---|---|---|---|
| 24 hr Ca (mmol) | 224 | 3.64 ± 0.12 | 282 | 4.15 ± 0.12 | <0.01 |
| 24 hr Cr (mmol) | 224 | 9.93 ± 0.16 | 282 | 10.10 ± 0.12 | ns |
| 24 hr Ca/Cr | 224 | 0.38 ± 0.011 | 282 | 0.42 ± 0.0012 | <0.02 |
| P | | <0.001 | | <0.001 | |
| Fasting Ca/Cr | 77[a] | 0.18 ± 0.013 | 157[b] | 0.27 ± 0.011 | <0.001 |

[a]FSH ≤45 U/L
[b]FSH >45 U/L

Assuming a calcium/sodium slope of 1/100, reducing the sodium intake by 50 mmol would return this to the premenopausal value of 21 mmol (840 mg). Increasing the sodium intake by 50 mmol would increase the postmenopausal requirement to 32 mmol (1280 mg). The figures are lower than the estimate of Heaney *et al.* (1978) of 37.5 mmol (1500 mg) which did not take into account skin losses, and lower than that of Hasling *et al.* (1990) in osteoporotic women of 34.5 mmol (1380 mg), which did include skin losses. The main reason for these differences is rather higher absorption coefficients in our data set.

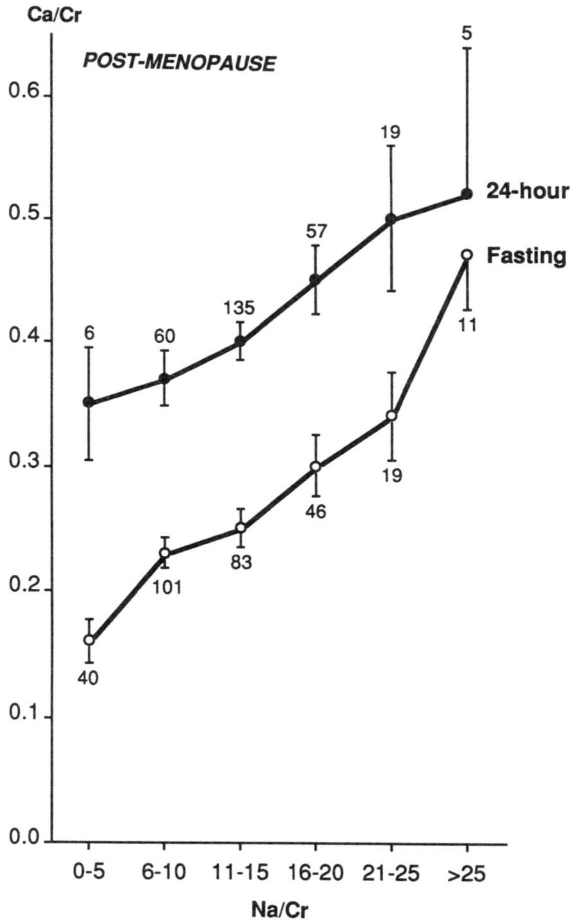


Figure 6. The relation between urine calcium and sodium in 24-hr and fasting urine samples from postmenopausal women. The numbers in each group are indicated.

As shown in Figure 6, urine calcium and sodium are also related in postmenopausal women but urine calcium is again higher for any given urine sodium in the 24-hr collections than in the fasting samples. This is because of the higher intercept on the calcium axis in the 24-hour collections (due presumably to absorbed dietary calcium) rather than to any difference in slope between the fasting and non-fasting samples (Table 6).

Table 6. Regression of Ca/Cr on Na/Cr in 24-hr and Fasting Urine Samples in Postmenopausal Women

| Collection | n | Slope | Intercept | r | P |
|---|---|---|---|---|---|
| 24 hour | 282 | 0.0093 | 0.29 | 0.23 | <0.001 |
| Fasting | 157[a] | 0.0090 | 0.15 | 0.42 | <0.001 |
| 't' | | 0.13 | 3.8 | | |
| P | | ns | <0.001 | | |

[a]FSH >45 U/L

When the calcium/sodium relationships in premenopausal and postmenopausal women are compared (Tables 2 and 6), the slope of Ca/Cr on Na/Cr appears steeper and the Ca/Cr intercept higher in post- than premenopausal women, but neither difference is in fact significant. However, when urine calcium is corrected for urine sodium by covariance analysis, the corrected Ca/Cr is significantly higher in postmenopausal than premenopausal women both in 24-hr collections (0.42 vs 0.37: P<0.001) and in fasting specimens (0.26 vs 0.19: P<0.001).

Table 7. Effect of Calcium Intake on the Regression of 24-hr Urine Calcium on Sodium in Normal Postmenopausal Women

| Calcium intake | n | Slope | Intercept | r | P |
|---|---|---|---|---|---|
| ≤1000 mg | 274 | 0.015 | 2.03 | 0.38 | <0.001 |
| >1000 mg | 178 | 0.0089 | 3.04 | 0.21 | <0.005 |

The influence of calcium intake on the calcium/sodium relationship in urine (referred to earlier) can be seen by dividing the whole Adelaide postmenopausal set (Nordin and Polley, 1987) into those women on calcium intakes up to 1000 and over 1000 mg. This is shown in Table 7. In those women on the lower calcium intakes, the correlation between 24-hr calcium and sodium is much higher than in those on the higher intakes, though both are significant.

Not only is calcium excretion higher relative to sodium in postmenopausal than premenopausal women, but there may also be an actual rise in overnight sodium excretion as well after the menopause. This is suggested by the Na/Cr ratios in both the Australian and New Zealand data (Table 8).

Table 8. 24-hr and Fasting Urine Sodium (mean ± SE) in Australian and New Zealand Surveys (numbers in parentheses)

| Source | 24 hour specimens (mmol sodium/day) | | P |
|---|---|---|---|
| | Premenopausal | Postmenopausal | |
| Need & Nordin (unpubl.) | 133 ± 3.2 (224) | 138 ± 3.2 (282) | ns |
| Goulding (1981)[(a)] | 142 ± 6.2 (82) | 146 ± 4.8 (85) | ns |
| | Fasting specimens (molar Na/Cr) | | |
| | Premenopausal | Postmenopausal | |
| Nordin *et al.*, (1990) | 10.8 ± 0.80 (77)[(b)] | 12.8 ± 0.53 (157)[(c)] | <0.05 |
| Goulding (1981) | 11.2 ± 0.61 (146) | 14.8 ± 0.89 (171) | <0.005 |

[(a)]Assuming 30-39 year olds are premenopausal and 60-69 year olds are postmenopausal
[(b)]FSH ≤45 U/L
[(c)]FSH >45 U/L

This might be due to a decline in creatinine output after menopause. This was not seen in the Australian study in which 24-hr urine creatinine was 9.93 ± 0.16 mmol in the premenopausal and 10.10 ± 0.16 mmol in the postmenopausal women. In the New Zealand series, however, 24-hr creatinine was 11.7 ± 0.35 mmol in the 30–39 year olds and 10.1 ± 0.34 mmol in the 60–69 year olds, a reduction which was significant (P<0.001) and which would account for about half of the rise in Na/Cr. There is therefore a suggestion that the pattern of sodium excretion changes at the menopause, i.e., that there is a delay in the excretion of the previous day's sodium such as has been described to occur with aging (Epstein and Hollenberg, 1976). This would have the effect of taking calcium out of the body during the night and early morning when absorption of dietary calcium has ceased and could be regarded as delaying the overnight calcium adaptation referred to earlier. Thus, whereas in premenopausal women the fasting Ca/Cr is about 50% of the 24-hr value, in postmenopausal women it is about 65% of the 24-hr value (Table 4). Some of this apparent delay in calcium excretion (or failure of adaptation) may be due to a corresponding delay in sodium excretion, but the difference in fasting Ca/Cr between pre- and post-menopausal women remains after correction for the sodium effect by covariance analysis (sodium-corrected Ca/Cr in premenopausal women $0.19 \pm 0.014$; $0.26 \pm 0.0098$ in postmenopausal women) ($P<0.001$). The alternative explanation that the extra sodium comes from bone is incompatible with the low sodium content of bone (ratio to calcium 1:40 by weight (McLean and Urist, 1968).

The effect of sodium load on bone resorption can be gauged from its effect on urine hydroxyproline, which is derived from the collagen component of bone and generally measured as the molar hydroxyproline/creatinine ratio in the fasting urine (Nordin, 1978). It is well established that this ratio rises at the menopause (Nordin and Polley, 1987; Goulding, 1981; Stepan *et al.*, 1987; Kelly *et al.*, 1989; Nordin *et al.*, 1990) (Table 9) and that it is restored to the premenopausal level by hormone treatment (Gallagher and Nordin, 1975; Christiansen *et al.*, 1982).

Table 9. Mean Fasting Molar Hydroxyproline/Creatinine Ratios (x 1000) ± SE in Pre- and Postmenopausal Women

| Source | n | Premenopausal | n | Postmenopausal | P |
|---|---|---|---|---|---|
| Nordin *et al.* (1990) | 77[a] | 14.1 ± 0.48 | 157[b] | 18.4 ± 0.57 | <.001 |
| Goulding(1981)[c] | 146 | 22.0 ± 0.1 | 171 | 29.0 ± 0.2 | <.005 |

[a] FSH ≤45 U/L

[b] FSH >45 U/L

[c] Assuming 30-39 year olds are premenopausal and 60-69 year olds are postmenopausal

The relation between urine sodium and urine hydroxyproline is shown in Table 10 and illustrated in Figure 7. In premenopausal women, hydroxyproline is not significantly related to urine sodium but in postmenopausal women there is a significant correlation between them. However, Goulding and Lim (1983) did find a significant correlation between urine sodium and hydroxyproline even in premenopausal women.

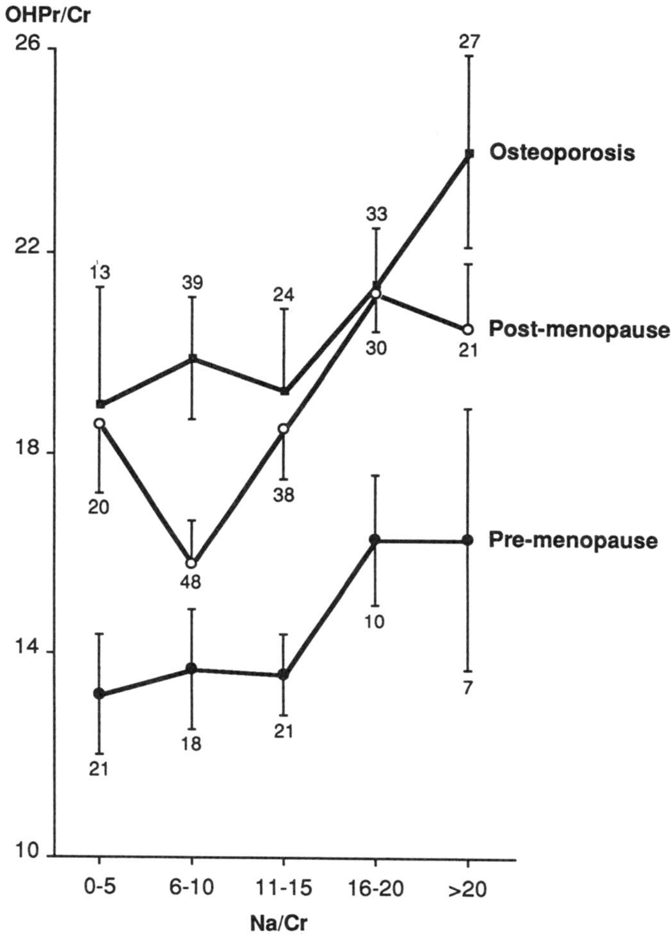


Figure 7. The relation between urine hydroxyproline and urine sodium in fasting urine samples from premenopausal and osteoporotic women. The numbers in each group are indicated

The figure suggests that the effect of sodium on bone resorption is not seen until the sodium/creatinine ratio exceeds about 15. Above that level, there is an apparent rise in hydroxyproline even in premenopausal women, but it is not statistically significant. As far as is known, the dependence of OHPr/Cr on Na/Cr is mediated via the urine calcium; when urine hydroxyproline is regressed on urine sodium and calcium, the former determinant loses its statistical significance. The chain of events therefore appears to be:

sodium load → urine calcium loss → increased urine hydroxyproline.

Table 10. Regression of Molar OHPr/Cr (x 1000) on Molar Na/Cr in Fasting Urine Samples from Premenopausal and Normal and Osteoporotic Postmenopausal Women [unpublished data]

| Group | n | Slope | Intercept | r | P |
|---|---|---|---|---|---|
| Premenopausal[a] | 77 | 0.13 | 12.7 | 0.11 | ns |
| Postmenopausal[b] | 157 | 0.23 | 15.5 | 0.26 | <.001 |
| Osteoporotic | 136 | 0.20 | 17.9 | 0.21 | <.02 |

[a] FSH ≤45 U/L
[b] FSH >45 U/L

Table 10 and Figure 7 also include a series of "osteoporotic" postmenopausal women, i.e., postmenopausal women with vertebral compression. In these women, the mean fasting urinary sodium, calcium and hydroxyproline are all higher than they are in age-matched controls (Morris *et al.,* 1991). The slope of urine hydroxyproline on urine sodium is similar to that in normal postmenopausal women but the intercept on the hydroxyproline axis is higher. (This is probably due to that component of their bone resorption which is a response to impaired calcium absorption (Morris *et al.,* 1991)).

## 6. Clinical Intervention

There have been two clinical studies of the effect of manipulating salt intake on urine hydroxyproline in postmenopausal women. In the first (McParland *et al.,* 1989), a dietary salt supplement of 100 mmol raised the 24-hr urine sodium from 70 to 167 mmol and the hydroxyproline from 158 to 217 mmol ($P<0.05$). In the second (Need *et al.,* 1991), moderate salt restriction reduced the hydroxyproline/creatinine ratio within a few days but the effect was significant ($P<0.001$) only in those subjects who started from a sodium/creatinine ratio over 15 (Figure. 8). The reason for this is perhaps apparent from inspection of Figure 7, which suggests that the effect of sodium load on bone resorption is most clearly seen at Na/Cr ratios over 15.

Unpublished intervention data from Adelaide are shown in Table 11.

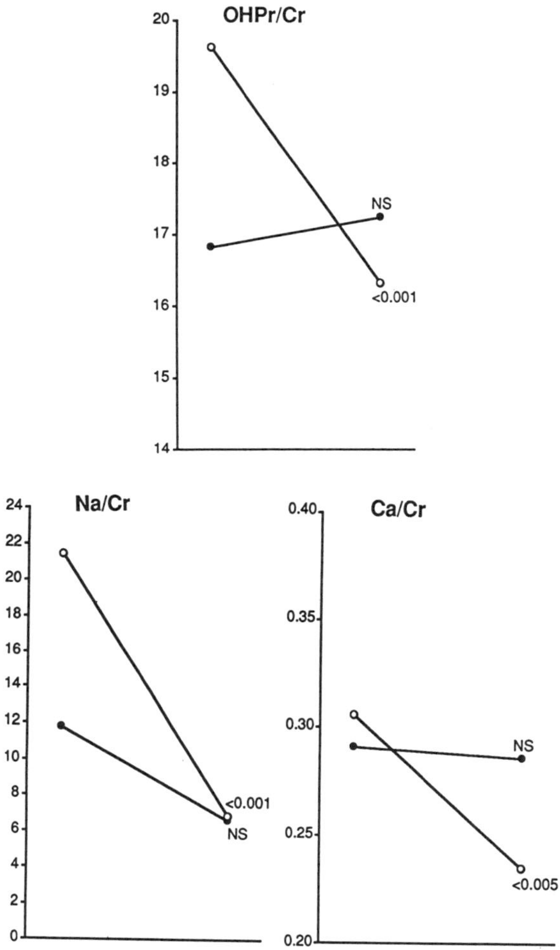


Figure 8. The effect of 2–7 days' salt restriction on fasting urinary sodium, calcium and hydroxyproline in postmenopausal women. The subjects have been divided into two groups—those starting at sodium/creatinine ratios up to 15 (31) [ o ] and those with initial sodium/creatinine ratios over 15 (28) [•]

In this study, 24-hr urine calcium and sodium and fasting urine calcium, sodium and hydroxyproline were determined in 30 normal postmenopausal women on their own basal diets. The measurements were then repeated on salt-supplemented (+ 30 mmol) and salt-restricted (- 30 mmol) intakes, to which the subjects were allocated in random order. As Table 11 shows, 24-hr urine sodium excretion clearly reflected salt intake but there was little change in 24-hr urine calcium. In the fasting urines, sodium excretion again varied with the intake, and the hydroxyproline excretion rose and fell with urine sodium although urine calcium changed very little. The explanation for this apparent anomaly is not clear, but it seems that sodium-induced changes in urine calcium do not necessarily follow the sodium/calcium relationship seen in cross-sectional studies, although variations in dietary sodium have the predicted effect on urine hydroxyproline.

Table 11. Effect of Salt-Supplementation and Salt Restriction on Urine Calcium, Sodium and Hydroxyproline (mean ± SE) in 30 Normal Postmenopausal Women

| | | Salt Intake | | |
|---|---|---|---|---|
| | | Low | Basal | High |
| 24-hour Urine | Ca/Cr | 0.38 ± 0.042 | 0.42 ± 0.039 | 0.42 ± 0.052 |
| | Na/Cr | 12.5 ± 2.2 | 14.5 ± 1.4 | 17.6 ± 0.98 |
| Fasting Urine | Ca/Cr | 0.26 ± 0.032 | 0.27 ± 0.030 | 0.28 ± 0.024 |
| | Na/Cr | 8.0 ± 0.81 | 11.7 ± 1.3 | 16.0 ± 1.6 |
| | OHPr/Cr [a] | 13.4 ± 1.0 | 15.7 ± 0.87 | 17.1 ± 1.5 |

[a] Molar ratio x 1000

## 7. Mechanisms and Conclusions

The above observations show that urine calcium varies with urine sodium, particularly when calcium intake is low or the patient is fasting. In postmenopausal women in particular, this effect of sodium load on calcium excretion may make a significant contribution to the rate of bone resorption. The potential importance of this has been underlined by intervention studies, of which more are needed. In particular, it needs to be established whether there is a change in renal handling of calcium after menopause.

The way in which the sodium-load signal is transmitted to the bone is not clear. In theory, a sodium-dependent rise in urine calcium would tend to lower plasma ionized calcium and stimulate PTH secretion; high serum PTH levels have in fact been reported in four women with osteoporosis and hypercalciuria (Sakhaee *et al.*, 1985) and rises in serum PTH and $1,25(OH)_2D_3$ in response to salt-loading have also been described in a study of 11 normal subjects (Breslau *et al.,* 1982). Experimental studies have also demonstrated increased calcium excretion and reduced bone mass in salt-loaded rats (Goulding, 1980), particularly on a restricted calcium intake (Goulding and Campbell, 1984) and after oophorectomy (Goulding and Campbell, 1983). These changes were associated with increased cyclic AMP and hydroxyproline excretion in intact but not in thyroparathyroidectomised animals, suggesting that the bone loss in the intact animals was PTH-mediated. However, we have not seen in our own large series of postmenopausal women any positive correlation between urine sodium or calcium and serum PTH and we are therefore doubtful whether the sodium effect on urine calcium and bone resorption is PTH-mediated. Nor is it clear why sodium-dependent calcium excretion should—apparently—stimulate bone resorption in postmenopausal but not in premenopausal women, which is what our urine hydroxyproline data suggest. Does the normal response to a sodium-dependent calciuria involve activation of parathyroid hormone and $1{,}25(OH)_2D_3$ production and a compensatory increase in intestinal calcium absorption? If so, is this response lacking in postmenopausal women as Breslau *et al.* (1985) have suggested? Or is the obligatory calcium loss so much greater in postmenopausal than premenopausal women, possibly due to reduced renal tubular reabsorption of calcium (Nordin *et al.,* 1991), that its further elevation by sodium raises it beyond the level at which available calcium stores can meet the demand without breaking down bone? Is there perhaps a blood/bone equilibrium which is set by parathyroid hormone but then sustains the plasma calcium without further PTH intervention (Nordin, 1960; Staub *et al.,* 1989; Parfitt, 1989)? Or is there even some way in which local cytokines can transmit a calcium signal? These questions are unanswered at the moment.

It is clear, however, that salt intake is an important determinant of urine calcium, particularly on low calcium intakes and in the postmenopausal state. It is also clear that this effect is due to the sodium ion since it cannot be reproduced with potassium chloride (Walser, 1961). Nor does it follow sodium administration as an alkali (Goulding *et al.,* 1984; Kurtz *et al.,* 1987), presumably because of the positive effect of alkali on tubular reabsorption of calcium (Sutton *et al.,* 1979). The clinical significance of sodium-dependent calcium appears to be borne out by the effect of salt loading and restriction on urine hydroxyproline in postmenopausal women (McParland *et al.,* 1989; Need *et al.,* 1991) but the only study of the effect of salt restriction on bone loss was inconclusive (Nordin and Polley, 1987), possibly because the subjects were selected for high rather than low calcium intakes. Future studies of the effect of salt intake on postmenopausal bone loss are awaited.

## References

Agus, Z.S., and Goldfarb, S., 1985, "Renal regulation of calcium balance," in: *The Kidney: Physiology and Pathophysiology* (D.W. Seldin and G. Giebisch, eds.), pp. 1323-1335, Raven Press, New York.

Berry, C.A., and Rector, F.C., 1991, "Renal transport of glucose, amino acids, sodium, chloride, and water," in: *The Kidney* (B.M. Brenner and F.C. Rector, eds.), pp. 245-282, W.B. Saunders, Philadelphia.

Bourdeau, J.E., and Burg, M.B., 1980, Effect of PTH on calcium transport across the cortical thick ascending limb of Henle's loop, *Am. J. Physiol.* **239:**F121.

Breslau, N.A., McGuire J.L., Zerwekh, J.E., and Pak, C.Y.C., 1982, The role of dietary sodium on renal excretion and intestinal absorption of calcium and on vitamin D metabolism, *J. Clin. Endocrinol. Metab.* **55:**369.

Breslau, N.A., Sakhaee, K., and Pak, C.Y.C., 1985, Impaired adaptation to salt-induced urinary calcium losses in postmenopausal osteoporosis, *Trans. Assoc. Am. Physician* **98:**107.

Brunette, M.G., Mailloux, J., and Lajeunesse, D., 1992, Calcium transport through the luminal membrane of the distal tubule: I. Interrelationship with sodium, *Kidney Int.* **41:**281.

Charles, P., Taagehoj, F., Jensen, L., Mosekilde, L., and Hansen, H.H., 1983, Calcium metabolism evaluated by $Ca^{45}$ kinetics: estimation of dermal calcium loss, *Clinical Science,* **65:**415.

Christiansen, C., Christensen, M., Larsen, N-E., and Transbol, I.B., 1982, Pathophysiological mechanisms of estrogen effect on bone metabolism. Dose-response relationships in early postmenopausal women, *J. Clin. Endocrinol. Metab.* **55:**1124.

Costanzo, L.S., 1984, Comparison of calcium and sodium transport in early and late rat distal tubules: effect of amiloride, *Am. J. Physiol.* **246:**F937.

Costanzo, L.S., and Weiner, I.M., 1976, Relationship between clearances of Ca and Na: effect of distal diuretics and PTH, *Am. J. Physiol.* **30:**67.

Costanzo, L.S., and Windhager, E.E., 1980, Effects of PTH, ADH, and cyclic AMP on distal tubular Ca and Na reabsorption, *Am. J. Physiol.* **239:**F478.

Edwards, B.R., Baer, P.G., Sutton, R.A.L., and J.H. Dirks, 1973, Micropuncture study of diuretic effects on sodium and calcium reabsorption in the dog nephron, *J. Clin. Invest.* **52:**2418.

Epstein, F.H., 1968, Calcium and the kidney, *Am. J. Med.* **45:**700.

Epstein, M., and Hollenberg, N.K. 1976, Age as a determinant of renal sodium conservation in normal man, *J. Lab. Clin Med.* **87:**411.

Gallagher, J.C., and Nordin, B.E.C., 1975, "Effects of oestrogen and progestogen therapy on calcium metabolism in postmenopausal women," in: *Estrogens in the Post-Menopause,* Frontiers of Hormone Research, vol. 3 (P.A. vanKeep and C. Lauritzen, eds.), pp. 150-176, Karger, Basel.

Goulding, A., 1980, Effects of dietary NaCl supplements on parathyroid function, bone turnover and bone composition in rats taking restricted amounts of calcium, *Mineral Electrolyte Metab.* **4:**203.

Goulding, A., 1981, Fasting urinary sodium/creatinine in relation to calcium/creatinine and hydroxyproline/creatinine in a general population of women, *New Zealand. Med J.* **93:**294.

Goulding, A., and Campbell, D., 1983, Dietary NaCl loads promote calciuria and bone loss in adult oophorectomized rats consuming a low calcium diet, *J. Nutr.* **113:**1409.

Goulding, A., and Campbell, D.R., 1984, Effects of oral loads of sodium chloride on bone composition in growing rats consuming ample dietary calcium, *Mineral Electrolyte Metab.* **10:**58.

Goulding, A., and Lim, P.E., 1983, Effects of varying dietary salt intake on the fasting urinary excretion of sodium, calcium and hydroxyproline in young women, *New Zealand Med. J.* **96**:853.

Goulding, A., McIntosh, J., and Campbell, D., 1984, Effects of sodium bicarbonate on calcium and phosphorus balances in the rat, *J. Nutr.* **114**:653.

Goulding, A., Everitt, H.E., Cooney, J.M., and Spears, G.F.S., 1986, "Sodium and osteoporosis," in: *Recent Advances in Clinical Nutrition 2* (M.L. Wahlqvist and A.S. Truswell, eds.), pp. 99-108, John Libbey, London.

Hanson, R.C., White, J.B., and Gomoll, A.W., 1982, Acute and chronic diuretic-induced alterations in urinary sodium and calcium excretion in the conscious rat, *Mineral Electrolyte Metab.* **8**:314.

Harden, R.McG., 1964, Calcium excretion in thermal sweat in thyrotoxicosis, *J. Endocrinol.* **28**:153.

Harinck, H.I.J., Bijvoet, O.L.M., Plantingh, A.S.T., Body, J-J., Elte, J.W.F., Sleeboom, H.P., Wildiers, J., and Neijt, J.P., 1987, Role of bone and kidney in tumor-induced hypercalcemia and its treatment with bisphosphonate and sodium chloride, *Am. J. Med.* **82**:1113.

Hasling, C., Charles, P., Jenson, F.T., and Mosekilde, L., 1990, Calcium metabolism in postmenopausal osteoporosis: the influence of dietary calcium and net absorbed calcium, *J. Bone Min. Res.* **5**:939.

Heaney, R.P., Recker, R.R., and Saville, P.D., 1978, Menopausal changes in calcium balance performance, *J. Lab. Clin. Med.* 92:953.

Kelly, P.J., Pocock, N.A., Sambrook, P.N., and Eisman, J.A., 1989, Age and menopause-related changes in indices of bone turnover, *J. Clin. Endocrinol. Metab.* **69**:1160.

King, J.S., Jackson, R., and B. Ashe, 1964, Relation of sodium intake to urinary calcium excretion, *Invest. Urology* **1**:555.

Kleeman, C.R., Bohannan, J., Bernstein, D., Ling, S., and Maxwell, M.H., 1964, Effect of variations in sodium intake on calcium excretion in normal humans, *Proc. Soc. Exp. Biol.Med.* **115**:29.

Kurtz, T.W., Al-Bander, H.A., and Morris, R.C., 1987, 'Salt-sensitive' essential hypertension in men, *N. Engl. J. Med.* **317**:1043.

Massry, S.G., Coburn, J.W., Chapman, L.W., and Kleeman, C.R., 1967, Effect of NaCl infusion on urinary $Ca^{++}$ and $Mg^{++}$ during reduction of their filtered loads, *Am. J. Physiol.* **213**:1218

McCarron, D.A., Rankin, L.I., Bennett, W.M., Krutzig, S., McClung, M.R., and Luft, F.C., 1981, Urinary calcium excretion at extremes of sodium intake in normal man, *Am. J. Nephrol.* **1**:84.

McLean F.C., and Urist, M.R., eds., 1968, *Bone: Fundamentals of the Physiology of Skeletal Tissue,* 3rd Edition, University of Chicago Press, Chicago.

McParland, B.E., Goulding, A., and Campbell, A.J., 1989, Dietary salt affects biochemical markers of resorption and formation of bone in elderly women, *Br. Med. J.* **299**:834.

Morris, H.A., Need, A.G., Horowitz, M., O'Loughlin, P.D., and Nordin, B.E.C., 1991, Calcium absorption in normal and osteoporotic postmenopausal women, *Calcif. Tissue Int.* **49**:240.

Muldowney, F.P., Freaney, R, and Moloney, M.F., 1982, Importance of dietary sodium in the hypercalciuria syndrome, *Kidney Int.* **22**:292.

Need, A.G., Morris, H.A., Cleghorn, D.B., DeNichilo, D., Horowitz, M., and Nordin, B.E.C., 1991, Effect of salt restriction on urine hydroxyproline excretion in postmenopausal women, *Arch. Intern. Med.* **151**:757.

Nordin, B.E.C., 1960, Osteomalacia, osteoporosis and calcium deficiency. *Clin. Orthopaedics* **17**:235.

Nordin, B.E.C., 1978, Diagnostic procedures in disorders of calcium metabolism. *Clinical Endocrinol.* **8**:55.

Nordin, B.E.C., and Marshall, D.H., 1988, "Dietary requirements for calcium," in: *Calcium in Human Biology* (B.E.C. Nordin, ed.), pp. 447-471, Springer, Berlin.

Nordin, B.E.C., and Polley, K.J., 1987, Metabolic consequences of the menopause: A cross-sectional, longitudinal, and intervention study on 557 normal postmenopausal women, *Calcif. Tissue Int.* **41**:S1.

Nordin, B.E.C., Need, A.G., Hartley, T.F., Philcox, J.C., Wilcox, M., and Thomas, D.W., 1989, Improved method for calculating calcium fractions in plasma: reference values and effect of menopause, *Clin. Chem.* 35:14.

Nordin, B.E.C., Morris, H.A., Need, A.G., Horowitz, M., and Robertson, W.G., 1990, Relationship between plasma calcium fractions, other bone-related variables, and serum follicle-stimulating hormone levels in premenopausal, perimenopausal, and postmenopausal women, *Am. J. Obstet. Gynecol.* **163**:140.

Nordin, B.E.C., Need, A.G., Morris, H.A., Horowitz, M., and Robertson, W.G., 1991, Evidence for a renal calcium leak in postmenopausal women, *J. Clin. Endocrinol. Metab.* **72**:401.

Parfitt, A.M., 1989, Plasma calcium control at quiescent bone surfaces: A new approach to the homeostatic function of bone lining cells, *Bone* **10**:87.

Phillips, M.J., and Cooke, J.N.C., 1967, Relation between urinary calcium and sodium in patients with idiopathic hypercalciuria, *Lancet* **i**:1354.

Reid, I.R., Schooler, B.A., Hannan, S.F., and Ibbertson, H.K., 1986, The acute biochemical effects of four proprietary calcium preparations. *Aust. NZ J. Med.* **16**:193.

Sabto, J., Powell, M.J., Breidahl, M.J., and Gurr, F.W., 1984, Influence of urinary sodium on calcium excretion in normal individuals. *Med. J. Aust.* **140**:354.

Sakhaee, K., Nicar, M.J., Glass, K., and Pak, C.Y., 1985, Postmenopausal osteoporosis as a manifestation of renal hypercalciuria with secondary hyperparathyroidism, *J. Clin. Endocrinol. Metab.* **61**(2):368.

Silver, J., Rubinger, D., Friedlaender, M.M., and Popovtzer, M.M., 1983, Sodium-dependent idiopathic hypercalciuria in renal-stone formers, *Lancet* **ii**:484.

Simpson, F.O., Nye, E.R., Bolli, P., Waal-Manning, H.J., Goulding, A.W., Phelan, E.L., deHamel, F.A., Stewart, .D.H., Spears, G.F.S., Leek, G.M., and Stewart, A.C., 1978, The Milton Survey: Part 1, General methods, height, weight and 24-hour excretion of sodium, potassium, calcium, magnesium and creatinine, *New Zealand Med. J.* **87**:379.

Staub, J.F., Tracqui, P., Lausson, S., Milhaud, G., and Perault-Staub, A.M., 1989, A physiological view of in vivo calcium dynamics: the regulation of a nonlinear self-organized system, *Bone* **10**:77.

Stepan, J.J., Pospichal, J., Presl, J., and Pacovsky, V., 1987, Bone loss and biochemical indices of bone remodeling in surgically induced postmenopausal women, *Bone* **8**:279.

Sutton, R.A.L., Wong, N.L.M., and Dirks, J.H., 1979, Effects of metabolic acidosis and alkalosis on sodium and calcium transport in the dog kidney, *Kidney Int.* **15**:520.

Walser, M., 1961, Calcium clearance as a function of sodium clearance in the dog, *Am. J. Physiol.* **200**:769.

Wong, N.L.M., and Quamme, G.A., 1990, Association of calcium and sodium handling in the rabbit nephron, *Renal Physiol. Biochem.* **13**:306.

Chapter 13

# Fluoride in the Prevention and Treatment of Osteoporosis

Jukka A. Inkovaara

## 1. Introduction

The halogen fluorine is the most electronegative and one of the most reactive of the elements. It does not exist free in nature but in numerous inorganic salts and countless organic compounds, substituting for hydrogen.

The skeletal effect of fluoride has been known since the time when Möller and Gudjonson (1932) first described skeletal fluorosis. Dental mottling or fluorosis was first observed in the early 1900s in persons residing in high fluoride areas. A relationship between the prevalence of dental caries and naturally occurring fluoride in drinking water was shown by Dean (1938, 1942).

Fluoridation of public water supplies was initiated in 1945 in the US and in 1985 it was estimated that 121 million Americans received fluoridated drinking water (Kaminsky *et al.*, 1990).

## 2. Pharmacology

Fluoride absorbs from the duodenum and absorption is almost total, 90% of the dose being absorbed across the gastrointestinal tract. There are numerous factors that inhibit this process, in particular calcium and phosphorus.

Jukka A. Inkovaara • University Central Hospital, PL 2000, 33521 Tampere, Finland

*Advances in Nutritional Research*, Vol. 9
Edited by Harold H. Draper
Plenum Press, New York, 1994

In clinical practice two fluoride salts are used: sodium fluoride and sodium monofluorophosphate. The plasma half-life is 4–5 hours and approximately 60% of ingested fluoride is deposited in the skeleton and other calcified tissues such as the teeth. The remainder is excreted by the kidney (Ekstrand *et al.,* 1990).

In bone and dental enamel fluoride ion may substitute for the hydroxyl ion of calcium hydroxyapatite to form calcium fluorapatite, which has an improved chemical stability (Levine, 1990; Blinkhorn, 1991). When fluoride exposure increases in the skeleton the number of osteoblasts and the rate of bone formation increase, and the serum activity of skeletal alkaline phosphatase rises (Gruber and Baylink, 1991).

Fluoride concentrations in plasma and bone are variable and increase with age. In subjects studied in artificially fluoridated drinking water areas, concentrations increase about 80% from 2 to 87 years of age (Hanhijärvi, 1974). The corresponding increase in a non-fluoridated drinking water area was about 40%, which was highly significantly less than in the fluoridated drinking water community. In ashed bone, fluoride concentrations increased from about 400 ppm to 2500 ppm in individuals living in fluoride-poor regions. This means that there was an approximately sixfold increase in bone fluorides while the age increased by 65 years (Hanhijärvi, 1974).

Diseases with potential renal complications, for example diabetes and systemic lupus erythmatosis, cause an increase in plasma fluoride concentrations; likewise oedematous diseases like heart insufficiency, cor pulmonale and liver cirrhosis (Hanhijärvi, 1974).

There are many sources of fluoride other than drinking water, e.g., fluoride-containing dentifrices, mouth rinses, fluoride tablets and dental sealants. Marier and Rose (1972) have made a review of selected literature aimed at identifying how much humans nowadays are exposed to environmental fluoride, the total ingestion from various sources, and the types of fluoride present in air, foods, beverages and other commodities.

## 3. Epidemiology

The results of epidemiological studies of fracture risk in populations in areas where the drinking water has a high or low fluoride content are contradictory.

### 3.1. Vertebral Fractures

Bernstein *et al.* (1966) found that reduced bone density and collapsed vertebrae were substantially more frequent in a low fluoride area. Vertebral fracture prevalence rose with age, being 35% among women 65 years or older in the low fluoride region compared to 10% in the high fluoride area. Rates among men in the low and high fluoride area were 45% and 53% ($P > 0.05$).

In another study among people living in a high fluoride area, Leone *et al.* (1955) reported less frequent osteoporosis and vertebral fractures than in people living in a low fluoride control area. On the other hand, in an area of endemic fluorosis in India, the inhabitants showed evidence of spinal osteosclerosis and extensive osteoporotic changes in the bones of the extremities (Krishnamachari *et al.*, 1973).

In an English study (Ansell *et al.*, 1965) there were no more lumbar collapses in a low fluoride town than in a city with fluoridated drinking water. In two New York towns with and without fluoridated drinking water no difference was found in vertebral fractures under X-ray (Melton *et al.*, 1989).

## 3.2. Peripheral Fractures

A history of fractures was found to be more common among women in a community with high fluoride (4 mg/L) and low calcium (15 mg/L) drinking water than with a high calcium (375 mg/L) and low fluoride (1 mg/L) water supply or with a low calcium (60 mg/L) and low fluoride (1 mg/L) supply (Sowers *et al.,* 1986). Among older women (55–85 years) there were more total fractures (46%, 35% and 36%) and more hip, wrist and spine fractures (16%, 9% and 12%, respectively).

In a later study in the same area and by the same workers (Sowers *et al.*, 1991) there was no difference in the 5-year relative risk of any fracture in the higher calcium community versus the low calcium, low fluoride control community. However, the relative risk was 2.1 in women in the high fluoride community compared with women in the control community. There was no difference in the 5-year risk of wrist, hip or spine fracture in the higher calcium areas vs. control areas. The 5-year relative risk for women in the higher fluoride community compared with women in the control community was 2.2.

Similar findings have been recorded for long bone fractures among younger women (Avorn *et al.*, 1986). Fracture rates were slightly greater in high fluoride counties. The high fluoride counties were all urban and the low fluoride areas were all rural. The hip fracture incidence appears to be greater in urban areas and urbanization increases the number of hip fractures in both sexes (Anderson, 1992).

There are results eliciting no difference in fracture rate from low and high fluoride areas. No change in hip fracture incidence was seen before and after drinking water fluoridation (Goggin *et al.*, 1965). Korns (1969) arrived at similar results in a study on two New York communities.

No difference was found in hip fracture rates among three Swedish communities with low or intermediate fluoride levels in their drinking water (Alffram *et al.*, 1969). No correlation between the fluoride content of drinking water and hip fracture rates was found in studies by Madans *et al.* (1983) and Cooper *et al.* (1991).

In a Finnish study (Simonen and Laitinen, 1985), the incidence of femoral neck fracture between 1967 and 1978 was compared in two townships. The age-standardized incidence in men was 2.5 in Kuopio and 7.0 in Jyväskylä (64% lower) and in women 6.0 in Kuopio and 9.0 in Jyväskylä (33% lower). Kuopio has fluoridated its drinking water since 1959, whereas Jyväskylä has only trace amounts of fluoride in its drinking water. The differences in incidence in men were greatest in the age group 50–59 and smallest in the age group 80 and over. There may be some reason other than osteoporosis for the difference in the younger male age group because older men do not have osteoporosis and accelerated rates of hip fractures until the mid-80s onward, as in women (Anderson, 1992). There may have been a difference in duration of urbanization. In the study of Arnala *et al.* (1986) in the same region but at a later period (1972–1981) there was no difference in the incidence of hip fracture in three areas.

## 4. Fluoride and Bone Strength

Van Gastel *et al.* (1980) observed that after two years' therapy with sodium fluoride and dihydrotachysterol, histological findings in a first and second bone biopsy indicated an increased amount of bone but that the mineralization of newly formed bone was incomplete. Stein and Granik (1980), in mechanical tests of vertebral bone cylinders with various exposures to fluoride, found that higher fluoride hard tissue was weaker in a static test than that with less fluoride. They concluded that water fluoridation would not appear to alter static bone strength.

Alhava *et al.* (1980) studied the fluoride content of bones from Kuopio, where drinking water has been fluoridated since 1959, and found it to increase significantly with age. The highest fluoride content in bone ash was observed in women with severe osteoporosis. Cancellous bone strength measured by a strain transducer was significantly higher in women from Kuopio suffering from chronic immobilizing disease. No statistically significant differences were found in bone mineral density between the samples from the fluoridated and non-fluoridated areas.

Carter and Beaupre (1990) introduced a conceptual framework for understanding how fluoride treatment alters the tissue loading control of bone architecture and can cause a systemic increase in bone mass. Due to possible adverse influences of fluoride, an increase in bone mass does not necessarily result in an increase in bone strength.

Aaron *et al.* (1991) compared bone biopsies taken before and after two years' treatment with sodium fluoride. They observed a marked increase in bone volume, which was attributable to an increase in trabecular thickness rather than number. These data suggest that the restoration of skeletal mass with fluoride may not lead to a comparable reduction in fracture risk.

## 5. Fluoride Therapy

Fluoride has been used in the treatment of osteoporosis since 1961 (Rich and Ensink, 1961) because it increases trabecular bone mass in the spine and may be effective in the treatment of spinal osteoporosis.. Fluoride treatment is still controversial because of side effects, a high rate of non-responders, a possible osteomalacic effect on bone, deleterious effects on cortical bone, and especially in view of the uncertain effect on the fracture rate (Laroche and Maziéres, 1991).

At present, fluoride therapy is highly questionable in the prophylaxis and treatment of osteoporosis.

### 5.1 Therapeutic Window

The therapeutic window of fluoride is narrow. The serum fluoride level must be at least 95 ng/ml before a beneficial effect on the skeleton is obtained, and it should be kept below 190 ng/ml if toxic effects are to be avoided (Pak, 1989). Fluoride retention depends on fluoride dose, body size, renal function and intestinal absorption of fluoride and calcium. A high dosage of sodium fluoride (45.8 mg/day) may cause hyperostosis in patients with high retention of fluoride in bone, even when serum fluoride is not significantly increased (Budden *et al.*, 1988).

### 5.2 Adverse Effects, Stress Fractures and Non-Responders

In clinical trials with fluoride the rate of adverse effects (20–40%) and that of non-responders (15–37%) have been moderately high. Furthermore, the number of dropouts has been considerable. Approximately 20% of patients do not respond to fluoride therapy with a change in bone mass. As yet there is no effective means of identifying responders at the commencement of therapy (Hodsman and Drost, 1989; Franke, 1988).

Hodsman and Drost (1989) prospectively studied 48 women who suffered from postmenopausal osteoporosis. During treatment 25 patients developed significant side effects due to fluoride, and 18 (37%) were ultimately intolerant of the drug and discontinued treatment after an average of 17.3 months. Moreover, an additional 15% required temporary interruption of therapy because of side effects. Fifteen patients were classified as non-responders. When comparing patients tolerant of fluoride therapy with those who discontinued treatment because of drug intolerance, there was no significant difference in dose or serum fluoride levels. During fluoride therapy they reported a total of 25 side effects, 11 gastrointestinal (1 gastrointestinal hemorrhage), 10 arthralgias, 2 tendonitis and 2 stress fractures).

In another study with 158 patients (Franke, 1988) there were 15–20% non-responders, 34.2% had gastrointestinal disturbances, 7 (14.4%) gastric ulcers, 34.2% arthralgias and 10 (6.3%) stress fractures (7 femoral neck). Fifty-one percent of patients experienced side effects in a study by Hasling *et al.* (1987), of which 37% were joint-related and 25% gastrointestinal. Six percent of the patients withdrew from treatment due to side effects. In 8 patients receiving daily doses of 1.09 mg/kg sodium fluoride, elemental calcium and vitamin D for osteoporosis, Schnitzler and Solomon (1985) found 17 episodes of periarticular pain and swelling in the lower limbs. Radiographs 4–6 weeks later showed features of healing stress fractures in the periarticular cancellous bone: 2 in the distal femur, 3 in the proximal tibia, 6 in the distal tibia and 6 in the calcaneum. Bone scintigraphy was positive on 5 occasions.

In another study, Schnitzler *et al.* (1990) studied 18 patients who developed 41 stress fractures: 26 periarticular, 6 femoral neck, 5 long bone shaft, 1 greater trochanter and 3 pubic rami fractures. The sodium fluoride dose was 0.99 mg/kg per day, calcium 1 g/day, and vitamin D. The authors concluded that stress fracture patients had more severe trabecular and cortical osteoporosis and possibly a poorer bone-forming capacity before therapy than patients without stress fractures.

Compston *et al.* (1980) reported a case treated with sodium fluoride and large doses of vitamin D in which osteomalacia developed during the treatment. Bone biopsy showed moderately severe osteomalacia and secondary hyperparathyroidism. Orcel *et al.* (1990) reported 18 patients who experienced stress fractures during fluoride therapy; there were 13 cases of tibial metaphysis, 10 femoral neck fractures and 4 calcaneus fractures. In bone histomorphometry trabecular resorption was increased and features of osteomalacia were found in two patients with decreased renal function. In a study by Bayley *et al.* (1990) on 61 patients treated with sodium fluoride (40–60 mg/day) in combination with calcium and vitamin D, four patients had traumatic stress fractures. Seven had 10 upper femur fractures, of which 5 were stress fractures. Nine patients developed lower extremity pain syndrome. Patients with femur fractures were older and had significantly higher bone fluoride retention. Thus fluoride therapy may be implicated in the pathogenesis of hip fractures, which may occur despite a rapid and marked increase in bone mass.

Table I. Controlled Studies on Fractures During Sodium Fluoride Therapy

| Study | Patients Controls | | NaF dose | Fractures (n or %) | | |
|---|---|---|---|---|---|---|
| | | | | Vertebral | Peripheral | Hip |
| Inkovaara *et al.* (1984) | Pat | 146 | 25-27 mg | 2 | 9 + 3 | 11 |
| | Contr | 169 | | | 6 | 5 |
| Gutteridge *et al.* (1984) | Pat | 16 | 40-80 mg | 9 | 11 | 11 |
| | Contr | 8 | | 8 | 0 | 0 |
| Power *et al.* (1986) | Pat | 25 | 40-60 mg | 23.3% | 7 | 7 |
| | Contr | 24 | | 25.0% | 1 | 1 |
| Mamelle *et al.* (1988) | Pat | 180 | 50 mg | 39.2% | 24 | 6 |
| | Contr | 136 | | 50.8% | 22 | 4 |
| Kleerekoper *et al.* (1989) | Pat | 42 | 75 mg | 602[a] | | |
| | Contr | 33 | | 491[a] | | |
| Hedlund *et al.* (1989) | Pat | 35 | 50 mg | 4.1% | 8 | 6 |
| | Contr | 43 | | 4.3% | 0 | 0 |
| Riggs *et al.* (1990) | Pat | 66 | 75 mg | 163 | 72 | 13 |
| | Contr | 69 | | 136 | 24 | 4 |

[a] Calculated numbers 718 and 529 for Pat and Contr, respectively.

## 6. Fractures

The best basis for measuring the effectiveness of fluoride therapy is fracture incidence. However, only a few controlled studies have been published hitherto (Table 1).

In an early trial by Inkovaara *et al.* (1975), NaF (25 mg/day) was administered for 5 months and thereafter 25 mg twice a week for a further 3 months in a double-blind test. Eleven fractures were detected in the fluoride group (n = 237), 2 compressions of the vertebrae and another 3 fractures during the following month (11 hip fractures). In the control group (n = 233), 6 fractures (5 hip fractures) occurred ($P < 0.1$). No vertebral compressions occurred. The fact that there were more fractures in the fluoride group did not encourage the use of fluoride therapy in the care of geriatric patients.

Gutteridge *et al.* (1984) gave sodium fluoride up to 80 mg/day, 1000 mg calcium/day and vitamin $D_2$ at an initial dosage of 1.25 mg/day to 16 patients for 4 years. The vitamin D dosage subsequently was reduced to maintain normal urine Ca levels and plasma Ca and 25-OHD levels. The NaF dosage was adjusted to maintain a plasma fluoride ion concentration of 0.18–0.30 mg/L. The control group was comprised of 8 patients. In the treated group 5 patients developed 11 spontaneous femoral fractures during or soon after therapy. Vertebral fractures decreased in number in the treatment group. These data suggest that fluoride has no value in countering the risk of femoral fractures. There were subsequent femoral neck or shaft fractures, 6 bilateral, in 7 patients who continued treatment (Gutteridge *et al.*, 1990). In all, there were 19 spontaneous fractures, 5 of which were asymptomatic and 5 were stress fractures. When compared with another identically treated group of 14 patients without spontaneous femoral fracture, the bone fluoride content was higher in 6 biopsied patients in the fracture group than in 4 in the non-fracture group. In post-treatment biopsies in the fracture group there were features of thinning, increased porosity and advanced tunnelling resorption.

For three years Dambacher *et al.* (1986) monitored 15 patients with postmenopausal osteoporosis subjected to continuous therapy with 80 mg NaF/day and 14 patients without this treatment. The fracture rate was significantly different in the first year, with 0.3 new vertebral fractures in the untreated group and 2.9 new fractures in the treated group ($P < 0.01$). During the second and third year the mean number of new fractures was approximately equal in both groups. Osteoarticular side effects were observed in 47% of the treated patients. In 27%, scintigraphy of the ankle was abnormal, alkaline phosphatase was increased and radiologic signs of healing stress fractures were present.

In a study by Power *et al.* (1986), the incidence of vertebral fracture was reduced ($P < 0.01$) only in 8 fluoride-treated patients who developed radiographic fluorosis. Six sodium fluoride-treated patients sustained a total of 7 femoral fractures, 6 fractures of the femoral neck and 1 of the femoral shaft.

Mamelle *et al.* (1988) studied 257 patients with primary osteoporosis receiving 25 mg sodium fluoride twice a day combined with calcium (1 g/day) and a vitamin $D_2$ supplement. Two hundred and nine patients received one of the alternative therapies usually prescribed in France. After 24 months there was a significantly lower rate of new vertebral fractures (39.2%) in the NaF group than in the non-NaF group (50.8%, $P < 0.05$). Twenty-four nonvertebral fractures were noted in the NaF group (6 hip) and 22 in the non-NaF group (4 hip).

Hedlund and Gallagher (1989) compared the incidence of hip fracture in four groups of osteoporotic women: 22 treated with placebo, 17 with fluoride and calcium, 18 treated with fluoride and calcitriol, and 21 with calcitriol alone. After 2.1 years 6 hip fractures occurred in the fluoride groups. In the non-fluoride group there were none ($P = < 0.01$). Kleerekoper *et al.* (1989) found that continuous sodium fluoride treatment did not appear to be more effective than calcium carbonate in reducing the vertebral fracture rate and height loss in post-menopausal white women.

To measure the effect of fluoride treatment on fracture rate Riggs *et al.* (1990) carried out a four-year prospective clinical trial with 202 postmenopausal women with osteoporosis who were randomly assigned to receive sodium fluoride (75 mg/day) or placebo. All were given calcium supplementation (1500 mg/day). The number of new vertebral fractures was similar in the treatment (163) and placebo (136) groups ($P > 0.05$). In the treatment group there were 72 nonvertebral fractures and in the placebo group 24 fractures ($P < 0.01$). The authors concluded that under the conditions of this study the fluoride-calcium regimen was not an effective treatment for postmenopausal osteoporosis. In a more recent study (Kanis *et al.*, 1992) the use of fluoride proved to be associated with a high risk of hip fracture. After adjustment for other risk factors the relative risk decreased.

Among the controlled studies cited above, vertebral fractures were lower in only two (Gutteridge *et al.*, 1984; Mamelle *et al.*, 1988) and there was either no difference or a higher fracture rate in the remainder. These findings do not support the positive results of uncontrolled studies (Pak *et al.*, 1989; Farley *et al.*, 1990).

Many studies employing densitometry have yielded fine but misleading results. Densitometry is not the best method of measuring the effect of fluoride therapy on skeletal strength because while bone crystallinity may be increased, bone strength is not.

## 7. Conclusions

Fluoride is important for developing bone tissue and tooth enamel, but no one can say what is the need of the adult skeleton or whether there is any. Fluoridation of drinking water is safe but it may be that nowadays other sources of fluoride like toothpastes, food and beverages, etc., are sufficient for the needs of the skeleton and teeth. Fluoridation of drinking water may reduce the prevalence of vertebral fractures but there is no evidence that fluoridation prevents peripheral fractures, and the use of fluoride in the prevention of osteoporosis is not recommended.

The use of fluoride in the treatment of osteoporosis seems highly questionable in that the therapeutic window is extremely narrow: 15-37% of patients do not respond to fluoride, and furthermore there are frequent gastrointestinal disturbances and arthralgias. Vertebral fractures may be reduced, but patients receiving fluoride have experienced more nonvertebral (especially hip) fractures, as well as spontaneous hip fractures, than control patients. The high morbidy and mortality of osteoporotic patients with hip fractures do not justify the use of fluoride in the treatment of osteoporosis.

## References

Aaron, J.E., de-Vernejoul, M.C., and Kanis, J.A., 1991, The effect of sodium fluoride on trabecular architecture, *Bone* **12**:307.

Alffram, P.A., Hernborg, J., and Nilsson, B.E.R., 1969, The influence of high fluoride content in the drinking water on the bone mineral mass in man, *Acta Orthop. Scand.* **40**:137.

Alhava, E.M., Olkkonen, H., Kauranen, P., and Kari, T., 1980, The effect of drinking water fluoridation on the fluoride content, strength and mineral density of human bone, *Acta Orthop. Scand.* **51**:413.

Anderson, D., 1992, Osteoporosis in men, *Br. Med. J.* **305**:489 (Editorial).

Ansell, B.M., and Lawrence, J.S., 1965, Fluoridation and rheumatic diseases: A comparison of rheumatism in Watford and Leigh, *Ann. Rheum. Dis.* **25**:67.

Arnala, I., Alhava, E.M., Kivivuori, R., and Kauranen, P., 1986, Hip fracture incidence not affected by fluoridation. Osteofluorosis studied in Finland, *Acta Orthop. Scand.* **57**:344.

Avorn, J., and Niessen, L.C., 1986, Relationship between long bone fractures and water fluoridation, *Gerodontics* **2**:175.

Bayley, T.A., Harrison, J.E., Murray, T., Josse, R.G., Sturtridge, W., Pritzker, K.P.H., Strauss, A., Veith, R., and Goodwin, S., 1990, Fluoride-induced fractures: Relation to osteogenic effect, *J. Bone Min. Res.* **5** (Suppl.1):S217.

Bernstein, D.S., Sadowsky, N., Hegsted, D.M., Guri, C.D., and Stare, F.J., 1966, Prevalence of osteoporosis in high- and low-fluoride areas in North Dakota, *JAMA* **198**:499.

Blinkhorn, A.S., 1991, Fluoride and caries prevention: 2, *Dent. Update* **18**:146.

Budden, F.H., Bayley, T.A., Harrison, J.E., Josse, R.G., Murray, T.M., Sturtridge, W.C., Kandel, R., Veith, R., Strauss, A.L., and Goodwin, S., 1988, The effect of fluoride on bone histology in postmenopausal osteoporosis depends on adequate fluoride absorption and retention, *J. Bone Min. Res.* **3**:127.

Carter, D.R., and Beaupre, G.S., 1990, Effects of fluoride on bone strength, *J. Bone Min. Res.* **5** (Suppl. 1): S177.

Compston, J.E., Chadha, S., and Merrett, A.L., 1980, Osteomalacia developing during treatment of osteoporosis with sodium fluoride and vitamin D, *Br. Med. J.* **281**:910.

Cooper, C., Wickham, C., Lacey, R.F., and Barker, D.J.P., 1990, Water fluoride concentration and fracture of the proximal femur, *J. Epidemiol. Commun. Health* **44**:17.

Dambacher, M.A., Ittner, J., and Ruegsegger, P., 1986, Longterm fluoride therapy of postmenopausal osteoporosis, *Bone* **7**:199.

Dean, H.T., 1938, Endemic fluorosis and its relation to dental caries, *Public Health Rep.* **53**:1443.

Dean, H.T., Arnold, F.A., and Elvone, E., 1942, Domestic water and dental caries. V. Additional studies of the relation of fluoride domestic waters to dental caries experience in 4425 white children aged 12 to 14 years of 13 cities in 4 states, *Public Health Rep.* **57**:1155.

Duursma, S.A., Raymakers, J.A., De-Raadt, M.E., Karsdorp, N.J., van Dijk, A., and Glerum, J., 1990, Urinary fluoride excretion in responders and nonresponders after fluoride therapy in osteoporosis, *J. Bone Min. Res.* **5** (Suppl.1):S43.

Ekstrand, J., and Spak, C.J., 1990, Fluoride pharmacokinetics: its implications in the fluoride treatment of osteoporosis, *J. Bone Min. Res.* **5** (Suppl. 1):S53.

Farley, S.M., Wergedal, J.E., Farley, J.R., Javier, G.N., Schulz, E.E., Libanati, C.R., Talbot, J.R., Mohan, S.S., Perkel, V.S., and Baylink, D.J., 1990, "Fluoride decreases spinal fracture rate: A study of over 500 patients," in: *Third International Syposium on Osteoporosis* (K. Overgaard and C. Christiansen, eds.), p. 141, Handelstrykkeriet Aalborg ApS, Aalborg, Denmark, (abstract).

Franke, J., 1988, Fluoride and osteoporosis, *Ann. Chir. Gynaecol.* **77**:235.

van Gastel, C., Becker, H., Visser, W.J., Bonne, A., Megens, J.G.N., and Duursma, S.A., 1980, Sodium fluoride in osteoporosis, *Neth. J. Med.* **23**:155.

Goggin, J.E., Haddon, W.J., Hambly, G.S., and Hoveland, J.R., 1965, Incidence of femoral fractures in postmenopausal women, *Public Health Rep.* **80**:1005.

Gruber, H.E., and Baylink, D.J., 1991, The effects of fluoride on bone, *Clin. Orthop. Rel. Res.* **267**:264.

Gutteridge, D.H., Price, R.I., Nicholson, G.C., Kent, G.N., Retallack, R.W., Devlin, R.D., Worth, G.K., Glancy, J.J., Michell, P., and Gruber, H., 1984, "Fluoride in osteoporotic vertebral fractures—trabecular increase, vertebral protection, femoral fractures," in: *Osteoporosis: International Symposium on Osteoporosis* (C. Christiansen, C.D. Arnaud, B.E.C. Nordin, A. M. Parfitt, W.A. Peck, and B. L. Riggs, eds.) Aalborg Stiftsbigrykkeri, Copenhagen, **2**:705-707.

Gutteridge, D.H., Price, R.I., Kent, G.N., Prince, R.L., and Michell, P.A., 1990, Spontaneous hip fractures in fluoride-treated patients: Potential causative factors, *J. Bone Min. Res.* **5** (Suppl 1):S205.

Hanhijärvi, H., 1974, Comparison of free ionized fluoride concentration of plasma and renal clearence in patients of artificially fluoridated and non-fluoridated drinking water areas, *Proc. Finn. Dent. Soc.* **70**:1.

Hasling, C., Nielsen, H.E., Melsen, F., and Mosekilde, L., 1987, Safety of osteoporosis treatment with sodium fluoride, calcium phosphate and vitamin D, *Min. Electrolyte Metab.* **13**:96.

Hedlund, L.R., and Gallagher, J.C., 1989, Increased incidence of hip fracture in osteoporotic women treated with sodium fluoride, *J.Bone Min. Res.* **4**:223.

Hodsman, A.B., and Drost, D.J., 1989, The response of vertebral bone mineral density during the treatment of osteoporosis with sodium fluoride, *J. Clin. Endocrinol. Metab.* **69**:932.

Inkovaara, J., Heikinheimo, R., Järvinen, K., Kasurinen, U., Hanhijärvi, H., and Iisalo, E., 1975, Prophylactic fluoride treatment and aged bones, *Br. Med. J.* **3**:73.

Kaminsky, L.S., Mahoney, M.C., Leach, J., Melius, J., and Miller, M.J., 1990, Fluoride: Benefits and risks of exposure, *Oral Biol. Med.* **1**:261.

Kanis, J.A., Johnell, O., Gullberg, B., Allander, E., Dilsen, G., Gennari, C., Vaz, A.A.L., Lyritis, G.P., Mazzuoli, G., Miravet, L., Passeri, M., Cano, R.P., Rapado, A., and Ribot, C., 1992, Evidence for efficacy of drugs affecting bone metabolism in preventing hip fractures, *Br. Med. J.* **305**:1124.

Kleerekoper, M., Peterson, E., Phillips, D., Nelson, D., Tilley, B., and Parfitt, A., 1989, Continuous sodium fluoride therapy does not reduce vertebral fracture rate in postmenopausal osteoporosis, *J. Bone Min. Res.* **4** (Suppl. 1):376.

Korns, R.F., 1969, Relationship of water fluoridation to bone density in two N.Y. towns, *Public Health Rep.* **84**:815.

Krishnamachari, K.A., and Krishnaswamy, K., 1973, Genu valgum and osteoporosis in an area of endemic fluorosis, *Lancet* **ii**:877.

Laroche, M., and Mazieres, B., 1991, Side-effects of fluoride therapy, *Bailliere's Clin. Rheumatol.* **5**:61.

Leone, N.C., Stevenson, C.A., Hilbish, T.F., and Sosman, M.C., 1955, A roentgenologic study of a human population exposed to high-fluoride domestic water. A ten-year study, *Am. J. Roentgenol.* **74**:874.

Levine, R.S., 1990, Fluoride and caries prevention: 1. Scientific rationale, *Dent. Update* **18**:105.

Madans, J., Kleinman, J.C., and Cornoni-Huntley, J., 1983, The relationship between hip fracture and water fluoridation: an analysis of national data, *Am. J. Public Health* **73**:296.

Mamelle, N., Meunier, P.J., Dusan, R., Guillaume, M., Martin, J.L., Gaucher, A., Prost, A., Ziegler, G., and Netter, P., 1988, Risk-benefit ratio of sodium fluoride treatment in primary vertebral osteoporosis, *Lancet* **ii**:361

Marier, J.R., and Rose, D., 1972, "Environmental Fluoride," *NRC Publication No.12,226,* Enviromental Secretariat, Division of Biology, National Research Council of Canada, Ottawa, pp 1–32.

Melton, L.J., Kan, S.H., Frye, M.A., Wahner, H.W., O'Fallon, W.M., and Riggs, B.L., 1989, Epidemiology of vertebral fractures in women, *Am. J. Epidemiol.* **129**:1000.

Möller, P.F., and Gudjonsson, S.V., 1932, Massive fluorosis of bones and ligaments, *Acta Radiol. Scand.* **13**:269.

Orcel, P., de-Vernejoul, M.C., Prier, A., Miravet, L., Kuntz, D., and Kaplan, G., 1990, Stress fractures of the lower limbs in osteoporotic patients treated with fluoride, *J. Bone Min. Res.* **5** (Suppl. 1):S191.

Pak, C.Y., 1989, Fluoride and osteoporosis, *Proc. Soc. Exp. Biol. Med.* **191**:278.

Pak, C.Y., Sakhaee, K., Zerwekh, J.E., Parcel, C., Peterson, R., and Johnson, K., 1989, Safe and effective treatment of osteoporosis with intermittent slow release sodium fluoride: augmentation of vertebral bone mass and inhibition of fractures, *J. Clin. Endocrinol. Metab.* **68**:150.

Power, G.R.I., and Gay, J.D.L., 1986, Sodium fluoride in the treatment of osteoporosis, *Clin. Invest. Med.* **9**:41.

Rich, C., and Ensink, F., 1961, Effect of sodium fluoride on calcium metabolism of human beings, *Nature* **191**:184.

Riggs, B.L., Hodgson, S.F., O'Fallon, W.M., Chao, E.Y.S., Wahner, H.W., Muhs, J.M., Cedel, S.L., and Melton, L.J., III, 1990, Effect of fluoride treatment on the fracture rate in postmenopausal women with osteoporosis, *N. Engl. J. Med.* **322**:802.

Schnitzler, C.M., and Solomon, L., 1985, Trabecular stress fractures during fluoride therapy for osteoporosis, *Skeletal Radiol.* **14**:276.

Schnitzler, C.M., Wing, J.R., Mesquita, J.M., Gear, A.K., Robson, H.J., and Smyth, A.E., 1990, Risk factors for the development of stress fractures during fluoride therapy for osteoporosis, *J. Bone Min. Res.* **5** (Suppl. 1):S195.

Simonen, O., and Laitinen, O., 1985, Does fluoridation of drinking water prevent bone fragility and osteoporosis? *Lancet* **ii**:432.

Sowers, M.R., Wallace, R.B., and Lemke, J.H., 1986, The relationship of bone mass and fracture history to fluoride and calcium intake: a study of three communities, *Am. J. Clin. Nutr.* **44**:889.

Sowers, M.F., Clark, M.K., Jannausch, M.L., and Wallace, R.B., 1991, A prospective study of bone mineral content and fracture in communities with differential fluoride exposure, *Am. J. Epidemiol.* **133**:649.

Stein, I.D., and Granik, G., 1980, Human vertebral bone: relation of strength, porosity, and mineralization to fluoride content, *Calcif. Tissue Int.* **32**:189.

Chapter 14

# Bone Mineral Content in Postmenopausal Vegetarians and Omnivores

Isabelle F. Hunt

## 1. Introduction

There is evidence that the consumption of diets containing large amounts of animal protein is associated with low bone mineral content (BMC). Mazess and Mather (1974, 1975) observed that in Eskimos, whose diets consisted chiefly of meat and provided about 200 g of animal protein per day, bone loss occurred earlier in life and was of greater magnitude than in Caucasians in the U.S.A. An explanation for these observations is based on findings that the higher content of sulfur-containing amino acids in animal as compared to plant proteins results in a greater production of metabolic acid (Halperin and Jungas, 1983; Breslau *et al.*, 1988) and higher net excretion of urinary acid. This increased acid load is buffered by calcium which in individuals with marginal or negative calcium balance is made available by the dissolution of bone (Allen *et al.*, 1979; Schuette *et al.*, 1980; Bushinsky, 1989). Because foods that have a high protein content also tend to be high in phosphorus, the calciuretic effect of a diet high in protein may be ameliorated by the concomitant presence of phosphorus. There is evidence that increased phosporus intake depresses serum calcium levels and stimulates the synthesis of parathyroid hormone which leads to increased reabsorption of calcium by the renal tubules (Yuen *et al.*, 1984).

Additional evidence for a role of dietary protein in the promotion of bone mineral loss is to be found in epidemiologic studies of hip fracture rates. A

Isabelle F. Hunt • School of Public Health, UCLA, Los Angeles, CA 90024

*Advances in Nutritional Research,* Vol. 9
Edited by Harold H. Draper
Plenum Press, New York, 1994

positive cross-cultural association between dietary protein intake and hip fracture was first reported by Hegsted (1986). In a recent epidemiologic study of 16 countries, Abelow *et al.* (1992) reported a strong, positive cross-cultural association between the incidence of age-adjusted hip fracture in women over 50 years of age and the intake of animal protein, but not the intake of calcium or energy.

The association of a high intake of animal protein with an increased loss of bone mineral and a higher rate of hip fracture raises the question whether vegetarianism protects against this loss. Before reviewing the studies of BMC in vegetarians, vegetarian diet patterns and health habits that may affect BMC will be discussed. It is recognized that many variables in addition to diet may influence BMC including genetics, endocrine status, and health and exercise habits (Cummings *et al.*, 1985). Weight-bearing exercise has been shown to have a positive effect on BMC (Riggs and Melton, 1992) but there is little published information about the exercise habits of vegetarians (Fraser, 1988).

## 2. Description of the Types of Vegetarian Diets

A Roper Report (1978) stated that about four percent of adults in the U.S.A. consumed vegetarian diets. Over the past few decades there has been a rapid increase in the consumption of vegetarian diets in affluent Western countries (Dwyer, 1983). Most vegetarians have chosen this lifestyle for religious, philosophical or health reasons. Studies indicate that many health benefits are associated with vegetarianism, including lower blood pressure (Beilin *et al.*, 1987), lower body weight (Levin *et al.*, 1986a) and lower incidence of certain chronic diseases such as diabetes, heart disease and some types of cancer (Snowdon and Phillips, 1985; Fraser, 1988; Phillips, 1980).

Traditional vegetarian diets differ in the severity of restriction of animal products (Havala and Dwyer, 1988). The pure vegetarian or vegan diet completely excludes meat, fish, fowl, dairy products, eggs and other animal products such as honey, whereas the ovovegetarian diet includes eggs and the lactoovovegetarian diet includes eggs and dairy products. Some of these vegetarian regimens also allow special foods such as meat analogs and fortified soy products, and permit or encourage the use of dietary supplements such as vitamin $B_{12}$, iron and essential fatty acids. Alcohol and caffeinated beverages are commonly excluded or restricted by vegans, ovovegetarians and lactoovovegetarians. Traditional vegetarians are usually careful to follow good health habits (Havala and Dwyer, 1988).

In addition to the three fairly well defined types of vegetarian diets, many other diets that exclude various kinds and amounts of animal foods are commonly termed "vegetarian" by their adherents. These diets are often more liberal than the lactoovovegetarian diet but some, such as the macrobiotic vegan diets, are more restrictive than the vegan diet described above. This review deals primarily with the three most usual types of vegetarian diets, especially with the lactoovovegetarian diet that is followed by about 90 percent of vegetarians in the U.S.A. (Roper Report, 1978).

Herbert (1990) states that the Recommended Dietary Allowance for protein should be about 25% higher for adult vegans than for omnivores because the proteins derived from plant sources are less digestible than those from animal sources. However, most vegans who meet their energy requirement probably also meet their protein requirement, but to meet their need for the essential amino acids they must select foods that have a complementary amino acid content. For example, every day they should eat a combination of legumes and grains, legumes, nuts and seeds, or nuts, seeds and grains. It is not difficult to include these combinations in the vegan diet; the procedure is described fully by Herbert (1990). Other authors also have given practical advice in planning vegetarian diets. Mutch (1988) reviewed and evaluated food guides for nutritious diets for adult and pregnant vegetarians, particularly vegans. Johnston (1988) provided suggestions for counseling pregnant vegetarians.

## 3. Nutritional Assessment of Vegetarian Diets

Recent studies of the dietary intakes of vegetarians living in Britain (Bull and Barber, 1984; Davies *et al.*, 1985), France (Millet *et al.*, 1989), Israel (Levin *et al.*, 1986), and the U.S.A. (Kelsay *et al.*, 1988; Nieman *et al.*, 1989; Rider *et al.*, 1984; Havala and Dwyer, 1988) show that vegetarian diets contain less protein and more fiber than nonvegetarian diets. Vegetarian diets are usually lower in total fat, saturated fat and cholesterol and have a higher ratio of polyunsaturated to saturated fats than omnivorous diets.

Vitamin $B_{12}$ is deficient in the unsupplemented diets of vegans but not in vegetarian diets that include adequate amounts of eggs or dairy foods. Iron intakes tend to be low in diets that exclude meat and calcium is likely to be low in diets that exclude dairy products. There is some concern that because of the high fiber content of vegetarian diets, the bioavailability of minerals and trace elements may be lower than in omnivorous diets.

Differences in the non-nutrient content of vegetarian and omnivorous diets have also received some attention. In addition to higher fiber, a lower cholesterol content of vegetarian diets has been found repeatedly in dietary surveys. Nair *et al.* (1984) showed that the plant sterols $\beta$-sitosterol and stigmasterol are present in larger amounts in vegetarian diets. $\beta$-sitosterol, which is not well absorbed from the gastro-intestinal tract, has been reported to cause a reduction in the incidence of colon tumors in experimental animals (Raicht *et al.*, 1980). These non-nutrient components of the vegetarian diet may contribute to the lower cancer risk attributed to the vegetarian diet.

## 3.1. Diets of Elderly Vegetarians

There have been only a few studies of dietary intakes of elderly vegetarians (Hunt *et al.*, 1988, 1989; Tylavsky and Anderson, 1988). In a study in which the reported 24-hr dietary intakes of 146 postmenopausal omnivores aged 65.7 ± 8.5 years (mean ± SD) and 144 Seventh-day Adventist (SDA) lactoovovegetarians aged 66.6 ± 10.7 years in California were compared, Hunt *et al.* (1988) found that although, as expected, the omnivores consumed significantly more foods from the meat group and the vegetarians significantly more from the bean and nut group, there were no significant differences between the omnivores and the vegetarians in their daily intake of foods from the milk, vegetable and fruit, or bread and cereal groups. The nutrient density of the vegetarian diets was somewhat higher than that of the omnivorous diets for folate, thiamin, and vitamins A and C. The vegetarian diets were significantly lower in total fat, saturated fatty acids, cholesterol and zinc, and higher in dietary fiber and carbohydrates, than the omnivorous diets. There was no significant difference in the reported intake of calcium, iron, magnesium and copper, or in the energy value of the diets of the two groups of women (Hunt *et al.*, 1989). Similar data on the dietary intakes of elderly vegetarians living in North Carolina were reported by Tylavsky and Anderson (1988).

The diets of the postmenopausal vegetarians in the California study were compared with the reported intakes of a younger sample (mean age 46 ± 18 years) of 31 SDA lactoovovegetarians living in Oregon (Shultz and Leklem, 1983). The vegetarians in Oregon had a significantly higher caloric intake than the older vegetarians in California but both groups appeared to have nutritious diets with a higher calcium content and a lower percent of kilocalories from fat and saturated fatty acids than the women included in the NHANES II (National Center for Health Statistics, 1983).

## 3.2. Diets of Infants and Children

Vegan diets may not be appropriate for infants older than six months or for young children because these groups may not be able to tolerate the large volume of vegan foods that is needed to meet their energy and nutrient requirements (Jacobs and Dwyer, 1988). Lactovegetarian and lactoovovegetarian diets are more likely to be nutritionally adequate provided that they are planned in accordance with accepted pediatric guidelines.

The effect of vegetarian diets on the growth of children has been studied. In one large study (O'Connell *et al.*, 1989), the height and weight of 404 vegetarian children aged 4 months to 10 years who lived on "The Farm," a collective community in Tennessee, were compared to the U.S. reference standards for height for age, weight for age, and weight for height. Until 1983, a year before the study began, the children were fed a vegan diet with soybeans as the principal source of protein. Supplements of vitamins A, D, and $B_{12}$ were added to the soy milk and other vitamin supplements were also permitted. In 1983, some members of The Farm began to add eggs and dairy products to their diets. The investigators found that the vegetarian children had grown normally, though modestly slower than the reference population.

## 4. Biochemical Assessment of Nutritional Status of Vegetarians

Because of differences in their diets, differences could be expected in the nutrient and non-nutrient content of body fluids in omnivores and vegetarians. Indeed, there is general agreement that in vegetarians plasma cholesterol and arachidonic acid levels tend to be lower and linoleic acid values higher than in omnivores. Plasma levels of palmitoleic, vaccenic and docosahexaenoic acids are much lower in vegetarians (Kritchevsky *et al.*, 1984; Fisher *et al.*, 1986; Melchert *et al.*, 1987). It has been suggested that the apparent lower risk for coronary heart disease in vegetarians may be related to differences in their plasma lipid profile (Fisher *et al.*, 1986).

Because of the possibility that the bioavailability of minerals may be decreased by high fiber diets, blood levels of minerals and trace elements have been measured in vegetarians. Levin *et al.* (1986b) studied the dietary intakes and blood levels of zinc, iron, and magnesium in 113 omnivores and 92 lacto-ovovegetarians living in Israel. The reported mean intakes of iron and magnesium were significantly higher in the vegetarians than in the omnivores, but the intake of zinc was not significantly different. The intakes of these three minerals met or exceeded the U.S. National Research Council recommended dietary allowances (1989). The fiber intake of the vegetarians in the Israel study came mainly from fruits and vegetables; it was almost twice the intake of the omnivores in the study and about five times the estimated mean dietary fiber intake of 12 g/day in the population of the U.S.A. (Lanza *et al.*, 1987). Levin *et al.* (1986b) found no significant difference in blood levels of zinc, iron, and magnesium between vegetarians and omnivores. Mean total iron-binding capacity was significantly lower in the vegetarians, but in both groups the mean blood levels of zinc, iron, and magnesium and the mean iron-binding capacity were within normal ranges. The authors concluded that long-term adherence to a lactoovovegetarian diet does not lead to deficiencies of these minerals.

In contrast, other investigations have indicated that vegetarian subjects were at higher risk for marginal iron and zinc status than omnivorous subjects (Anderson *et al.*, 1981; Gibson *et al.*, 1983). Srikumar *et al.* (1992) found that three months after healthy subjects switched from an omnivorous to a lacto-ovovegetarian diet, plasma and hair concentrations of zinc, copper, and selenium decreased, hair concentrations of mercury, lead and cadmium decreased, and plasma and hair concentrations of magnesium increased. The reduced concentrations of mercury and lead in the hair of the vegetarians may have been due to their lower consumption of fish and alcohol, but there was no explanation given for the decline in cadmium levels. Johnston *et al.* (1987) reported that plasma copper levels and the activity of superoxide dismutase in red cells were similar in a sample of postmenopausal vegetarians and omnivores who had similar copper intakes. Schultz and Leklem (1987) found no significant differences between vegetarians and omnivores in plasma pyridoxal phosphate, urinary 4-pyridoxic acid or urinary vitamin $B_6$ levels, suggesting that there was no adverse effect of fiber on the bioavailability of vitamin $B_6$. There have been several reports of low plasma vitamin $B_{12}$ concentrations in vegetarians and of clinical symptoms of vitamin $B_{12}$ deficiency in infants born to vegetarians (McPhee *et al.*, 1988; Wighton *et al.*, 1979; Higginbottom *et al.*, 1978; Specker *et al.*, 1990). Plasma carotenoid levels are usually much higher in vegetarians than in omnivores (Rider *et al.*, 1984) and the antioxidant properties of carotenoids may be a factor in the apparent decreased risk of certain types of cancers in vegetarians.

In summary, studies of nutrient status support the premise that, except for vitamin $B_{12}$, an adequate nutrient intake can be obtained from well planned vegetarian diets.

## 5. Studies of Bone Mineral Content in Postmenopausal Vegetarians And Omnivores

The mean BMC of small numbers of vegetarians over the age of 50 years has been reported to be higher than that of omnivores (Ellis *et al.*, 1972; Marsh, 1980, 1983). In these studies the BMC of the radius or the bones of the fingers was measured. No data were given in these reports on nutrient intakes or other lifestyle habits that might have affected BMC. In the late 1980s, two cross-sectional studies of postmenopausal lactoovovegetarians were reported, one with 88 subjects living in North Carolina (Tylavsky and Anderson, 1988) and the other with 144 subjects living in California (Hunt *et al.*, 1989). The purpose of these studies was to investigate relationships between BMC and dietary factors, primarily vegetarianism, with emphasis on intakes of protein and calcium. Reference groups consisted of elderly omnivorous women (278 and 146, respectively). In the North Carolina study, the mean age of the vegetarians was about 73 years and in California it was about 65 years. In both studies, mean calcium and caloric intakes were similar but mean protein intakes were significantly lower in the vegetarians than in the omnivores. More details are given in section 3.1.

Lifestyle habits such as smoking and the use of medications that might affect bone density were comparable in the vegetarians and omnivores in both studies. A single photon bone densitometer (Norland-Cameron) was used to measure BMC at the mid-radius of the nondominant arm. In both studies, no significant difference was found between the omnivores and the vegetarians in mean age-adjusted BMC. Additionally, in the California study, the data for BMC in women at each decade of life were in agreement with those obtained in a cross-sectional study of 1552 German women whose mean BMC was 0.68 g/cm$^2$ at age 50 years and 0.52 g/cm$^2$ at age 80 years (Runge *et al.*, 1980).

The data obtained in the California study also indicated that during the late premenopausal period the BMC of vegetarians and omnivores was similar. In a recent study (Lloyd *et al.*, 1991), no significant difference was found in the density of the lumbar spine, measured by dual photon absorptiometry, in 27 vegetarian and 37 omnivorous premenopausal women with a mean age of 35 years. This is evidence that during the late premenopausal as well as the post-menopausal period, the dietary practices of vegetarians do not have significant effects on BMC.

A recent study (Tesar *et al.*, 1992) of 28 matched pairs of postmenopausal Caucasian vegetarians and omnivores confirmed that cortical bone density was not significantly different in the vegetarians and omnivores. They also found no difference in trabecular bone density between the two groups.

In another aspect of bone metabolism, the mean age-adjusted plasma levels of osteocalcin, a putative marker for bone resorption, was not significantly different in the postmenopausal omnivores and vegetarians in the California study (Hunt *et al.*, 1989). Mean age-adjusted BMC was significantly lower in both omnivores and vegetarians who had high plasma osteocalcin levels (Johnston, 1987).

In the North Carolina study, a multiple regression model including data from both vegetarian and omnivorous subjects indicated that bone indices were *positively* influenced by dietary protein over the range of reported intakes. The authors suggest that since protein intakes were within the normally accepted range, the need for protein in an elderly population may be higher than currently recommended (Tylavsky and Anderson, 1988).

The studies of Mazess and Mather (1974, 1975) indicating that there is a greater rate of bone loss in Eskimos consuming about 200 g of animal protein per day provide evidence that extremely high protein intakes may have deleterious effects on bone dynamics. In a study of 300 white omnivorous women aged 20 to 39 years, protein intake was not a significant predictor of BMC (Mazess and Barden, 1991). The mean protein intake of these women was about 65 ± 20 grams at each 5 year interval for age. At present it is not known what levels of protein intake or types of dietary protein are required for optimal bone growth and maintenance.

In the North Carolina study, no significant beneficial effect of a high ongoing calcium intake on BMC in vegetarian and omnivorous postmenopausal women was found using a multiple regression model. A low lifetime calcium intake was found to be deleterious only for the BMC of the midradius. Likewise, in the California study, no significant effect of the current intake of calcium or phosphorus, or of their dietary ratio, on age-adjusted BMC was found in either vegetarians or omnivores. Also, no significant difference in age-adjusted BMC was found between women of either group who reported calcium intakes >1200 mg/day during the early period of life (from 15 to 30 years) and those who consumed less calcium during that period. Calcium intakes early in life were predictive of current BMC only in subjects (omnivores and vegetarians) who had adequate estrogen replacement therapy (ERT). Adequate ERT was defined as a dose >0.625 mg conjugated equine estrogen per day for >20 days per month beginning within 3 years after menopause and continuing for >3 years. In subjects with adequate ERT and early calcium intakes >500 mg/day, BMC could be predicted to fall below the purported "fracture threshold" of 0.58 g/cm$^2$ at about 78 years of age, about 17 years later than for all the other subjects in the study (i.e., subjects who did not receive ERT irrespective of early calcium intake and subjects who had adequate ERT but early calcium intakes <500 mg/day). Current calcium intakes were not associated with BMC in either the subjects who received adequate ERT or in those who received no ERT (Hunt *et al.*, 1990).

## 6. Conclusion

Most published studies indicate that there is no significant difference in the BMC of postmenopausal vegetarians and omnivores when calcium intakes are similar and when protein intakes in the omnivores are within the range usually reported for women in the U.S.A.

## References

Abelow, B.J., Holford, T.R., and Insogna, K.L., 1992, Cross-cultural association between dietary animal protein and hip fracture: a hypothesis, *Calcif. Tissue Int.* **50**:14.

Allen, L.H., Bartlett, R.S., and Block, G.D., 1979, Reduction of calcium absorption in man by consumption of dietary protein, *J. Nutr.* **109**:1345.

Anderson, B.M., Gibson, R.S., and Sabry, J.H., 1981, The iron and zinc status of long-term vegetarian women, *Am. J. Clin. Nutr.* **34**:1042.

Beilin, L.J., Armstrong, B.K., Margetts, B.M., Rouse, I.L., and Vandongen, R., 1987, Vegetarian diet and blood pressure, *Nephron* **47** (Suppl.):37.

Breslau, N.A., Brinkley, L., Hill, K.D., and Pak, C.Y.C., 1988, Relationship of animal protein-rich diet to kidney stone formation and calcium metabolism, *J. Clin. Endocrinol. Metab.* **66**:140.

Bull, N.L., and Barber, S.A., 1984, Food and nutrient intakes of vegetarians in Britain, *Human Nutr.: Applied Nutr.* **38A**:288.

Bushinsky, D.A., 1989, "Internal exchanges of hydrogen ions: bone," in: *The Regulation of Acid Base Balance,* (D. W. Seldin and G. Giebisch, eds.), pp. 69-88, Raven Press, New York.

Cummings, S.R., Kelsey, J.L., Nevitt, M.C., and O'Dowd, K.J., 1985, Epidemiology of osteoporosis and osteoporotic fractures, *Epidemiol. Rev.* **7**:178.

Davies, G.J., Crowder, M., and Dickerson, J.W.T., 1985, Dietary fiber intakes of individuals with different eating patterns, *Human Nutr.: Applied Nutr.* **39A**:139.

Dwyer, J., 1983, "Nutritional status and alternative life style diets with special reference to vegetarianism in the U.S.," in: *CRC Handbook of Nutritional Supplements, Human Use,* Vol. 1 (M. Reichaig, ed.), pp. 343-410, CRC Press, Boca Raton, FL.

Ellis, F.R., Holesh, S., and Ellis, J.W., 1972, Incidence of osteoporosis in vegetarians and omnivores, *Am. J. Clin. Nutr.* **25**:555.

Fisher, M., Levine, P.H., Weiner, B., Ockene, I.S., Johnson, B., Johnson, M.H., Natale, A.M., Vaudreuil, C.H., and Hoogasian, M.A., 1986, The effect of vegetarian diets on plasma lipid and platelet levels, *Arch. Intern. Med.* **146**:1193.

Fraser, G.E., 1988, Determinants of ischemic heart disease in Seventh-day Adventists: a review, *Am. J. Clin. Nutr.* **48**:833.

Gibson, R.S., Anderson, B.M., and Sabry, J.H., 1983, The trace metal status of a group of post-menopausal vegetarians, *J. Am. Diet. Assoc.* **82**:246.

Halperin, J.L., and Jungas, R.L., 1983, Metabolic production and renal disposal of hydrogen ions, *Kidney Int.* **24**:709.

Havala, S., and Dwyer, J., 1988, Position of the American Dietetic Association: vegetarian diets–technical support paper, *J. Am. Diet. Assoc.* **88**:352.

Hegsted, D.M., 1986, Calcium and osteoporosis, *J. Nutr.* **116**:2316.

Herbert, V., 1990, "Vegetarianism," in: *The Mount Sinai School of Medicine Complete Book of Nutrition* (V. Herbert and G.J. Subak-Sharpe, eds., and D.A. Hammock, assoc. ed.), pp. 415-427, St. Martin's Press, New York.

Higginbottom, M.C., Sweetman, L. and Nyhan, W.L., 1978, A syndrome of methylmalonic aciduria, homocystinuria, megaloblastic anemia and neurologic abnormalities in a vitamin B-12 deficient breast-fed infant of a strict vegetarian, *N. Engl. J. Med.* **229**:317.

Hunt, I.F., Murphy, N.J., and Henderson, C., 1988, Food and nutrient intake of Seventh-day Adventist women, *Am. J. Clin. Nutr.* **48**:850.

Hunt, I.F., Murphy, N.J., Henderson, C., Clark, V.A., Jacobs, R.M., Johnston, P.K., and Coulson, A.H., 1989, Bone mineral content in postmenopausal women: comparison of omnivores and vegetarians, *Am. J. Clin. Nutr.* **50**:517.

Hunt, I.F., Murphy, M.J., Clark, V.A., Judd, H.L., Cedars, M.I., and Browdy, B.L., 1990, Bone mineral content in postmenopausal women, calcium intake early in life, and estrogen therapy, *Nutr. Res.* **10**:1061.

Jacobs, C., and Dwyer, J.T., 1988, Vegetarian children: appropriate and inappropriate diets, *Am. J. Clin. Nutr.* **48**:811.

Johnston, P.K., 1987, Relationship between osteocalcin and other indicators of bone metabolism, bone density, copper status, and dietary intake in postmenopausal vegetarian women. Unpublished dissertation.

Johnston, P.K., 1988, Counseling the pregnant vegetarian, *Am. J. Clin. Nutr.* **48**:901.

Johnston, P.K., Hunt, I.F., Murphy, N., and Swendseid, M.E., 1988, Copper status of post-menopausal vegetarian women, Proc. First Int. Congr. on Vegetarian Nutr., *Am. J. Clin.. Nutr.* **48**:925 (Abstr.).

Kelsay, J.L., Frazier, C.W., Prather, E.S., Canary, J.L., Clark, W.M., and Powell, A.S., 1988, Impact of variation in carbohydrate intake on mineral utilization by vegetarians, *Am. J. Clin. Nutr.* **48**:875.

Kritchevsky, D., Tepper, S.A., and Goodman, G., 1984, Relationship of diet to serum lipids, *Am. J. Clin. Nutr.* **40**:921.

Lanza, E., Jones, D.Y., Block, G., and Kessler, L., 1987, Dietary fiber intake in the U.S. population. *Am. J. Clin. Nutr.* **46**: 790.

Levin, N., Rattan, J., and Gilat, T., 1986a, Energy intake and body weight in ovolacto-vegetarians, *J. Clin. Gastroenterol.* **8**:451.

Levin, N., Rattan, J., and Gilat, T., 1986b, Mineral intake and blood levels in vegetarians, *Isr. J. Med. Sci.* **22**:105.

Lloyd, T., Schaffer, J.M., Walker, M.A., and Demers, M., 1991, Urinary hormonal concentrations and spinal bone densities of premenopausal vegetarian and nonvegetarian women, *Am. J. Clin. Nutr.* **54**:1005.

Marsh, A.G., Sanchez, T.V., Mickelsen, O., Keiser, J., and Mayor, G., 1980, Cortical bone density of adult lacto-ovo-vegetarian and omnivorous women, *J. Am. Diet. Assoc.* **76**:148.

Marsh, A.G., Sanchez, T.V., Chaffee, F.L., Mayor, G.H., and Mickelsen, 0., 1983, Bone mineral mass in adult lacto-ovo-vegetarian and omnivorous males. *Am. J. Clin. Nutr.* **37**:453.

Mazess, R.B., and Barden, H.S., 1991, Bone density in premenopausal women: effect of age, dietary intake, physical activity, smoking, and birth-control pills, *Am J. Clin. Nutr.* **53**:132.

Mazess, R.B., and Mather, W.E., 1974, Bone mineral content of North Alaskan Eskimos, *Am. J. Clin. Nutr.* **27**:916.

Mazess, R.B., and Mather, W.E., 1975, Bone mineral content in Canadian Eskimos, *Hum. Biol.* **47**:45.

McPhee A.J., Davidson, G.P., Leahy, M., and Beare, T., 1988, Vitamin $B_{12}$ deficiency in a breast-fed infant, *Arch. Dis. Child.* **63**:921.

Melchert, H.-U., Limsathayourat, N., Mihajlović, H., Eichberg, J., Thefeld, W., and Rottka, H., 1987, Fatty acid patterns in triglycerides, diglycerides, free fatty acids, cholesteryl esters and phosphatidylcholine in serum from vegetarians and non-vegetarians, *Atherosclerosis* **65**:159.

Millet, P., Guilland, J.C., Fuchs, F., and Kippling, J., 1989, Nutrient intake and vitamin status of healthy French vegetarians and nonvegetarians, *Am. J. Clin. Nutr.* **50**:718.

Mutch, P.B., 1988, Food guides for the vegetarian, *Am. J. Clin. Nutr.* **48**:913.

Nair, P.P., Turjman, N., Kessie, G., Calkins, B., Goodman, G.T., Davidovitz, H., and Nimmagadda, G., 1984, Dietary cholesterol, $\beta$-sitosterol, and stigmasterol, *Am. J. Clin. Nutr.* **40**:927.

National Center for Health Statistics, 1983, *Dietary Intake Source Data: United States, 1976-80*, pp. 83-1681 (Vital and health statistics, series 11: data from the National Health and Nutrition Survey #231 D.H.H.S.), Washington, D.C.

National Research Council (U.S.), Subcommittee on the Tenth Edition of the RDAs, 1989, *Recommended Dietary Allowances*, Food and Nutrition Board, Commission on Life Sciences, National Research Council, p. 285, National Academy Press, Washington.

Nieman, D.C., Underwood, B.C., Sherman, K.M., Arabatzis, K., Barbosa, J.C., Johnson, M., and Shultz, T.D., 1989, Dietary status of Seventh-Day Adventist vegetarian and non-vegetarian elderly women, *J. Am. Diet. Assoc.* **89**:1763.

O'Connell, J.M., Dibley, M.J., Sierra, J., Wallace, B., Marks, J.S., and Yip, R., 1989, Growth of vegetarian children: The Farm study, *Pediatrics* **84**:475.

Phillips, R.L., 1980, Cancer among Seventh-day Adventists, *J. Envir. Pathol. Toxicol.* **3**:157.

Raicht, R.F., Cohen, B.I., Fazzini, E.P., Sarwal, A.N., and Takahashi, M., 1980, Protective effect of plant sterols against chemically induced colon tumors in rats, *Cancer Res.* **40**:403.

Rider, A.A., Arthur, R.S., Calkins, B.M., and Nair, P.P., 1984a, Selected biochemical parameters in blood and urine, *Am. J. Clin. Nutr.* **40**:917.

Rider, A.A., Calkins, B.M., Arthur, R.S., and Nair, P.P., 1984b, Concordance of nutrient information obtained by different methods, *Am. J. Clin. Nutr.* **40**:906.

Riggs, B.L., and Melton, L.J. III, 1992, The prevention and treatment of osteoporosis, *N. Engl. J. Med.* **327**:620.

Roper Report 78-10, Oct./Nov. 1978. Roper Center for Public Opinion Research, University of Connecticut, Storrs.

Runge, H., Fengler, F., Franke, J., and Koall, W., 1980, Evaluation of the mineral content of peripheral bones (radius) by photon- absorption technique in normals as well as in patients with various types of bone diseases, *Radiologe* **20**:505.

Schuette, S.A., Zemel, M.B., and Linkswiler, H.M., 1980, Studies on the mechanism of protein-induced hypercalciuria in older men and women, *J. Nutr.* **110**:305.

Shultz, T.D., and Leklem, J.E., 1983, Dietary studies of Seventh-day Adventists and non-vegetarians, *J. Am. Diet. Assoc.* **83**:27.

Schultz, T.D., and Leklem, J.E., 1987, Vitamin $B_6$ status and bioavailability in vegetarian women, *Am. J. Clin. Nutr.* **46**:647.

Snowdon, D.A., and Phillips, R.L., 1985, Does a vegetarian diet reduce the occurrence of diabetes? *Am. J. Public Health* **75**:507.

Specker, B.L., Black, A., Allen, L., and Morrow, F., 1990, Vitamin $B_{12}$: low milk concentrations are related to low serum concentrations in vegetarian women and to methylmalonic aciduria in their infants, *Am. J. Clin. Nutr.* **52**:1073.

Srikumar, T.S., Johansson, G.K., Öckerman, P.A., Gustafsson, J.A., and Åkesson, B., 1992, Trace element status in healthy subjects switching from a mixed to a lactovegetarian diet for 12 mo., *Am. J. Clin. Nutr.* **55**:885.

Tesar, R., Notelovitz, M., Shim, E., Kauwell, G., and Brown, J., 1992, Axial and peripheral bone density and nutrient intakes of postmenopausal vegetarian and omnivorous women, *Am. J. Clin. Nutr.* 56:699.

Tylavsky, F.A., and Anderson, J.J.B., 1988, Dietary factors in bone health of elderly lactoovovegetarian and omnivorous women, *Am. J. Clin. Nutr.* **48**:842.

Yuen, D.E., Draper, H.H., and Trilok, G., 1984, Effect of dietary protein on calcium metabolism in man, *Nutr. Abst. and Rev. in Clin. Nutr.* **54**:447.

Wighton, M.C., Manson, J.I., Speed, I., Robertson, E., and Chapman, E., 1979, Brain damage in infancy and dietary vitamin B-12 deficiency, *Med. J. Aust.* **2**:1.

Chapter 15

# The Effect of Obesity on Post-menopausal Bone Loss and the Risk of Osteoporosis

Claude Ribot, Florence Trémollières and Jean-Michel Pouillès

## 1. Introduction

Primary osteoporosis is the major cause of spine, wrist, rib and hip fractures in postmenopausal women and in the elderly (Consensus Development Conference, 1991; Kanis *et al.*, 1992). Such fractures are a major health problem in industrialized countries and represent a high socio-economic cost since the incidence and morbidity are high. The aging population in industrialized countries and the increase in age-specific incidence noted in several epidemiological studies (Lewis *et al.*, 1981; Obrant *et al.*, 1989; Maggi *et al.*, 1991; Cooper *et al.*, 1992) indicate that there will be a significant increase in such fractures. This justifies preventive measures and improved characterization of the predisposing factors.

Osteoporosis has recently been redefined (Consensus Development Conference, 1991) as a disease characterized by low bone mass, microarchitectural deterioration of bone tissue leading to enhanced bone fragility and a subsequent increase in fracture risk. A variety of diseases may enhance the risk of secondary osteoporosis but the common form is involutional osteoporosis, which Riggs and

Claude Ribot, Florence Trémollières and Jean-Michel Pouillès • Endocrinology Department, Metabolic Bone Diseases Unit, C.H.U. Purpan 31059 Toulouse, France

*Advances in Nutritional Research*, Vol. 9
Edited by Harold H. Draper
Plenum Press, New York, 1994

Melton (1983, 1986) have suggested may be divided into two separate syndromes. Type I (postmenopausal) osteoporosis affects women for 15 to 20 years after menopause and is characterized by fractures in skeletal sites that contain a large amount of cancellous bone (vertebrae, distal forearm). Type II (senile or age-related) osteoporosis occurs in both men and women over the age of 70 and concerns mainly hip fractures, i.e., sites containing substantial amounts of both cortical and trabecular bone. Individualization of these two forms remains controversial since the conditions promoting the occurrence of osteoporosis in these different sites might be similar.

After reaching a peak in the early twenties, perhaps before (Gilsanz *et al.*, 1988; Katzman *et al.*, 1991; Bonjour *et al.*, 1991; McCormick *et al.*, 1991; Thomas *et al.*, 1991), bone mass declines throughout life in both sexes as a universal aging phenomenon (Mazess, 1982). In women, bone mass decreases most rapidly for several years after menopause. Low bone mass at maturity, the accelerated rate of postmenopausal bone loss and aging-related bone loss are the main factors predisposing to osteoporosis. The quality of bone tissue and its architectural structure certainly contribute to its capacity for resistance to trauma, although in proportions that are almost impossible to determine, since bone rarefaction and architectural abnormalities are frequently associated. In fact, bone mineral density is responsible for 70 to 80% of bone resistance and remains the best predictive factor for fracture risk, as recently published longitudinal studies have shown. Following adjustment for age, this risk is multiplied by 1.5 to 2 for each standard deviation (10–12%) decrease in bone mass measured at either the radius or the calcaneum (Hui *et al.*, 1989; Cummings *et al.*, 1990; Porter *et al.*, 1990). Furthermore, from still preliminary results it would seem that direct measurement at the potential fracture site provides more reliable risk data than peripheral measurements (Melton, 1992).

Peak bone mass and bone loss are modulated by genetic, environmental and hormonal factors. Furthermore, there are conditions that are likely to have a negative influence on bone mass and thus are considered as osteoporotic risk factors. Such established and putative risk factors include age, sex, ethnic origin, early onset of menopause, sedentary lifestyle, tobacco and alcohol consumption and low dietary intake of calcium (Cooper *et al.*, 1991; Compston *et al.*, 1992; Pouilles *et al.*, 1991). In fact, the predictive value of these factors with respect to low bone mass and fracture risk is too slight (with the exception of some particular circumstances such as early menopause), to be valid for use in screening (Kleerekopper *et al.*, 1989; Slemenda *et al.*, 1990; Ribot *et al.*, 1992). On the other hand, "non-risk" factors for osteoporosis are even less well known. Since low body weight is a clinical condition often associated with osteoporosis, it has been suggested that obesity might be a protective factor.

This paper aims to review the influence of body weight on bone mass and on the risk of osteoporosis and to discuss the underlying mechanisms.

## 2. Obesity: Definition, Assessment

The term obesity needs to be interpreted carefully. Obesity is an excess of body fat. "Overweight" is defined as a body weight in excess of ideal weight. Some overweight subjects, especially those who exercise regularly, are not overfat. However, in women who do not exercise regularly, and particularly after the menopause, it can reasonably be assumed that excess weight corresponds to excess fat.

There are many methods of measuring body fat. Most are not generally available for clinical purposes and either lack precision and accuracy or are costly. The most popular body fat index relies on skin fold measurement made with calipers. The body mass index (BMI) (weight/height$^2$) is far easier to use. According to some authors, it provides only a rough estimate of fat mass, whereas for others it corresponds closely to the measurement of body density under water, which is the most accurate method of determining body fat. It has been stated that the BMI does not differentiate sex-related forms of obesity (i.e., android or upper body obesity and gynoid or lower body obesity) and that the higher BMI in postmenopausal than in premenopausal women reflects a greater fat mass (Ley *et al.,* 1992). Furthermore, BMI may be overestimated in cases where vertebral crush fractures have reduced the patient's original height. Schematically, it is possible to distinguish between overweight ($25 < \text{BMI} < 30$) and obesity ($\text{BMI} > 30$). New methods of assessment of body fat such as bioelectric impedance and dual photon X-ray energy absorptiometry (DEXA) have been used only in the most recent studies (Svendsen *et al.,* 1991; Ley *et al.,* 1992).

## 3. Protective Effect of Obesity

Three types of data support the hypothesis that obesity has a protective effect against osteoporosis.

### 3.1. Obesity and Incidence of Fracture

Many epidemiological studies have shown the existence of an inverse relationship between the risk of fracture and body weight. The relation between excess body weight and risk of fracture has been documented primarily for fractures of the femoral neck; data relating to fracture of the radius and vertebrae are more limited.

Kreiger *et al.* (1982) and Hutchinson *et al.* (1979) have shown that BMI is significantly lower in women having suffered a fracture of the femoral neck than in age-matched control subjects. Whickam *et al.* (1989) reported that the risk of hip fracture was strongly related to BMI with a 5-fold increase in risk among women in the lowest third of the BMI distribution. Paganini-Hill *et al.* (1981) found a 42% reduction in the relative risk of sustaining a hip fracture in women with a BMI above 33. The Framingham study confirms a protective effect of excess body weight on hip fracture and also on fracture of the radius, which fell significantly with increasing increments of weight. There was a clear decrease to 0.33 in the relative risk of fracture when the weight index as assessed by Metropolitan Relative Weight was higher than 138 (Kiel *et al.*, 1987). Finally, in the MEDOS (Mediterranean Osteoporosis Study), whose aim was to assess the incidence of hip fracture in six southern European countries, the relative risk was reduced to 0.35 for a weight increase of 2 SD from normal values (personal data). Concerning vertebral fractures, the scarcity of epidemiological data available does not make it possible to specifically determine the influence of excess body weight on risk of fracture. However, indirect arguments indicate that overweight could also be a protective factor. Hassager *et al.* (1991) found a decrease in both weight and fat mass (assessed by X-ray absorptiometry) in women with vertebral osteoporosis.

Overall, these data suggest that risk of osteoporotic fractures, whatever the type, is inversely correlated with body weight and that excess body weight is a protective factor.

### 3.2. Obesity and Bone Mass

Studies on the determinants of bone mass in normal subjects have all shown that age, weight and menopausal status are the major factors. Mazess *et al.* (1990) found that in men each 10 kg gain in weight was associated with a 4% increase in vertebral bone density and a 3% femoral bone mass increase assessed by dual photon absorptiometry (DPA). In a cross-sectional study on 510 healthy white French women aged between 25 and 70 years Ribot *et al.* (1988) found a positive correlation between weight and vertebral bone mineral density (BMD). It is interesting to note that in the above study, which aimed to establish normal values for bone mass according to age group, there was a difference of almost 10% in relation to U.S. population reference values (Mazess *et al.*, 1987). It is likely that the lower density in our sample was influenced by our exclusion of overweight women, which gave a mean weight of 55 kg as compared to 62 kg in the U.S. study. Liell *et al.* (1988) reported spinal BMD values in U.S. women similar to the values for French women when overweight subjects were excluded. In all these studies, weight was a predictive factor not only for vertebral bone mass but also for femoral and radial bone mass.

A few studies have been carried out specifically on bone mineral density in overweight women; in general increased bone mass was found in all measured sites in obese women. It is possible that a moderate overestimation of vertebral BMD due to arthritis, which is frequent in obese subjects, was obtained using DPA. However, this artifact does not seem to interfere in measurements made on other peripheral sites.

Dawson-Hughes *et al.* (1987) reported a significant relationship between both BMD and bone mineral content (BMC) at different sites (lumbar spine, femoral neck, radius) and the percent ideal body weight (% IBW). In this study, postmenopausal women 115% over IBW had higher mean BMD and BMC than women of normal weight. Bagur *et al.* (1992) have reported that even in cases of premature menopause, obese women (BMI > 30) had significantly increased vertebral and femoral BMD when compared with nonobese women, but not with age-matched controls. Liel *et al.* (1988) measured the BMD of the radius, hip and spine in obese white and black premenopausal women who were more than 30% over their IBW. Vertebral and femoral bone densities (but not radial bone densities) were found to be significantly higher in obese women than in age- and race-matched women of normal weight. In a cross-sectional study on 176 women, Ribot *et al.* (1987) found that the protective effect of obesity on bone mass was particularly evident in postmenopausal women. Spinal BMD in postmenopausal obese women was higher than in women of normal weight, and was comparable to that of younger perimenopausal women. This difference in BMD was not found in premenopausal women and this suggests that obesity is more likely to play a protective role in postmenopausal bone loss. Kin *et al.* (1991) also found that the protective effect of obesity on vertebral bone mass became evident only after age 40–49. Women with higher BMI ($\geq$ 25) had higher vertebral BMD than thin and normal women after the fifth decade of life. Thus, even though the methodology of these cross-sectional studies may have partially accounted for the differences in BMD observed between pre- and postmenopausal women, the results strongly support a role of obesity in the rate of bone loss after menopause.

In fact, the influence of obesity on the acquisition of peak bone mass remains poorly understood. Bone mass and bone density increase in both sexes during puberty and reach a plateau at the end of adolescence. Variations in bone density remain difficult to interpret because of concomitant modifications in the geometry of the vertebral bodies (size, volume). Furthermore, increase in bone mass is closely associated with anthropometric growth variables (height, weight, pubertal status) whose respective roles remain difficult to assess. In the only study available, McCormick *et al.* (1991) have shown that BMD per kilogram body weight was much reduced in overweight children and adolescents compared to normal control subjects. This suggests that obesity does not positively affect vertebral bone mass during growth.

### 3.3. Obesity and Rate of Bone Loss

As shown previously, the difference in bone mass between obese women and women of normal weight is more evident after menopause than before, and this suggests that they have a slower rate of bone loss. In this context, it is interesting to note that bone remodeling, assessed by serum osteocalcin levels and the urinary excretion of calcium, was found to be significantly reduced in overweight postmenopausal women (Ribot *et al.*, 1987). A negative correlation between body weight and the urinary excretion of calcium has been reported by Frumar *et al.* (1980), who suggested that excess weight was associated with lower bone resorption. More interestingly, Ribot *et al.* (unpublished results) extended their study on obese women by longitudinally examining the influence of excess weight on changes in vertebral BMD. One hundred and fifteen postmenopausal women of normal weight (BMI < 25) and 40 postmenopausal obese women (BMI ≥ 25) were followed over an average period of 31 months. The rate of bone loss in the two groups was calculated using linear regression analysis of BMD over time. For the same age and interval since menopause, postmenopausal obese women had a significantly lower annual rate of vertebral bone loss than nonobese controls (-0.54% vs -1.5%, $p<0.05$). A negative relationship was found between the rate of bone loss and BMI, but not body weight. Since direct measurement of body fat has been shown to correlate better with BMI than with body weight, these results suggest that it is the excess adipose tissue that is more likely to play a protective role in postmenopausal bone loss. In a recently published longitudinal study, Harris *et al.* (1992) also observed a protective effect of weight on bone loss in postmenopausal women with up to 106% IBW. Mean femoral and radial BMD changes tended to be lower in the heaviest women but the difference was not significant. However, Aloia *et al.* (1991) found no relation between total body fat (indirectly determined from evaluation of total body potassium), total body calcium and the BMD of the radius, spine and femoral neck, and no evidence that excess adipose tissue plays a major role in protecting against bone loss.

## 4. Mechanisms of the Protective Effect of Obesity

The protective effect of obesity may be explained by a combination of hormonal and mechanical factors.

### 4.1. Hormonal Mechanisms

Obesity is associated with several endocrine changes (insulin secretion, increased cortisol production, decreased growth hormone secretion and reduced secretory capacity of glucagon) which could affect bone metabolism either directly or indirectly. So far, the mechanism best understood is the increase in endogenous estrogens, which is due to increased conversion of androgens in adipose tissue and to decreased sex hormone binding globulin (SHBG) levels. Alterations in the vitamin D endocrine system and in the production of some growth factors have been mentioned but insufficiently studied. This protective hormonal effect mainly, if not exclusively, concerns bone loss.

Sex steroids are necessary for the skeleton's integrity. There is no doubt that estrogen deficiency is the major determining factor for type I osteoporosis and could also promote type II (Davidson *et al.*, 1982). Estrogen deficiency leads to activation of bone remodeling with resorption activity preponderant over formation (Parfitt *et al.*, 1982). The result is altered microarchitecture and mineral bone loss which, in addition to the effects of aging, is more pronounced in trabecular rather than in cortical bone. It is estimated that menopause-related bone loss is responsible, by the age of 70 years, for more than half the trabecular bone loss and approximately one-third the cortical bone loss (Genant *et al.*, 1989).

Postmenopausal bone loss rate varies according to individuals and may be modulated by genetic (Kelly *et al.*, 1991) and nutritional (Dawson-Hughes, 1991) factors as well as by regular exercise (Smith *et al.*, 1991). It depends mostly on the extent of estrogen deficiency, which is itself partly related to the type of menopause and body mass.

Several studies have shown a positive correlation between plasma concentrations of estradiol or estrone and perimenopausal (Johnston *et al.*, 1985; Slemenda *et al.*, 1987) and postmenopausal (Riis *et al.*, 1984; Cauley *et al.*, 1986) bone loss. Women classified as rapid losers have significantly lower estrogen concentrations than those classified as slow losers. Early menopause is a recognised risk factor. Recently it was shown that, whereas the natural menopause bone loss rate falls significantly after 5 years, in early menopause it remains linear for more than 20 years (Pouilles *et al.*, 1993). Similarly, in cases of surgical menopause, the relative hyperresorption phase, assessed by the increase in plasma tartrate resistant acid phosphatase activity and in urinary hydroxyproline excretion, can last more than 12 years (Stepan *et al.*, 1987). Finally, administration of gonadotrophin releasing hormone (GnRH) agonists which achieve a hormonal castration within a few weeks leads to a vertebral bone loss of 4 to 6% in a few months, i.e., a level three times as high as that of natural menopause (Johansen *et al.*, 1988).

On the other hand, after menopause, body mass becomes the main determinant of endogenous estrogen activity (Meldrum *et al.*, 1981). It has frequently been demonstrated that obese women are not as deficient in estrogen after menopause; this could explain the higher frequency of cancer of the endometrium (Rizkallah *et al.*, 1975) and of the breast (Poortman *et al.*, 1973) in obese women. It could also explain the decrease in bone resorption and in the consequently higher bone mass found in postmenopausal overweight women. Moreover, several studies have shown that there is a positive correlation after menopause between endometrium status and bone mineral content on one hand and the concentration of estrogens and androgens on the other (Brody *et al.*, 1981; Longcope *et al.*, 1984).

Prior to menopause, plasma concentrations of estradiol and estrone essentially reflect ovarian secretion. After menopause, although the ovaries may continue to produce small quantities of hormones (Longcope *et al.*, 1980), most of the circulating estrogens are a result of peripheral conversion from androstenedione and of the interconversion between androgens (androstenedione, testosterone) and estrogens (estrone, estradiol) (Longcope, 1976). Adipose tissue has the enzymes required for this metabolism (aromatase, 17$\beta$-dehydrogenase) and, as numerous *in vivo* and *in vitro* studies have shown, it is an important site for aromatization and conversion (Schindler *et al.*, 1972; Bleau *et al.*, 1974; Longcope *et al.*, 1978; Perel *et al.*, 1979). Along with MacDonald *et al.* (1978) and Rizkalla *et al.* (1975), Longcope *et al.* (1986) have shown that peripheral aromatization of androstenedione is positively correlated ($r = 0.45$, $p<0.001$) with BMI. Furthermore, production of androgens, especially in cases of extreme obesity, is likely to be higher than in normal weight women (Samojlik *et al.*, 1984). Finally, Kirshner *et al.* (1990) have reported that sex hormone production and metabolism differ according to obesity phenotype: women with upper body obesity have higher androgen production rates and higher free testosterone and estradiol levels, whereas women with lower body obesity produce increased amounts of estrone (EI) from peripheral aromatization.

The fact that SHBG is generally lower in obese subjects (Haffner *et al.*, 1991), thereby resulting in an increased free fraction of biologically active sex hormones, could be an inhibiting factor in bone loss. Significantly high SHBG levels have been reported in women with type I (Van Hemert *et al.*, 1989) or type II (Davidson *et al.*, 1982) osteoporosis. A negative relationship between SHBG binding capacity and vertebral (Wild *et al.*, 1987) or forearm (Brody *et al.*, 1987) bone mass has been demonstrated in postmenopausal women. In a wide population-based study aiming to assess the role of endogenous estrogens and of SHBG on bone mass, bone loss and fracture risk, Van Hermet *et al.* (1989) found that the level of SHBG was a better determining factor of bone mass and bone loss than estrogen level. This parameter was also the only variable independently linked to fracture risk. This last point, together with the absence of correlation between the degree of obesity and the SHBG level noted by Brody *et al.* (1987), suggests that the relationship between SHBG and bone mass may be based on different mechanisms or on mechanisms not exclusively related to variations in biological activity of sex hormones. For example, the fact that SHBG is increased in states associated with a decrease in define (IGF-I) levels (growth factors involved in the autocrine modulation of bone remodeling) opens new research paths (Von Schoultz *et al.*, 1989). The role of SHBG in osteoporosis pathogeny is not yet properly understood.

An increase in the plasma concentrations of immunoreactive parathyroid hormone (iPTH), 1,25 dihydroxy vitamin $D_3$ and osteocalcin and a decrease in urinary excretion of calcium have all been reported in young extremely obese adults and in black subjects (Andersen *et al.*, 1986; Bell *et al.*, 1985). These changes in the vitamin D endocrine system were interpreted as factors that could explain the high bone mass noted in these two population groups. In moderately obese postmenopausal women there is a decrease in bone remodeling parameters (especially in plasma osteocalcin levels) that makes the secondary hyperparathyroidism hypothesis unlikely.

## 4.2. Mechanical Effects

Another possibility, of course, is a purely mechanical effect of excess weight on bone density. Numerous *in vitro* and *in vivo* studies have demonstrated that biomechanical factors play a role in bone density and in the resistance of bone to fractures (Hayes *et al.*, 1985). It is currently well known that certain types of physical exercise can lead to an increase in bone density, particularly at sites affected by mechanical loading. The clear increase in the bone density of the dominant radius in tennis players is a classic example (Huddleson *et al.*, 1980). Various studies have shown that fairly prolonged weightbearing training exercise of moderate intensity can lead to a significant increase in vertebral (Dalsky *et al.*, 1988; Michel *et al.*, 1989), femoral and radial (Pruitt *et al.*, 1992) bone mass and the mass of the total skeleton (Chow *et al.*, 1987). Inversely, in women, walking (Cavanaugh *et al.*, 1988), stretching (Chow *et al.*, 1987), cycling (Ismail *et al.*, 1989), swimming (Orwoll *et al.*, 1989) and endurance sports (Nelson *et al.*, 1988) do not appear to have a positive effect on bone mass. It is true that the increase in bone mass in obese subjects is most noticeable in the weight- bearing bones (vertebrae, femur) (Liel *et al.*, 1988; DeSimone *et al.*, 1989); however, an increase in peripheral bone mass (radius), although less noticeable (Aloia *et al.*, 1991; Dawson-Hughes *et al.*, 1987) suggests that the phenomenon cannot be explained as a purely mechanical effect. Such an effect very probably plays a role in marked obesity, especially in cases of primary obesity that had its origin in childhood or adolescence, when mechanical loading is most likely to exert a positive effect on the skeleton. It seems unlikely in postmenopausal women with only a moderate excess of weight due primarily to an increase in lower body fat (Haarbo *et al.*, 1991) and in whom muscular mass decreases with age.

Finally, for hip fractures, in which the precipitating event is usually a fall, it is possible that excess adipose tissue surrounding the hip decreases the impact of the fall. It is also possible that elderly obese women have a slower gait that makes a fall on the hip less likely (Cummings *et al.*, 1987).

## 5. Summary

There are many data indicating that osteoporotic fractures, and particularly hip fractures, are less frequent in obese subjects. Overweight and obese women have a higher bone mass after menopause than women of the same age who are not overweight, and thus in all probability have a slower bone loss. This protective effect appears to be related both to mechanical factors and to estrogen synthesis in adipose tissue.

## References

Aloia, J.F., McGowan, D.M., Vaswani, A.N., Ross, P., and Cohn, S.H., 1991, Relationship of menopause to skeletal and muscle mass, *Am. J. Clin. Nutr.* **53**:1378.

Andersen, T., McNair, P., Fogh-Andersen, N., Nielsen, T.T., Hyldstrup, L., and Transbol, I., 1986, Increased parathyroid hormone as a consequence of changed complex. Binding of plasma calcium in morbid obesity, *Metabolism* **35**:147.

Bagur, A.C., and Mautalen, C.A., 1992, Risk for developing osteoporosis in untreated premature menopause, *Calcif. Tissue Int.* **51**:4.

Bell, N.H., Epstein, S., Greene, A., Shary, J., Oesmann, M. J., and Shaw, S., 1985, Evidence for alteration of vitamin D endocrine system in obese subjects. *J. Clin. Invest.* **76**:370.

Bleau, G., Roberts, K.D., and Chapdelaine, A., 1974, The in vitro and in vivo uptake and metabolism of steroids in human adipose tissue, *J. Clin. Endocrinol. Metab.* **39**:236.

Bonjour, J.P., Theintz, G., Buchs, B., Slosman, D., and Rizzoli, R., 1991, Critical years and stages of puberty for spinal and femoral bone mass accumulation during adolescence, *J. Clin. Endocrinol. Metab.* **73**:555.

Brody, S., Carlström, K., Lagrelius, A., Lunell, N.O., and Rosenberg, L., 1981, Adrenal steroids, bone mineral content and endometrial pathology in postmenopausal women, *Acta Obstet. Gynecol. Scand.* **60**:325.

Brody, S., Carlström, K., Lagrelius, A., Lunell, N.O., Möllerström, G., and Pouselle, A., 1987, Serum sex hormone binding globulin (SHBG), testosterone/SHBG index, endometrial pathology and bone mineral density in postmenopausal women, *Acta Obstet. Gynecol. Scand.* **66**:357.

Cauley, J.A., Gutal, J.P., Sandler, R.B., Laporte, R.E., Kuller, L.H., and Sashin, D., 1986, The relationship of endogenous estrogen to bone density and bone area in normal postmenopausal women, *Am. J. Epidemiol.* **124**:752.

Cavanaugh, D.J., and Cann, C.E., 1988, Brisk walking does not stop bone loss in postmenopausal women, *Bone* **9**:201.

Chow, R.K., Harrison, J.E., amd Notarius, C., 1987, Effect of two randomised exercise programmes on bone mass of healthy postmenopausal women, *Br. Med. J.* **292**:607.

Compston, J.E., 1992, Risk factors for osteoporosis, *Clin. Endocrinol.* **36**:223.

Consensus development conference: prophylaxis and treatment of osteoporosis, 1991, *Osteoporosis Int.* **1**:114.

Cooper, C., Shah, S.N., Hand, D.J., Adams, J., Compston, J.E., Davie, M., and Woolf, A., 1991, Screening for vertebral osteoporosis using individual risk factors, *Osteoporosis Int.* **2**:48.

Cooper, C., Atkinson, E.J., O'Fallon, W.M., and Melton, L.J., 1992, Incidence of clinically diagnosed vertebral fractures: a population-based study in Rochester, Minnesota, 1985-1989, *J. Bone Min. Res.* **7**:221.

Cummings, S.R., 1987, "Epidemiology of hip fractures," in: *Osteoporosis* (Christiansen, C., Johansen, J.S., Riis, B.J., eds.) pp. 40-44, Osteopress, Copenhagen.

Cummings, S.R., Black, D.M., Nevitt, M.C., Browner, W.S., Cauley, J.A., Genant, H.K., Scott, J.C., Seeley, D.G., Steiger, P., and Vogt, T.M., 1990, Appendicular bone density and age predict hip fracture in women, *JAMA* **263**:665.

Dalsky, G.P., Stocke, K.S., Ehsani, A.A., Slatopolsky, E., Lee, W.C., and Birge, S.J., 1988, Weight-bearing exercise training and lumbar bone mineral content in postmenopausal women, *Ann. Intern. Med.* **108**:824.

Davidson, B.J., Ross, R.K., Paganini-Hill, A., Hammon, G.L., Siiteri, P.K., and Judd, H.L., 1982, Total and free estrogens and androgens in postmenopausal women with hip fractures, *J. Clin. Endocrinol. Metab.* **54**:115.

Dawson-Hughes, B., 1991, Calcium supplementation and bone loss: a review of controlled clinical trials, *Am. J. Clin. Nutr.* **54**:2745.

Dawson-Hughes, B., Shipp, C., Sadowski, L., and Dallal, G., 1987, Bone density of the radius, spine, and hip in relation to percent of ideal body weight in postmenopausal women, *Calcif. Tissue Int.* **40**:310.

DeSimone, D.P., Stevens, J., Edwards, J., Shary, J., Gordon, L., and Bell, N.H., 1989, Influence of body habitus and race on bone mineral density of the midradius, hip, and spine in aging women, *J. Bone Min. Res.* **4**:827.

Frumar, A.M., Meldrum, D.R., and Geola, F., 1980, Relationship of fasting urinary calcium to circulating estrogen and body weight in postmenopausal women, *J. Clin. Endocrinol. Metab.* **50**:70.

Genant, H.K., Bayling, D.J., and Gallagher, J.C., 1989, Estrogens in the prevention of osteoporosis in postmenopausal women, *Am. J. Obstet. Gynecol.* **161**:1842.

Gilsanz, V., Gibbens, D.T., Roe, T.F., Carlson, M., Senac, M.O., Boechat, M.I., Huang, H.K., Schulz, E.E., Libanati, C.R., and Cann, C.C., 1988, Vertebral bone density in children: effect of puberty, *Radiology* **166**:847.

Haarbo, J., Marslew, U., Gotfredsen, A., and Christiansen, C., 1991, Postmenopausal hormone replacement therapy prevents central distribution of body fat after menopause, *Metabolism* **40**:1323.

Haffner, S.M., Katz, M.S., and Dunn, J.F., 1991, Increased upper body and overall adiposity is associated with decreased sex hormone binding globulin in postmenopausal women, *Int. J. Obesity* **15**:471.

Harris, S., Dallal, G.E., and Dawson-Hughes, B., 1992, Influence of body weight on rates of change in bone density of the spine, hip, and radius in postmenopausal women, *Calcif. Tissue Int.* **50**:19.

Hassager, C., and Christiansen, C., 1989, Influence of soft tissue body composition on bone mass and metabolism, *Bone* **10**:415.

Hayes, W.C., and Gerhart, T.N., 1985, "Biomechanics of bone: applications for assessment of bone strength," in: *Bone and Mineral Research 3*, (Peck, W.A., ed.), pp. 259-294, Elsevier Science Publishers, Amsterdam.

Huddleson, A.L., Rockwell, D., Kulund, D.L., and Harrison, R.B., 1980, Bone mass in lifetime tennis athletes, *JAMA* **244**:1107.

Hui, S.L., Slemenda, C.W., and Johnston, C.C., 1989, Baseline measurement of bone mass predicts fracture in white women, *Ann. Int. Med.* **111**:355.

Hutchinson, T.A., Polansky, S.M., and Feinstein, A.R., 1979, Postmenopausal oestrogens protect against fractures of hip and distal radius, *Lancet* **2**:705.

Ismail, F., Epstein, S., Gorman, K., Posner, J., Windsor, L.A., Makler, T., and Movsowitz, C., 1989, The influence of exercise on bone mineral metabolism in the elderly, *J.Bone Min. Res.* **4**(S):S231 (abstract).

Johansen, J.S., Riis, B.J., Hassager, C., Moen, M., Jacobson, J., and Christiansen, C., 1988, The effect of gonadotrophin releasing hormone agonist analog (Nafarelin) on bone metabolism, *J. Clin. Endocrinol. Metab.* **67**:701.

Johnston, C.C., Hui, S.L., Witt, R.M., Appledorn, R., Baker, R., and Longcope, C., 1985, Early menopausal changes in bone mass and sex steroids, *J. Clin. Endocrinol. Metab.* **61**:905.

Kanis, J.A., and Pitt, F.A., 1992, Epidemiology of osteoporosis, *Bone* **13**:57.

Katzman, D.K., Bachrach, L.K., Carter, D.R., and Marcus, R., 1991, Clinical and anthropometric correlates of bone mineral acquisition in healthy adolescent girls, *J. Clin. Endocrinol. Metab.* **73**:1332.

Kelly, P.J., Hopper, J.L., Macaskill, G.T., Pocock, N.A., Sanbrook, P.N., and Eisman, J.A., 1991, Genetic factors in bone turnover, *J. Clin. Endocrinol. Metab.* 72:808.

Kiel, D.P., Felson, D.T., Anderson, J.J., Wilson, P.W.F., and Moskowitz, M.A., 1987, Hip fracture and the use of estrogens in postmenopausal women, *N. Engl. J. Med.* **317**:1169.

Kin, K., Kushida, K., Yamazaki, K., Okamoto, S., and Inoue, T., 1991, Bone mineral density of the spine in normal Japanese subjects using dual-energy X-ray absorptiometry: effect of obesity and menopausal status, *Calcif. Tissue Int.* **49**:101.

Kirshner, M.A., Samojlik, E., Drejka, M., Szmal, E., Schneider, G., and Ertel, N.H., 1990, Androgen-estrogen metabolism in women with upper body versus lower body obesity, *J .Clin. Endocrinol. Metab.* **70**:473.

Kleerekopper, M., Peterson, E., Nelson, D., Tilley, B., Phillips, E., Schork, M.A., and Kuder, J., 1989, Identification of women at risk for developing postmenopausal osteoporosis with vertebral fractures: role of history and single photon absorptiometry, *Bone and Mineral* **7**:171.

Kreiger, N., Kelsey, J.L., Holfors, T.R., and O'Connor, T., 1982, An epidemiologic study of hip fracture in postmenopausal women, *Am. J. Epidemiol.* **116**:141.

Lewis, A., 1981, Fracture of the neck of the femur: changing incidence, *Br. Med. J.* **283**:1217.

Ley, C.J., Lees, B., and Stevenson, J.C., 1992, Sex- and menopause-associated changes in body-fat distribution, *Am. J. Clin. Nutr.* **55**:950.

Liel, Y., Edwards, J., Shary, J., Spicer, K.M., Gordon, L., and Bell, N.H., 1988, The effects of race and body habitus on bone mineral density of the radius, hip, and spine in premenopausal women, *J. Clin. Endocrinol. Metab.*, **66**:1247.

Longcope, C., 1976, "The significance of steroid production by peripheral tissue," in: *Endocrinology of the Ovaries* (Scholler, E.R. ed.), pp. 23-34, Paris.

Longcope, C., Pratt, J.H., and Schneider, S.H., 1978, Aromatization of androgens by muscle and adipose tissue in vivo, *J. Clin. Endocrinol. Metab.* **46**:146.

Longcope, C., Hunter, R., and Franz, C.H., 1980, Steroid secretion by the postmenopausal ovary, *Am. J. Obstet. Gynecol.* **138**:564.

Longcope, C., Baker, R.S., Hui, S.L., and Johnston, C.C., 1984, Estrogen dynamics in women with vertebral crush fractures, *Maturitas* **6**:309.

Longcope, C., Baker, R., and Johnston, C.C., 1986, Androgen and estrogen metabolism: relationship to obesity, *Metabolism* **35**:235.

McCormick, D.P., Ponder, S.W., Fawcett, H.D., and Palmer, J.L., 1991, Spinal bone density in 335 normal and obese children and adolescents: evidence for ethnic and sex differences, *J. Bone Min. Res.* **6**:507.

MacDonald, P.C., Edman, C.D., and Hemsell, D.L., 1978, Effect of obesity on conversion of androstenedione to estrone in postmenopausal women with and without endometrial cancer, *Am. J. Obstet. Gynecol.* **130**:448.

Maggi, S., Kelsey, J.L., Litvak, J., and Heyse, S.P., 1991, Incidence of hip fracture in the elderly: a cross-national analysis, *Osteoporosis Int.* **1**:232.

Mazess, R.B., 1982, On aging bone loss, *Clin. Orthop.* **165**:239.

Mazess, R.B., Barden, H.S., Ettinger, M., Johnston, C., Dawson-Hughes, B., Baran, D., Powell, M., and Notelevitz, M., 1987, Spine and femur density using dual-photon absorptiometry in US white women, *Bone and Mineral* **2**:211.

Mazess, R.B., Barden, H.S., Drinca, P.J., Bauwens, S.F., Orwoll, E.S., and Bell, N.H., 1990, Influence of age and body weight on spine and femur bone mineral density in US white men, *J. Bone Min. Res.* **5**:645.

Meldrum, D.R., Davidson, B.J., Taturyn, I.V., and Judd, H.L., 1981, Changes in circulating steroids with aging in postmenopausal women, *Obstet. Gynecol.* **57**:624

Melton, L.J., 1992, Predicting osteoporotic fractures: the state of the art, *Bone and Mineral* 17(S):72 (abstract).

Michel, B.A., Bloch, D.A., and Fries, J.F., 1989, Weight-bearing exercise, overexercise, and lumbar bone density over age 50 years, *Arch. Intern. Med.* **149**:2325.

Nelson, M.E., Meredith, C.N., Dawson-Hughes, B., and Evans, W.J., 1988, Hormone and bone mineral status in endurance-trained and sedentary postmenopausal women, *J. Clin. Endocrinol. Metab.* **66**:927.

Obrant, K.J., Bengner, U., Johnell, O., Nilsson, B.E., and Sernbo, I., 1989, Increasing age-adjusted risk of fragility fractures: a sign of increasing osteoporosis in successive generations? *Calcif. Tissue Int.* **44**:157.

Orwoll, E.S., Ferar, J., Oviatt, S.K., McClung, M.R., and Huntington, K., 1989, The relationship of swimming exercise to bone mass in men and women, *Arch. Intern. Med.* **149**:2197.

Paganini-Hill, A., Ross, R.K., Gerkins, V.R., Henderson, B.E., Arthur, M., and Mack, T.M., 1981, Menopausal estrogen therapy and hip fractures, *Ann. Int. Med.* **95**:28.

Parfitt, A.M., 1982, The coupling of bone formation to bone resorption: a critical analysis for the concept and its relevance to the pathogenesis of osteoporosis, *Metab. Bone Dis. Relat. Res.* **4**:1.

Perel, E., and Killinger, D.W., 1979, The interconversion and aromatization of androgens by human adipose tissue, *J. Steroid Biochem.* **10**:623.

Poortman, J., Thijssen, H.H., and Schwartz, F., 1973, Androgen production and conversion to estrogen in normal postmenopausal women and in selected breast cancer patients, *J. Clin. Endocrinol. Metab.* **37**:101.

Pouilles, J.M., Ribot, C., Tremollieres, F., Bonneu, M., and Brun, S., 1991, Facteurs de risque de l'ostéoporose vertébrale. Résultats d'une étude chez 2,279 femmes adressées à une consultation de ménopause, *Rev. Rhum. Mal. Osteoartic.* **3**:169.

Pouilles, J.M., Tremollieres, F., Vellas, B., Albarede, J.L., and Ribot, C., 1992, Fracture de l'extrémité supérieure du fémur chez la femme agée: rôle respectif de la chute et de la déminéralisation osseuse, *Rev. Rhum. Mal. Osteoartic.* **4**:241.

Pruitt, L.A., Jackson, R.D., Bartels, R.L., and Lehnhard, H.J., 1992, Weight-training effects on bone mineral density in early postmenopausal women, *J. Bone Min. Res.* **7**:179.

Ribot, C., Tremollieres, F., Pouilles, J.M., Bonneu, M., Germain, F., and Louvet, J.P., 1987, Obesity and postmenopausal bone loss: the influence of obesity on vertebral density and bone turnover in postmenopausal women, *Bone* **8**:327.

Ribot, C., Tremollieres, F., Pouilles, J.M., Louvet, J.P., and Guiraud, R., 1988, Influence of the menopause and aging on spinal density in French women, *Bone and Mineral* **5**:89.

Ribot, C., Pouilles, J.M., Bonneu, M., and Tremollieres, F., 1992, Assessment of the risk of postmenopausal osteoporosis using clinical factors, *Clin. Endocrinol.* **36**:225.

Riggs, B.L., and Melton, L.J., 1983, Evidence for two distinct syndromes of involutional osteoporosis, *Am. J. Med.* **75**:889.

Riggs, B.L., and Melton, L.J., 1986, Involutional osteoporosis, *N. Engl. J. Med.* **314**:1676.

Riis, B.J., Christiansen, C., Deftos, L.J., and Catherwood, B.D., 1984, "The role of serum concentrations of estrogen on postmenopausal osteoporosis and bone turnover," in: *Osteoporosis* (Christiansen, C., Arnaud, C.D., Nordin, B.E.C., Parfitt, A.M., Peck, W.A., and Riggs, B.L., eds.), pp. 87-88, Glostrup: Department of Clinical Chemistry, Glostrup Hospital.

Rizkallah, T.H., Tovell, H.M.M., and Kely, W.G., 1975, Production of estrone and fractional conversion of circulating androstenedione to estrone in women with endometrial carcinoma, *J. Clin. Endocrinol. Metab.* **40**:1045.

Samojlik, E., Kirshner, M.A., Silber, D., Schneider, G., and Ertel, N.H., 1984, Elevated production and metabolic clearance rates of androgen in morbidly obese women, *J. Clin. Endocrinol. Metab.* **49**:949.

Schindler, A.E., Ebert, A., and Friedrich, E., 1972, Conversion of androstenedione to estrone by human fat tissue, *J. Clin. Endocrinol. Metab.* **35**:627.

Slemenda, C., Hui, S.L., Longcope, C., and Johnston, C.C., 1987, Sex steroids and bone mass: a study of changes about the time of menopause, *J. Clin. Invest.* **80**:1261.

Slemenda, C., Hui, S.L., Longcope, C., Wellman, H., and Johnston, C.C., 1990, Predictors of bone mass in perimenopausal women. A prospective study of clinical data using photon absorptiometry, *Ann. Intern. Med.* **112**:96.

Smith, E.L., and Gilligan, C., 1991, Physical activity effects on bone metabolism, *Calcif. Tissue Int.* **49**:550.

Stepan, J.J., Pospichal, J., Presl, J., and Pacovsky, V., 1987, Bone loss and biochemical indices of bone remodeling in surgically induced postmenopausal women, *Bone* **8**:279.

Svendsen, O.L., Haarbo, J., Heitmann, B.L., Gotfredsen, A., Christiansen, C., 1991, Measurement of body fat in elderly subjects by dual-energy X-ray absorptiometry, bioelectrical impedance and anthropometry, *Am. J. Clin. Nutr.* **53**:1117.

Thomas, K.A., Cook, S.D., Bennett, J.T., Whitecloud, T.S., and Rice, J.C., 1991, Femoral neck and lumbar spine bone mineral densities in a normal population 3–20 years of age, *J. Pediatr. Orthop.* **11**:48.

Trevisan, C., Ortolani, S., Bianchi, M.L., Caraceni, M.P., Ulivieri, F.M., Gandolini, G., and Polli, E.E., 1991, Age, time since menopause, and body parameters as determinants of female spinal bone mass: a mathematical model, *Calcif. Tissue Int.* **49**:1.

Van Hemert, A.M., Birkenhäger, J.C., De Jong, F.H., Vandenbroucke, J.P., and Valkenburg, H.A., 1989, Sex hormone binding globulin in postmenopausal women: a predictor of osteoporosis superior to endogenous oestrogens, *Clin. Endocrinol.* **31**:499.

Von Schoultz, B., and Carlström, K., 1989, On the regulation of sex hormone binding globulin; a challenge of an old dogma and outlines of an alternative mechanism, *J. Steroid Biochem.* **32**:327.

Wickham, C.A.C., Walsh, K., Cooper, C., Barker, D.J., Margetts, B.M., Morris, J., and Bruce, S.A., 1989, Dietary calcium, physical activity and risk of hip fracture: a prospective study, *Br. Med. J.* **299**:889.

Wild, R.A., Buchanan, J.R., Myers, C., Lloyd, T., and Demers, L.M., 1987, Adrenal androgens, sex hormone binding globulin and bone density in osteoporotic menopausal women, *Maturitas* **9**:55.

Chapter 16

# Exercise and Bone Loss

Everett L. Smith, Catherine Gilligan and Lorri J. Tommerup

## 1. Introduction

The skeleton has two major functions: it provides structural support and acts as a mineral reservoir. It responds dynamically to hormones and mechanical stress, homeostatic factors that control serum calcium and skeletal architecture, geometry and bone mineral content. In providing structural support, the skeleton permits movement and protects vital organs. In its function as a mineral reservoir, it responds to changes in hormone levels and helps maintain serum calcium at about 9.8 mg/dL. Bone is resorbed to maintain serum calcium when dietary calcium is inadequate. If the dietary inadequacy is chronic, calcium will be pulled continually from the bone reservoir, resulting in a negative calcium balance and a net loss of calcium and phosphorus. Structural integrity therefore may be threatened when the demands on the reservoir to maintain serum calcium homeostasis are too high.

Mechanical strain through weight bearing and muscle contraction during exercise is a major factor in maintaining skeletal structural integrity. Bone mineral content (BMC) changes in response to the magnitude and frequency of mechanical stressors induced by exercise. Both skeletal integrity and serum calcium are maintained by hormonal and mechanical homeostasis when these

Everett L. Smith, Catherine Gilligan and Lorri J. Tommerup • Department of Preventive Medicine, University of Wisconsin, Madison, WI 53706.

*Advances in Nutritional Research,* Vol. 9
Edited by Harold H. Draper
Plenum Press, New York, 1994

mechanisms are in balance. With aging, however, multiple factors precipitate bone involution. These factors include inadequate calcium intake, inadequate mechanical strain (because of decreased exercise and muscular strength) and hormonal changes.

The balance between bone formation and resorption in remodeling depends upon hormones, growth factors and mechanical loads. Serum calcium levels and skeletal homeostasis are affected by bone remodeling. Hormonal and local mechanical stimuli affect cells of osteoclast and osteoblast lineage that regulate bone remodeling. Precursor bone cells are recruited, replicate and differentiate in response to these stimuli. Skeletal integrity is maintained at a level appropriate to resist the mechanical loads applied during work, exercise and the other activities of daily living. Mechanical loads affect not only the quantity but also the quality and geometric organization of skeletal tissue, all of which are important determinants of skeletal strength and fracture resistance. Bone modeling and remodeling activities respond to both circulating hormones and local growth factors (e.g., insulin-like growth factor, transforming growth factors, platelet-derived growth factor, skeletal growth factor, prostaglandin $E_2$, interleukin-1, growth hormone, calcitonin and parathyroid hormone) (Canalis, 1990). Circulating hormones may directly stimulate bone cells and/or enhance growth factor function. The role of growth factors is still unclear; some are associated more with fracture healing and bone pathologies than with the normal physiological function of bone.

## 2. Bone Cells

The cells that make up the specialized connective tissue of the skeleton synthesize, maintain and organize the extracellular matrix. The five major cells involved in bone turnover and maintenance are pre-osteoblasts, osteoblasts, osteocytes, pre-osteoclasts and osteoclasts. Pre-osteoblasts determine the number and activity level of the bone-forming osteoblasts, which are involved in collagen synthesis and mineralization. Osteocytes are the resident cells, i.e., osteoblasts encased in bone matrix, and have low metabolic activity. The bone-resorbing osteoclasts are key cells in bone remodeling. When bone is remodeled, first the precursor cells are activated, then bone is resorbed and finally new bone is formed. The resorptive (osteoclastic) and formative (osteoblastic) activities are coupled. Osteoclasts form resorption cavities on bone surfaces and tunnel through cortical bone. Osteoblasts fill in the resorption cavities by forming a collagen matrix that later becomes mineralized. Bone formation generally fully replaces the bone lost to resorption in adults under 30 (Mundy, 1990). Net bone loss, resulting from incomplete replacement of mineralized matrix in the resorption cavities, is common after age 30–35 (Frost, 1991).

## 3. Mechanical Strain Homeostasis

Mechanical strain homeostasis acts to maintain the structural support function of the skeleton at a level appropriate to the ordinary demands made upon it. Mechanical loading results in a deformation of both hard and soft tissues. Deformation may be induced by the application of stress or pressure (both defined as force per unit area). The resultant deformation is called strain, defined as the change in the length of the structure, $\Delta L$, divided by the original length. Bone strain is most often expressed in microstrain or 1/1,000,000 of a strain. Bone adapts to strain history (the activities of daily living) at the cellular and architectural levels. The strain induced by activities of daily living tends to stabilize at a similar level regardless of the size of the organism. This was demonstrated by Rubin (1984), who showed that horses, dogs, pigs and turkeys, when trotting or running, all incurred similar levels of strain on weightbearing bones.

Lanyon (1989) hypothesized that the number of precursor cells activated and their metabolic activity in the modeling and remodeling process are directly proportional to the induced bone strain rather than to absolute load or stress (force per unit area). Using an *in vivo* isolated wing model, he showed that bone hypertrophy was related to both the magnitude of strain and the number of load cycles. He applied loads that resulted in 500 to 4000 microstrain. Strains below 1000 microstrain resulted in bone loss and strains above 1000 microstrain resulted in up to 40% cross-sectional area increases that were proportional to the strain applied (Rubin and Lanyon, 1985). When strain was held constant at a level that produced an osteogenic response (about 2000 microstrain), as few as four loading cycles per day (loads applied at 1 Hz) prevented disuse atrophy, and as few as 36 loadings per day resulted in a 30–40% increase in cross-sectional area. No further hypertrophy was observed if the number of loading cycles was increased to 360 or 1800 cycles per day at the same strain magnitude (Rubin and Lanyon, 1984).

Strain rate and magnitude both influence bone response (Lanyon, 1981). A stimulus must exceed a threshold magnitude and frequency to produce bone hypertrophy. Static loading, unlike dynamic loading, induces little or no hypertrophy (Lanyon and Rubin, 1984). Periosteal bone deposition increased slightly in sheep radii exposed to artificial stimulation of normal magnitude and higher than normal rate, and increased substantially when both strain magnitude and strain rate were higher than those resulting from the normal load induced by walking (Lanyon, 1981).

Recently, the biochemical and histological consequences of bone loading have been investigated. The mechanism by which strain produces an osteogenic cellular response is now being delineated. Osteoclastic function increases and osteoblastic function decreases with disuse. *In vivo* strain, however, does not appear to directly affect mature osteoblasts and osteoclasts. Osteocytes, numbering up to 20,000 per cubic millimeter (Mundy, 1990), appear to be excellent candidates for the transduction of mechanical strain to biochemical messages that stimulate other bone cells. Radioactive osteocyte RNA increased in core biopsy specimens loaded *in vitro* compared to non-loaded specimens (el-Haj *et al.,* 1990). In another study, a single period of loading stimulated the transformation of quiescent surface lining cells into active bone-forming osteoblasts within five days (Pead and Lanyon, 1989).

A few studies have addressed the effects of exercise on cellular activities that result in greater cortical and trabecular bone matrix. Sows (3–3.5 years old) trained on a motor-driven treadmill for 20 weeks exhibited significantly greater cellular activity at the femur midshaft than control sows (Raab *et al.,* 1991). Active periosteal surface was 27% greater in trained than in untrained sows. In addition, periosteal mineral apposition rate (MAR) was 76% higher and intracortical MAR was 23% higher. Overload also altered cellular activity in adult rats (Jee and Li, 1990). After 18 and 26 weeks of increased loading, the loaded animals had significantly higher trabecular number, thickness and density, as well as overall bone density, in the proximal tibia, than did controls. The MAR and bone formation rate at 26 weeks also were significantly higher (38% and 23%, respectively) in the loaded animals, while the percentage of eroded and labeled perimeter was lower.

### 3.1. Membrane Mechanosensor

The biochemical messages resulting from mechanical strain have only recently been recognized as regulators of bone modeling and remodeling in addition to local and systemic hormonal messages. The cell membranes of both plants and animals contain mechanosensitive areas. These mechanosensors transduce mechanical stimuli into electrical and/or biomechanical changes within the cell in proportion to the magnitude of the stimulus. The Venus flytrap and Characean algae are two examples of plants that generate action potentials in response to mechanical stimuli on cell membrane receptors (Wayne, 1993; Hodick and Sievers, 1989; Okihara *et al.,* 1991). In animals, capillary endothelial cells, muscle cells, fibroblasts, osteoblasts and osteocytes have been shown to respond to mechanical stimuli (Vandenburgh, 1992). Watson (1991) described the mechanosensitive area of the cell as a mechanotransducer that responds to mechanical stretch, pressure change and possibly electrical change.

Many cells respond to physical stimuli by altering their shape and internal biochemistry. In the past few years it has become clear that both the osteoblast and the osteocyte respond to mechanical stimuli by synthesizing prostaglandin $E_2$ ($PGE_2$), a mediator of bone modeling and remodeling (Reich and Frangos, 1993). In culture media, $PGE_2$ increases when osteoblasts and osteocytes are exposed to fluid shear stress (Reich and Frangos, 1993; Reich and Frangos, 1991; Hsieh, *et al.*, 1991; Walsh and Guzelsu, 1993) or to mechanical loading in an *in vitro* bone explant model (Lanyon 1992; Dodds *et al.,* 1993; Zaman *et al.,* 1992; Dallas *at al.,* 1993). Mechanical stimuli induce increases in $PGE_2$ and subsequently in adenosine 3′, 5′-cyclic monophosphate (cAMP) and insulin-like growth factor (IGF-I) in osteoblast cell culture. In cultures dosed with .01–1 $\mu$M of $PGE_2$ for 5 min, cAMP increased 8- to 54-fold, and within 6 hr prepro-IGF-I transcript levels rose about 4-fold (McCarthy *et al.,* 1991).

Hock *et al.* (1988) demonstrated that an increase in IGF-I stimulated DNA production in periosteal cells with a resultant increase in collagen synthesis. While further detailing of the response of the osteoblast and the osteocyte to mechanical loading is necessary, a number of biochemical events have been sequenced. Some of the events have been observed from within 10 sec to 96 hr after stimulation of cells either by increased flow shear stress or by mechanical loading of a bone explant. The events that occur after the mechanosensor has transduced the mechanical stimuli to a biochemical event are as yet not definitively established. Jones *et al.* (1991), however, observed changes in phosphoinositol, phospholipase C (PI-PLC) and inositol 1, 4, 5-triphosphate ($IP_3$) within $\mu$sec to 10 sec after a microstrain of 3000. $IP_3$ appeared in about 10 sec and intracellular calcium was released within the next $\mu$sec. They also reported that PI-PLC is activated by mechanical strain before any response of phospholipase $A_2$ ($PLA_2$). This is consistent with the research of Reich and Frangos (1993), who stated that protein kinase C (PKC) activates the phospholipase $A_2$ essential for the release of arachidonic acid (AA) required in the synthesis of $PGE_2$. A hypothetical sequence of events is summarized in Figure l. This sequence may be stimulated biochemically as well as mechanically. Regardless of the source of stimulus, $PGI_2$, $PGE_2$ and IGF-I are produced and are important in the mediation of bone modeling and remodeling.

Jee *et al.* (1993) used ovariectomized (OX) and sham ovariectomized 5.5-month-old rats to determine the osteogenic effects of $PGE_2$ on cortical bone mass decline in the proximal tibia. After 5 months, the OX rats had 78% less cortical bone in the proximal tibia than did controls. When the OX rats were treated with 6 mg/kg of $PGE_2$ for 75 days, they regained cortical bone to 47% of that of the controls. $PGE_2$ increased the ratio of bone formation to bone resorption from 0.6 to 5.8, as well as the bone formation rate and mineral appositional rate. When $PGE_2$ was withdrawn, bone resorption exceeded formation unless osteoclast activity was inhibited by a bisphosphonate (risedronate) (Tang *et al.,* 1992).

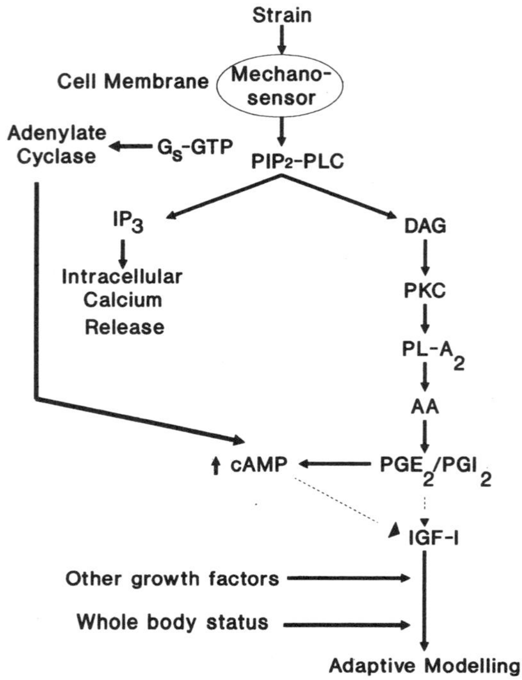


Figure 1. Proposed biochemical responses to mechanical stimuli (Mayer and Marshall, 1993; Lanyon, 1992.) PI-PLC = phosphoinositol specific phospholipase C (PI-PLC); $IP_3$ = inositol 1, 4, 5-triphosphate ($IP_3$); DAG = diacyl glycerol; PKC = protein kinase C; $PL\text{-}A_2$= phospholipase $A_2$; AA = arachidonic acid; $PGE_2$ = prostaglandin $E_2$; $PGI_2$ = prostacyclin; cAMP = adenosine 3′, 5′-cyclic monophosphate; IGF-I = insulin-like growth factor.

## 3.2. Effects of Exercise

Since weight bearing (gravity) and muscle contraction are the two major mechanical forces applied to bone, bone homeostasis is affected by physical activity. Immobilization or weightlessness decreases bone mass, while weight-bearing activity or muscle contraction produces bone hypertrophy. The degree of bone hypertrophy or atrophy is proportional to the change in magnitude and frequency of the mechanical stimulus from normal loading conditions. This is consistent with the mechanosensitive response of bone cells to threshold levels of bone strain. Weight-bearing segments (legs and spine) typically experience much higher loads than do the non-weight-bearing areas (ribs, arms and skull).

For example, muscle contractions in the arm during the activities of daily living produce much lower loads than does the impact of the heel during walking (1.2 to 1.5 times body weight). In weightlessness, therefore, less bone is lost from the radius than from the calcaneus because of a smaller decrease in loading and delta strain (Vogel and Whittle, 1976; Smith *et al.,* 1977).

To stimulate osteogenic bone response, changes in mechanical loading must exceed threshold levels. Bone neither atrophies nor hypertrophies within the normal range of stimuli specific to individual activities and genotype, because the strain-induced mechanical, stretch or pressure changes on the bone cell mechanosensor are not great enough to exceed membrane threshold. Bone hypertrophies at levels above this range and atrophies at levels below this range. An upper limit to stress-induced hypertrophy also seems to exist. If untrained persons increase their activity levels more rapidly than the bone can adapt to, or if soldiers and distance runners train at a higher speed, distance and/or frequency, fatigue fracture or microdamage can occur (Margulies *et al.,* 1986).

While various studies have shown that physical activity increases bone mass, research on the response to mechanical loading through physical activity in humans is still in its early stages. Studies on athletes have confirmed that the effect of physical activity is primarily local and proportional to the level of strain placed on the bone. Localized bone hypertrophy is seen in the dominant arm of young and old tennis players, whose humerus bone mineral density (BMD) is up to 35% greater than in the non-dominant arm (Huddleston *et al.,* 1980; Jones *et al.,* 1977; Montoye *et al.,* 1980). Athletes in sports that place the most strain on bone have the greatest hypertrophy: weightlifters have higher bone density in the spine and femur than aerobic athletes (Block *et al.,* 1986; Davee *et al.,* 1990; Heinrich *et al.,* 1990; Nilsson and Westlin, 1971), while swimmers have the least hypertrophy (Heinrich *et al.,* 1990; Nilsson and Westlin, 1971).

In intervention studies, physical activity has been found to increase bone mass or decrease bone loss in the spine, radius, calcaneus and tibia, but not to significantly affect BMD of the femur. Spinal bone density benefitted from aerobic weightbearing activities in both middle-aged postmenopausal women (Dalsky *et al.,* 1988; Nelson *et al.,* 1991) and women with osteoporosis (Krolner *et al.,* 1983). Bone loss in the radius and ulna was ameliorated by physical activity programs that included arm exercises (Smith *et al.,* 1981; Rikli and McManis, 1990; Simkin *et al.,* 1986; Smith *et al.,* 1989). The difference in bone loss between exercise and control groups was similar for premenopausal and postmenopausal women (Smith *et al.,* 1989). Bone density of the radius was not significantly affected by training that consisted primarily of weightbearing activity (Aloia *et al.,* 1978; Sandler *et al.,* 1987).

Despite the direct relationship demonstrated between strain and bone hypertrophy in animal models and weightlifters, intervention studies utilizing resistance training have not indicated that it is superior to other forms of training. In two studies in which groups of postmenopausal women participating in general aerobic training were compared to groups that added resistance (muscular strength) training, the calcium bone index (bone mass of the central third of the skeleton adjusted for body size) and density of the radius tended to increase more in the groups with both aerobic and resistance training (Rikli and McManis, 1990; Chow *et al.,* 1987). The results of intervention with resistance training alone, however, are conflicting. In one resistance-training program, vertebral BMD decreased significantly more in exercising premenopausal women than in controls (Rockwell *et al.,* 1990); in another, vertebral BMD change did not differ significantly between postmenopausal women who performed back extensions with light weights for two years and controls (Sinaki *et al.,* 1989). No explanation was proferred for the loss of vertebral BMD in the premenopausal subjects (Rockwell *et al.,* 1990). In contrast, two studies have indicated that the vertebral BMD of premenopausal and postmenopausal women participating in resistance training increased relative to controls (Gleeson *et al.,* 1990; Pruitt *et al.,* 1992), although calcaneus, femur, and distal forearm BMD were not significantly affected. In postmenopausal women concurrently receiving estrogen replacement therapy (ERT), spinal, total body and radial BMD increased significantly with resistance training, but only the change in radial BMD was significantly different from that in women receiving ERT alone (Notelovitz *et al.,* 1991).

Bone density gains induced by physical activity are not maintained long beyond the period of training. In a study of postmenopausal women participating in an exercise program for 9 or 22 months, followed by a detraining period of 13 months, bone density decreased rapidly during detraining and at the end of detraining was not significantly higher than the pre-training level (Dalsky *et al.,* 1988). Similarly, Michel *et al.* (1992) reported that decreased running heightened bone loss in men and women who were long-term runners (age 55 to 77). Runners who maintained their training level underwent a vertebral bone density decline of 3.8% in 5 years, compared to 13.4% in runners who decreased their training by 20%.

## 4. Exercise and Osteoporosis

The literature reviewed shows that exercise has a significant effect on bone. However, no studies have been carried out long enough to determine whether exercise can reduce the risk of hip fractures, a major public health problem in aging populations. Additional longitudinal research is necessary to determine the role of exercise in the prevention of osteoporotic fractures.

### 4.1. Hip Fracture Incidence and its Consequences

The incidence of hip fractures increases with age. After the age of 50, the incidence rate is about twice as high for white women as for white men (Wallace, 1983). The U.S. National Center for Health Statistics (1986) reported 247,000 hip fractures in adults over age 45 in 1985. Hip fractures are associated with an excess mortality of 12–20% and with greatly increased medical and assisted care expenses, generating costs of approximately 6–8 billion dollars a year in the U.S. Among women who suffer a hip fracture, 15-25% lose their independence in the first year after the fracture (Cummings, 1987).

### 4.2. Factors Associated with Hip Fractures

Three independent but interactive factors appear to contribute to the risk of hip fractures: the incidence of falling, the force of impact on the hip, and bone strength (Cummings, 1987). Bone density is an important component of bone strength and thus of fracture risk. The relative risk ratio for fracture increases by about two for every standard deviation below normal bone density (Mazess, 1990). However, other factors associated with the etiology of hip fractures must also be considered when developing strategies to prevent fractures (Lotz and Hayes, 1990). Ninety percent of hip fractures are associated with falls (Grisso *et al.,* 1984).

If the incidence and/or impact of falls were reduced, the frequency of fractures would be decreased. Most falls are multifactorial in nature (Prudham and Grimley, 1981). The incidence of recurrent falls in elderly subjects increases as the number of physiological and functional risk factors increases (Tinetti, 1986; Tinetti *et al.,* 1988; Nevitt *et al.,* 1989). Muscular strength of the legs is significantly correlated with balance and the risk of falls (Whipple *et al.,* 1987; Gehlsen and Whaley, 1990). Muscle strength and mass may also affect defensive responses and impact force if a fall occurs. Cummings and Nevitt (1989) observed that hip abductor strength declines about 27% from age 65 to age 80, indicating a similar decline in the capacity of muscles around the hip to absorb the energy of impact. The direction of falls, and thus the impact on the hip, is related to gait velocity (Cummings and Nevitt, 1989); those with the lowest velocity are at the greatest risk for hip fracture. The force of impact (reduced by protective responses) and the vector of force may be the dominant factors in whether a fall results in a hip fracture (Lotz and Hayes, 1990).

### 4.3. Benefits of Exercise

The potential of exercise to decrease fracture risk by reducing bone loss, increasing bone mass and/or reducing the frequency and severity of falls has received modest attention. However, exercise is the only preventive measure that is likely to reduce fracture risk not only through increasing bone mass, but also by preventing falls and lessening the force of impact (National Institutes of Health, 1984) . In addition to its effects on bone, exercise is known to improve muscular strength and flexibility. Several cross-sectional studies have indicated that the risk of falls and hip fractures is smaller in subjects who exercise (Campbell *et al.,* 1989; Boyce and Vessey, 1988; Astrom *et al.,* 1987; Wickham *et al.,* 1989). Virtually no studies have addressed the effects of exercise on femoral bone mass or other risk factors pertinent to hip fracture—reaction time, postural stability and gait. Exercise contributes to fracture prevention by strengthening the bones, reducing the incidence of falls and the force of impact, and by improving reaction time, gait and posture.

## References

Aloia, J.F., Cohn, S.H., Osuni, J., Cane, R. and Ellis, K., 1978, Prevention of involutional bone loss by exercise, *Ann. Intern. Med.* **89:**356.

Astrom, J., Ahnqvist, S., Beertema, J., and Jonsson, B., 1987, Physical activity in women sustaining fracture of the neck of the femur, *J. Bone Joint Surg.* **69B**:381.

Block, J., Genant, H.K., and Black, D., 1986, Greater vertebral bone mineral in exercising young men, *Western J. Med.* **145:**39.

Boyce, W.J., and Vessey, M.P., 1988, Habitual physical inertia and other factors in relation to risk of fracture of the proximal femur, *Age and Ageing* **17**:319.

Campbell, A.J., Borrie, M.J., and Spears, G.F., 1989, Risk factors for falls in a community-based prospective study of people 70 years and older, *J. Gerontol.* **44:**M112.

Canalis, E., 1990, "Regulation of bone remodeling," in: *Primer on the Metabolic Bone Diseases and Disorders of Mineral Metabolism,* (M. J. Favus, ed.), pp. 23-26, 1990, American Society for Bone and Mineral Research, Kelseyville, CA.

Chow, R.K., Harrison, J.E., and Notarius, C., 1987, Effect of two randomised exercise programmes on bone mass of healthy postmenopausal women, *Br. Med. J.* **292:**607.

Cummings, S.R., 1987, Epidemiology of osteoporotic fractures, *Osteoporosis Update 1987:*7

Cummings, S. R., and Nevitt, M. C., 1989, A hypothesis: The causes of hip fractures, *J. Gerontol.* **44:**M107.

Dallas, S.L., Zaman, G., Pead, M.J., and Lanyon, L.E., 1993, Early strain-related changes in cultured embryonic chick tibiotarsi parallel those associated with adaptive modeling in vivo, *J. Bone Min. Res.* **8**:251.

Dalsky, G.P., Stocke, K.S., Ehsani, A.A., Slatopolsky, E., Lee, W.C., and Birge, S.J., **1988**, Weight-bearing exercise training and lumbar bone mineral content in postmenopausal women, *Ann. Intern. Med.* **108:**824.

Davee, A.M., Rosen, C.J., and Adler, R.A., 1990, Exercise patterns and trabecular bone density in college women, *J. Bone Min. Res.* **5**:245.

Dodds, R.A., Ali, N., Pead, M.J., and Lanyon, L. E., 1993, Early loading-related changes in the activity of glucose 6-phosphate dehydrogenase and alkaline phosphatase in osteocytes and periosteal osteoblasts in rat fibulae in vivo, *J. Bone Min. Res.* **8**:261.

El Haj, A.J., Minter, S.L., Rawlinson, S.C.F., Suswillo, R., and Lanyon, L.E., 1990, Cellular responses to mechanical loading in vitro, *J. Bone Min. Res.* **5**:923.

Frost, H.M., 1991, A new direction for osteoporosis research: a review and proposal, *Bone* **12**:429.

Gehlsen, G.M., and Whaley, M.H., 1990, Falls in the elderly, Part II: Balance, strength, and flexibility, *Arch. Phys. Med. Rehab.* **71**:739.

Gleeson, P.B., Protas, E.J., LeBlanc, A.D., Schneider, V.S., and Evans, H.J., 1990, Effects of weight lifting on bone mineral density in premenopausal women, *J. Bone Min. Res.* **5**:153.

Grisso, J.A., Kelsey, J.L., Strom, B.L., Chiu, G.Y., Maislin, G., O'Brien, L.A., Hoffman, S., Kaplan, F., and the North East Hip Fracture Study Group, 1991, Risk factors for falls as a cause of hip fracture in women, *N. Engl. J. Med.* **324**:1326.

Heinrich, C.H., Going, S.B., Pamenter, R.W., Perry, C.D., Boyden, T.W., and Lohman, T.G., 1990, Bone mineral content of cyclically menstruating female resistance and endurance trained athletes. *Med. Sci. Sports Ex..* **22**:558.

Hock, J.M., Centrella, M., and Canalis, E, 1988, Insulin-like growth factor I (IGF-I) has independent effects on bone matrix formation and cell replication, *Endocrinol.* **122**:254.

Hodick, D., and Sievers, A., 1989, On the mechanism of trap closure of Venus flytrap (Dionaca Mucipula Ellis), *Planta* **179**:32.

Hsieh, H-J., Li, N-Q., and Frangos, J.A., 1991, Shear stress increases endothelial platelet-derived growth factor mRNA levels, *Am. J. Physiol.* **260**:H642.

Huddleston, A.L., Rockwell, D., Kulund, D.N., and Harrison, R.B., 1980, Bone mass in life-time tennis athletes, *JAMA* **244**:1107.

Jee, W.S.S., and Li, X.J., 1990, Adaptation of cancellous bone to overloading in the adult rat: a single photon absorptiometry and histomorphometry study, *Anat. Rec.* **227**:418.

Jee, W.S.S., Tang, L., Ke, H.Z., Setterberg, R.B., and Kimmel, D.B., 1993, Maintaining restored bone with bisphosphonate in the ovariectomized rat skeleton: dynamic histo-morphometry of changes in bone mass, *Bone* **14**:49

Jones, D.B., Nolte, H., Scholubbers, J-G., Turner, E., and Veltel, D., 1991, Biochemical signal transduction of mechanical strain in osteoblast-like cells, *Biomaterials* **12**:101.

Jones, H.H., Priest, J.D., and Hayes, W.C., 1977, Humeral hypertrophy in response to exercise, *J. Bone Joint Surg. (Am)* **59A**:204.

Krolner, B., Toft, B., Nielsen, S.P., and Tondevold, E., 1983, Physical exercise as prophylaxis against involutional vertebral bone loss: a controlled trial, *Clin. Sci.* **64**:541.

Lanyon, L.E., 1981, "Bone remodeling, mechanical stress and osteoporosis," in: *Osteoporosis: Recent Advances in Pathogenesis and Treatment* (H.F. DeLuca, H.M. Frost, W.S.S. Jee, C. C. Johnston, Jr., and A. M. Parfitt, eds.), University Park Press, Baltimore, p. 129.

Lanyon, L.E., 1989, Strain-related bone modeling and remodeling, *Topics Geriat. Rehab.* **4**:13.

Lanyon, L.E., 1992, Control of bone architecture by functional load bearing, *J. Bone Min. Res.* **7(suppl 2)**:S369.

Lanyon, L.E., and Rubin, C.T., 1984, Static vs. dynamic loads as a stimulus for bone remodeling, *J. Biomech.* **15**:767.

Lotz, J.C., Hayes, W.C., 1990, The use of quantitative computed tomography to estimate risk of fracture of the hip from falls, *J. Bone Joint Surg.* **72A**:689.

Margulies, J.Y., Simkin, A., Leichter, I., Bivas, A., Steinberg, R., Giladi, M., Stein, M., Kashtan, H., and Milgrom, C., 1986, Effect of intense physical activity on the bone-mineral content in the lower limbs of young adults, *J. Bone Joint Surg.* **68A**:1090.

Mayer, R.J., and Marshall, L.A., 1993, New insights on mammalian phospholipase $A_2$(s); comparison of arachidonoyl-selective and -nonselective enzymes, *FASEB J.* **7**:339.

Mazess, R.B., 1990, "Bone density for clinical diagnosis and monitoring," 1990, in: *Osteoporosis: Physiological Basis, Assessment and Treatment* (H.F. DeLuca and R.B. Mazess, eds.), pp. 63-85, Elsevier, New York.

McCarthy, T.L., Centrella, M., Raisz, L.G., and Canalis, E., 1991, Prostaglandin $E_2$ stimulates insulin-like growth factor I synthesis in osteoblast-enriched cultures from fetal rat bone, *Endocrinol.* **128**:2895.

Michel, B.A., Lane, N.E., Bjorkengren, A., Bloch, D.A., and Fries, J.F., 1992, Impact of running on lumbar bone density: a 5-year longitudinal study, *J. Rheumatol.* **19**:1759.

Montoye, H.J., Smith, E.L., Fardon, D.F., and Howley, E.T., 1980 , Bone mineral in senior tennis players. *Scand. J. Sports Sci.* **2**:26.

Mundy, G.R., 1990, "Bone resorbing cells," in: *Primer on the Metabolic Bone Diseases and Disorders of Mineral Metabolism* (M.J. Favus, ed.), pp. 18-22, American Society for Bone and Mineral Research, Kelseyville, CA.

Nelson, M.E., Fisher, E.C., Dilmanian, F.A., Dallal, G.E., and Evans, W. J., 1991, A 1-y walking program and increased dietary calcium in postmenopausal women: effects on bone, *Am. J. Clin. Nutr.* **53**:1304.

Nevitt, M.C., Cummings, S.R., Kidd, S., and Black, D., 1989, Risk factors for recurrent non-syncopal falls: a prospective study, *JAMA* **261**:2663.

NIH Osteoporosis Consensus Conference, 1984, *JAMA* **252**:799.

Nilsson, B.E., and Westlin, N.E., 1971, Bone density in athletes, *Clin. Orthop. Rel. Res.* **77**:179.

Notelovitz, M., Martin, D., Tesar, R., Khan, F.Y., Probart, C., and McKenzie, L., 1991, Estrogen therapy and variable-resistance weight training increase bone mineral in surgically menopausal women, *J. Bone Min. Res.* **6**:583.

Okihara, K., Ohkawa, T., Tsutsui, I., and Kasai, M., 1991, A $Ca^{2+}$ and voltage-dependent Cl-sensitive anion channel in the Chara Plasmalemma: a patch-clamp study, *Plant and Cell Physiol.* **32**:593.

Pead, M.J., and Lanyon, L.E., 1989, Indomethacin modulation of load-related stimulation of new bone formation in vivo, *Calcif. Tissue Int.* **45**:44.

Prudham, D., and Grimley, E.J., 1981, Factors associated with falls in the elderly: a community study, *Age Ageing* **10**:141.

Pruitt, L.A., Jackson, R.D., Bartels, R.L., and Lehnard, H.J., 1992, Weight-training effects on bone mineral density in early postmenopausal women, *J. Bone Min. Res.* **7**:179.

Raab, D.M., Crenshaw, T.D., Kimmel, D.B., and Smith, E.L., 1991, A histomorphometric study of cortical bone activity during increased weight-bearing exercise, *J. Bone Min. Res.* **6**:741.

Reich, K.M., and Frangos, J.A., 1991, Effect of flow on prostaglandin $E_2$ and inositol trisphosphate levels in osteoblasts, *Am. J. Physiol.* **261**:C428.

Reich, K.M., and Frangos, J.A., 1993, Protein kinase c mediates flow-induced prostaglandin $E_2$ production in osteoblasts, *Calcif. Tissue Int.* **52**:62.

Rikli, R.E., and McManis, B.G., 1990, Effects of exercise on bone mineral content in postmenopausal women, *Res. Quart. Ex. Sport* **61**:243.

Rockwell, J.C., Sorenson, A.M., Baker, S., Leahey, D., Stock, J.L., Michaels, J., and Baran, D.T., 1990, Weight training decreases vertebral bone density in premenopausal women: a prospective study, *J. Clin. Endocrinol. Metab.* **71**:988.

Rubin, C.T., 1984, Skeletal strain and the functional significance of bone architecture, *Calcif. Tissue Int.* **36**:S11.

Rubin, C.T., and Lanyon, L.E., 1984, Regulation of bone formation by applied dynamic loads, *J. Bone Joint Surg.* **66A**:397.

Rubin, C.T., and Lanyon, L.E., 1985, Regulation of bone mass by mechanical strain magnitude, *Calcif. Tissue Int.* **37**:411.

Sandler, R.B., Cauley, J.A., Hom, D.L., Sashin, D., and Kriska, A., 1987, The effects of walking on the cross-sectional dimensions of the radius in postmenopausal women, *Calcif. Tissue Int.* **41**:65.

Simkin, A., Ayalon, J., and Leichter, I., 1986, Increased trabecular bone density due to bone-loading exercises in postmenopausal osteoporotic women, *Calcif. Tissue Int.* **40**:59.

Sinaki, M., Wahner, H.W., Offord, K.P., and Hodgson, S.F., 1989, Efficacy of nonloading exercises in prevention of vertebral bone loss in postmenopausal women: a controlled trial, *Clin. Proc.* **64**:762.

Smith, E.L., Reddan, W., and Smith, P.E., 1981, Physical activity and calcium modalities for bone mineral increase in aged women, *Med. Sci. Sports Ex.* **13**:60.

Smith, E.L., Gilligan, C., Shea, M.M., Ensign, P., and Smith, P.E., 1989, Exercise reduces bone involution in middle-aged women, *Calcif. Tissue Int.* **44**:312.

Smith, M.C., Rambaut, P.C., Vogel, J.M., and Whittle, M.W., 1977, "Bone mineral measurement–experiment M078," in: *Biomedical Results from Skylab* (R.S. Johnston and L.F. Dietlein, eds.), p. 183, National Aeronautics and Space Administration, Washington, DC.

Tang, L.Y., Jee, W.S.S., Ke, H.Z., and Kimmel, D.B., 1992, Restoring and maintaining bone in osteopenic female rat skeleton: I. Changes in bone mass and structure, *J. Bone Min. Res.* **7**:1093.

Tinetti, M.E., 1986, Performance-oriented assessment of mobility problems in elderly patients, *J. Am. Geriat. Soc.* **34**:119.

Tinetti, M.E., Speechley, M., and Ginter, S.F., 1988, Risk factors for falls among elderly persons living in the community, *N. Engl. J. Med.* **319**:1701.

U.S. National Center for Health Statistics, 1986, "1985 Summary: National Hospital Discharge Survey, Advance Data from Vital and Health Statistics." *No. 127, DHHS Pub. No. [PHS] 86-1250.* Public Health Service, Hyattsville, MD.

Vandenburgh, H.H., 1992, Mechanical forces and their second messengers in stimulating cell growth in vitro, *Am. J. Physiol.* **262**:R350.

Vogel, J.M., and Whittle, M.W., 1976, Bone mineral changes; the second manned skylab mission, *Aviat. Space Environ. Med.* **47**:396.

Wallace, W., 1983, The increasing incidence of fractures of the proximal femur: an orthopedic epidemic, *Lancet* 2:1413.

Walsh, W. R., and Guzelsu, N., 1993, Ion concentration effects on bone streaming potentials and zeta potentials, *Biomaterials* **14**:331.

Watson, P.A., 1991, Function follows form: generation of intracellular signals by cell deformation, *FASEB J.* **5**:2013.

Wayne, R., 1993, Excitability in plant cells, *Am. Scientist* **81**:40.

Whipple, R.H., Wolfson, L.I., and Amerman, P.M., 1987, The relationship of knee and ankle weakness to falls in nursing home residents: an isokinetic study, *J. Am. Ger. Soc.* **35**:13.

Wickham C.A.C., Walsh, K., Cooper, C., Parker, D.J.P., Margetts, B.M., Morris, J., and Bruce, S.A., 1989, Dietary calcium, physical activity, and risk of hip fracture: a prospective study, *Br. Med. J.* **299**:889.

Zaman, G., Dallas, S.L. and Lanyon, L.E., 1992, Cultured embryonic bone shafts show osteogenic responses to mechanical loading, *Calcif. Tissue Int.* **51**:132.

Chapter 17

# The Menstrual Cycle: Effects on Bone in Menopausal Women

Susan I. Barr and Jerilynn C. Prior

## 1. Introduction

For many years it has been recognized that a significant loss of bone occurs in women at the time of menopause (Riggs and Melton, 1986). However, the importance to bone integrity of regular exposure to both estrogen and progesterone throughout a woman's reproductive life has received little recognition until recently. The purposes of this chapter are: (1) to demonstrate that disturbances of the menstrual cycle are not uncommon and are often clinically silent; (2) to demonstrate that these disturbances have significant effects on bone in premenopausal women; and (3) to discuss effects of exercise and nutritional parameters on the menstrual cycle, and through this mechanism, their potential to affect bone.

## 2. The Menstrual Cycle

Before describing disturbances of the menstrual cycle, it is first necessary to have a basic understanding of its normal physiology (Fritz and Speroff,

---

Susan I. Barr • School of Family and Nutritional Sciences, University of British Columbia, Vancouver, British Columbia, Canada V6T 1Z4 Jerilynn C. Prior • Department of Medicine, University of British Columbia, Vancouver, British Columbia, Canada V5Z 1M9

*Advances in Nutritional Research,* Vol. 9
Edited by Harold H. Draper
Plenum Press, New York, 1994

1983). The hormonal profile of an ovulatory menstrual cycle is shown in Figure 1. During menstrual flow, which starts on day one of the cycle, circulating levels of both estrogen and progesterone are low. These low levels of circulating hormones allow gonadotropin releasing hormone (GnRH) from the hypothalamus to stimulate release of low levels of both follicle stimulating hormone (FSH) and luteinizing hormone (LH) from the anterior pituitary. FSH stimulates the growth of a group of ovarian follicles, which begin to mature and synthesize estrogen. Estrogen levels rise gradually during the follicular phase of the cycle, and cause thickening of the endometrial lining of the uterus. The increasing levels of estrogen inhibit FSH secretion, and all but the most mature of the developing follicles undergo atresia. This dominant follicle produces large amounts of estrogen, called the midcycle estrogen peak, which stimulates a major surge in LH. The LH peak has at least three consequences: it inhibits estrogen production by follicular cells; it initiates changes which result in the rupture of the dominant follicle and release of the ovum (ovulation); and it also brings about the transformation of the ruptured follicle into the corpus luteum. The corpus luteum produces both estrogen and progesterone during the luteal phase of the cycle. The progesterone is important in preparing the endometrium to allow the fertilized ovum to implant in the uterine wall, should fertilization occur. If fertilization does not occur, estrogen and progesterone levels begin to fall and the corpus luteum undergoes luteolysis 10 to 16 days after its formation. The decline in estrogen and progesterone results in the shedding of the thickened endometrium as menstrual flow. The characteristics of the cycle are such that its normal length ranges from 21 to 35 days.

The regular ovulatory cycle described above does not occur without fail between menarche (the first menstrual flow) and menopause (cessation of menstrual flow). Deviations from the regular pattern are known to be more common during the periods following menarche and preceding menopause (Vollman, 1977); however, they can occur at any age. Abnormal patterns, ranging from the most to the least severe, include: (1) secondary amenorrhea (complete absence of menstrual flow for at least six months in a non-pregnant woman, characterized by marked deficiency of both estrogen and progesterone); (2) oligomenorrhea (irregular cycles lasting from 36 to 180 days, characterized by relative estrogen deficiency and, usually, complete progesterone deficiency); (3) anovulation (cycles of regular length in which ovulation does not occur, and in which there is therefore minimal progesterone secretion, although estrogen levels may be adequate); and (4) short luteal phase cycles (cycles in which the luteal phase lasts less than 10 days, and which are thus associated with relative progesterone deficiency, although, if cycle length is normal, estrogen levels may be adequate). It is important to note that the last two disturbances, anovulation and short luteal phase cycles, are clinically silent, as they commonly occur within cycles of normal length.

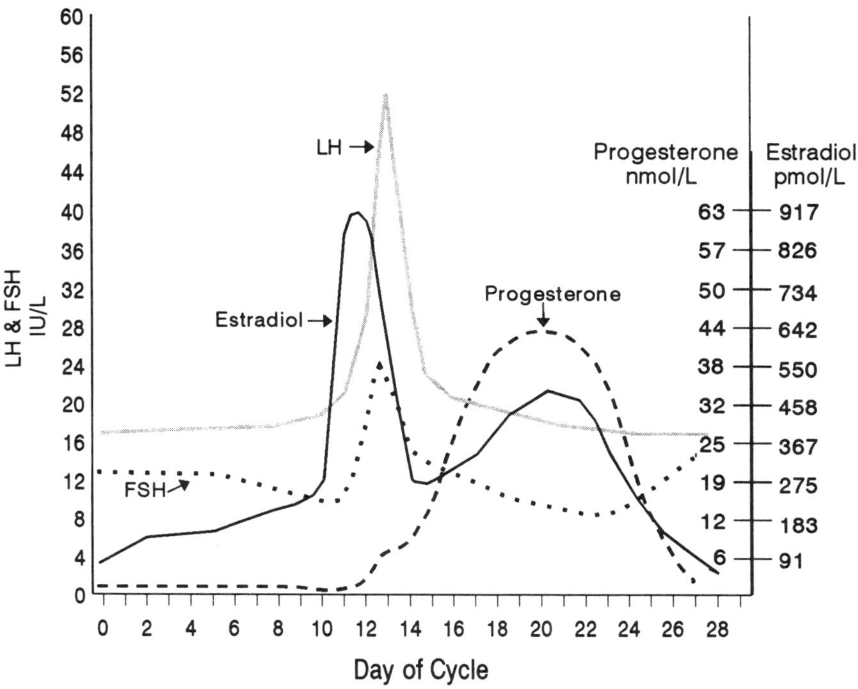


Figure 1. Hormonal characteristics of the ovulatory menstrual cycle. Modified from Fritz and Speroff (1983).

Although detection of amenorrhea and oligomenorrhea is straightforward, based on cycle interval, detection of anovulatory and short luteal phase cycles is more complex, and is often not performed. Instead, it is often simply assumed, both by the general public and by many researchers, that the presence of a regular length cycle reflects normal ovulatory function. This is clearly not the case; accordingly, it is critical that some method other than presence or absence of menses be used to assess the menstrual cycle. Available methods include:

a) Serial or single blood samples. Blood levels of estrogen and progesterone can be used to characterize the cycle and assess whether it follows the normal pattern shown in Figure 1. While daily or periodic samples are preferable because of variability in hormone levels, occasionally single samples are obtained between days 18 and 22 of the cycle (the mid-luteal phase of a normal length ovulatory cycle). If both estrogen and progesterone are present above baseline values, the cycle is assumed to be ovulatory. A single sample, however, cannot be used to assess the length of the luteal phase.

b) Serial or single salivary hormone measurements. Free progesterone equilibrates in body fluids, including saliva, in proportion to its presence in plasma. Measurement of progesterone in salivary samples obtained at 3–4-day intervals has been used to categorize cycles as ovulatory or anovulatory (Ellison, 1988). This method has the advantage of being noninvasive, and the sample can be collected by a woman at home. However, single samples have the disadvantage, like single blood samples, of being associated with considerable variability, and cannot be used to assess the adequacy of the luteal phase.

c) Urinary LH measurement. Assay of LH in urine samples collected at mid-cycle (days 12–16 in a typical cycle) detects likely ovulation, as the LH surge is a necessary prerequisite for ovulation. Occasionally, however, a surge in LH will occur without subsequent ovulation.

d) Basal body temperature measurement. The hypothalamus has a thermic response to progesterone, such that basal body temperature in the luteal phase of the cycle averages approximately 0.3°C higher than in the follicular phase. In the past, visual inspection was used to assess the presence or absence of an increase in temperature, and results were often unreliable. Recently, a computerized least square equation was developed and validated against the serum LH peak day (Prior *et al.,* 1990a). Daily monitoring of temperature allows cycles to be characterized as ovulatory or anovulatory, and also allows an estimate to be made of luteal phase length.

e) Endometrial biopsy. Cellular characteristics of the endometrium—such as glandular and stromal mitoses, glandular diameter, epithelial height and presence of vacuoles—change across the menstrual cycle in response to estrogen and progesterone. Histomorphometric examination of endometrial biopsy specimens can thus be used to indicate the phase of the cycle, and to characterize cycles as ovulatory or anovulatory (Bardawil, 1987).

f) Ultrasound visualization of the ovary. Visualization of the ovaries by ultrasound during the follicular phase can reveal the presence of developing follicles and the dominant follicle, and during the luteal phase, the presence of the corpus luteum (Shoupe *et al.,* 1989).

# 3. The Bone Life Cycle in Women

## 3.1. Measurement of Bone

Before describing the general patterns of change in bone mineral during a woman's life, it is first important to describe different *types* of bone and the techniques available to measure bone (Alhava, 1991). Cortical or compact bone is found in the shafts of the long bones. It remodels slowly, and likely reaches peak density values in the third or fourth decades of life. In contrast, cancellous or trabecular bone is found in areas such as the vertebral bodies and the hip, frequent locations of osteoporotic fractures. It turns over more rapidly than does cortical bone, appears to be more sensitive to hormonal influences, and likely reaches maximal values earlier in life.

Cortical bone can be evaluated using radiogrammetry of the metacarpals or single photon absorptiometry (SPA) of the radius. While these methods can be reasonably precise, they are not always correlated with measurements of trabecular bone at the sites where fractures are common. Quantitative computed tomography (QCT), in contrast, can be used to measure purely trabecular bone in areas such as the lumbar vertebrae. Its major disadvantage is the exposure to a high radiation dose. Dual photon absorptiometry (DPA) can be used for evaluating bone density of the spine, the hip or the total body. Dual energy X-ray absorptiometry (DXA) is also used to measure bone in the spine, the hip and the total body. Compared to DPA, it has the advantage of being faster, more precise, and associated with lower radiation doses.

The above methods give estimates of bone mass or density. The strength of a bone, however, is a product of both the amount of mineral it contains and the degree of connectedness (or "architecture") of the bone. The apparent velocity of ultrasound across a bone has recently been proposed as a method that reflects both the amount of substance that interferes with the transmission of ultrasound waves (i.e., bone mass) and the physical arrangement or architecture of that substance (Heaney *et al.,* 1989). Although this technique is not yet widely used, it may allow insights not possible with other techniques.

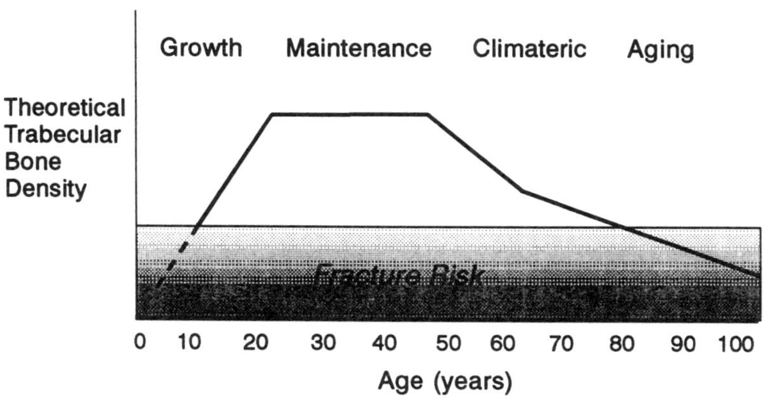


Figure 2. Life cycle of normal bone density in women, showing gain in trabecular bone density during childhood and adolescence, maintenance of bone density during adulthood, a period of rapid loss at the menopause, and continuing slow loss during aging.

### 3.2. General Changes over the Life Span

As shown schematically in Figure 2, the mineral content of bone increases during growth and maturation, is relatively stable during the middle adult years, undergoes a rapid decrease at the time of menopause, and decreases more gradually thereafter. This general pattern is believed to occur in all women irrespective of genetics, race, diet, exercise, or other lifestyle variables (Garn and Hawthorne, 1985). All these factors, however, potentially modify all aspects of this general pattern.

Until recently, research has focused on the accelerated bone loss that occurs at the menopause, and the potential to modify this loss through hormonal, nutritional, or exercise interventions. Because of the difficulty in treating osteoporosis once it is established, research is now underway to increase the understanding of variables that may influence the amount of bone that is gained during growth, and its maintenance during the "plateau" period. It is on this latter period, and the role of the menstrual cycle therein, that the remainder of the chapter will focus.

Based on cross-sectional studies, it has been assumed that only minor changes in bone density occur between the ages of 20 and 40 years (Rosenthal *et al.,* 1989). Cross-sectional studies, however, fail to reflect individual changes over time: "no change" with age in a cross-sectional study could actually reflect a situation in which clinically significant bone loss is occurring in some individuals and is being offset by gains in others. Longitudinal studies thus are necessary to truly understand the dynamics of this period of apparent stability in bone. In the next sections, the results of both longitudinal and cross-sectional studies that have addressed the effects on bone of disturbances of the menstrual cycle are described.

## 4. Amenorrhea and Oligomenorrhea

### 4.1. Definition and Prevalence

As indicated previously, secondary amenorrhea is defined as absence of menstrual flow for more than 180 days in women who have had at least one menstrual period and who are not pregnant or lactating, while oligomenorrhea is defined as menstrual periods lasting between 36 and 180 days. Amenorrhea has long been associated with rapid bone loss when it occurs at the time of menopause. In the past decade, amenorrhea in women athletes and in anorexia nervosa patients has also been associated with low bone density values (Drinkwater *et al.,* 1984; Rigotti *et al.,* 1984; Marcus *et al.,* 1985; Myerson *et al.,* 1992). Amenorrhea, however, is a clinical entity with diverse endocrinology

(Prior *et al.,* 1993). Although low levels of estrogen and progesterone are seen no matter what the underlying cause of amenorrhea (other than pregnancy), other hormones, such as prolactin, cortisol, and androgens, may be either normal or elevated, and the hormonal profile will modulate the effects on bone. For example, women with amenorrhea resulting from the polycystic ovary syndrome, who have higher androgen levels, had higher values for bone density than women with amenorrhea as a result of other causes (DiCarlo *et al.,* 1992).

Data on the prevalence of amenorrhea range from a one-year incidence of 0.7% in a population of Swedish women aged 18 to 45 years (Pettersson *et al.,* 1973), up to a 5% incidence in selected populations such as college students (Singh, 1981). Amenorrhea is a diagnostic criterion for anorexia nervosa (American Psychiatric Association, 1987), which occurs in approximately 1% of young women (Herzog and Copeland, 1985). Accurate information on the prevalence of oligomenorrhea is more difficult to obtain. Vollman (1977), who studied basal body temperature records of Swiss women during the 1950s and 1960s, reported that 8.2% of cycles in women aged 15 to 44 years were 36 days or longer (oligomenorrhea and amenorrhea). This rate varied considerably with age, decreasing from 27.5% of cycles in the first two years after menarche to approximately 15% of cycles in women in their early twenties, to < 5% of cycles in women aged approximately 30 to 45, and then increasing in the years before menopause. However, because these results are expressed in terms of the percentage of cycles that are long rather than in terms of the proportion of women who experience long cycles, they do not reveal the true prevalence of oligomenorrhea. Also, oligomenorrhea usually is used to refer to a *pattern* of long, irregular cycles, rather than one occasional long cycle. Nevertheless, taken together, these data suggest that irregular cycles are not uncommon, particularly in younger women.

### 4.2. Effects on Bone

The effects on bone of amenorrhea and oligomenorrhea have been documented mainly by cross-sectional studies, which revealed lower values for bone density than those for control populations (Drinkwater *et al.*, 1984; Marcus *et al.,* 1985; Myerson *et al.,* 1992; Snead *et al.,* 1992). It should be emphasized, however, that low bone density values are not found in all *individuals* with amenorrhea: in reports which show individual data, it is apparent that some amenorrheic women have normal or even above-normal bone density values (Marcus *et al.,* 1985; Biller *et al.,* 1991; Bachrach *et al.,* 1991). Very little prospective infomation is available on changes in bone with continued amenorrhea, but the available data suggest that the pattern of bone loss is similar to that seen postmenopausally; namely, a period of rapid loss followed by slow loss or relative stability. Biller *et al.* (1991), by monitoring spinal bone with QCT, found that a group of eight women who had been amenorrheic for more than five years experienced no significant change in bone density, whereas nine women who had been amenorrheic for fewer than five years lost trabecular bone at a mean annual rate of approximately 4%. Those who had been amenorrheic for a shorter period of time also had higher initial bone density values (147 ± 21 vs 124 ± 24 mg/cm$^3$) (mean ± SD). Cann *et al.* (1985) have also reported, in abstract form, a 4% annual rate of spinal bone loss observed by QCT in women who had been amenorrheic for three years or less, compared to no change in those amenorrheic for longer periods.

The question of the reversibility of bone loss—or the failure to gain bone—during periods of amenorrhea appears to be an open one at present, and additional studies are needed. Cross-sectional studies of bone density in women who were previously anorectic, but were of normal weight at the time their bone mineral density was assessed, suggest that many have normal bone density (Bachrach *et al.,* 1991; Treasure and Russell, 1987). For example, Bachrach *et al.* (1991) found that lumbar bone mineral density was within 1 SD of normal in 6 of 9 women, but was 2 SD below normal in 3 women. Other reports suggest that lumbar bone density in women who were formerly amenorrheic is lower than in women who were not (Drinkwater *et al.,* 1990).

Limited prospective data indicate that changes in bone density are highly correlated with change in body mass index (BMI) in anorexia nervosa patients (Bachrach *et al.,* 1991), but the extent of recovery is yet unknown. Athletes who resumed menses after being amenorrheic experienced significant gains (~6%) in bone density in the first year after resumption of menses (Drinkwater *et al.,* 1986; Lindberg *et al.,* 1987), but it appears that the rate of gain decreased and eventually stopped at values that were considerably below normal (Drinkwater *et al.,* 1990). Much work is still needed to clarify this area, but it seems reasonable to suggest that potential for recovery of bone density will vary according to the age of onset of amenorrhea and its duration, as well as to the extent of recovery of both body weight and the menstrual cycle. As will be shown subsequently, the fact that menses have resumed cannot be interpreted as full restoration of ovarian function.

## 5. Anovulation and Short Luteal Phase

### 5.1. Prevalence

The most comprehensive data on the prevalence of anovulatory and short luteal phase cycles are those of Vollman (1977), who analyzed basal body temperature records. Approximately 7% of all cycles were found to be anovulatory, and this was clearly related to gynecologic age, with the proportion decreasing from 55% of first menstrual cycles to 22% of cycles of women in their mid-late teens, to approximately 3% of cycles in mature women. An additional 14.6% of all cycles were short luteal phase cycles (< 10 days). Salivary progesterone measurements have also revealed higher frequencies of anovulatory cycles (Read *et al.,* 1984) and of short luteal phase cycles (Ellison *et al.,* 1987) during the early postmenarcheal years.

In recruiting subjects to participate in a study on habitual exercise and the menstrual cycle (Prior *et al.,* 1990b), a criterion for which was the presence of two consecutive ovulatory cycles with a normal length luteal phase (10–16 days), it was found that 32 of 113 women screened (28%) did not meet the criterion. During the subsequent 12-month study, only 13 of the 66 women who completed the study, and who had been prescreened as normally ovulatory in two cycles, had normally ovulatory cycles consistently. Of the remainder, 12 had one cycle with a short luteal phase, 28 had more than one short luteal phase cycle, and 13 had one or more anovulatory cycles. Looking at the data in another way, 29% of all cycles had luteal phase disturbances (short luteal phase or anovulatory cycle). Of note is that only 3% of all cycles were abnormal in total length. Thus, these data and those of others suggest that only a small proportion of women have normally ovulatory cycles consistently (20% in this study), and the majority experience ovulatory disturbances either occasionally (e.g., once a year) or more frequently. The results also indicate that it is not possible to predict characteristics of future cycles on the basis of the characteristics of any given cycle.

## 5.2. Effects on Bone

The effects of anovulatory or short luteal phase cycles on bone appear to be significant. In the study referred to above (Prior *et al.*, 1990b), cancellous spinal bone density was measured by QCT on two occasions one year apart. Mean bone density of the entire group decreased by 3.0 ± 4.8 mg/cm$^3$ (2%) from an initial value of 154.1 ± 2.7 mg/cm$^3$. The mean change in bone density varied depending on menstrual cycle characteristics: the 13 women with consistently normal cycles experienced a non-significant increase of 1.5 ± 3.9 mg/cm$^3$; those with one short luteal phase had a non-significant decrease of 0.7 ± 3.3 mg/cm$^3$; those with more than one short luteal phase lost 4.3 ± 4.2 mg/cm$^3$ ($P<0.001$) and those with one or more anovulatory cycles lost 6.4 ± 3.8 mg/cm$^3$ ($P<0.001$). When the luteal phase length was expressed as a ratio (luteal phase index = luteal phase length/total cycle length), the mean index for an individual was highly correlated with her one-year change in bone density ($r = 0.535$, $P < 0.001$). Few other variables were related to the change in bone density in these women: age, length of menstrual cycle, body mass index, skinfold thickness, calcium intake, exercise, and levels of serum estradiol, testosterone, triiodothyronine and cortisol were not associated with change in bone density. The mean daily energy intake was weakly related to one year bone change (r = 0.29, P = 0.02), as was family history of osteoporosis (chi-square = −4.27, P = 0.04).

A subsequent study was designed to assess the impact of progesterone replacement therapy in active women with ovulatory disturbances (Prior *et al.*, 1991). Women with abnormal cycles (short luteal phase, anovulation, oligomenorrhea, amenorrhea) were randomized by cycle type to receive either active or placebo medroxyprogesterone acetate (Provera$^R$, 10 mg/day on days 16–25 of the cycle, or 10 days per month in amenorrheic women), and active or placebo calcium (1000 mg/day), in a double-blind manner. Spinal bone density was evaluated using dual photon methods at enrollment and after one year of therapy. Among the 62 women who completed the study, initial bone density averaged 1.15 g/cm$^2$ and did not differ by treatment group. The one-year change in spinal bone density, however, differed significantly among treatment groups (P<0.01). Women receiving active Provera$^R$ and active calcium experienced a mean increase in spinal bone density, while those receiving both placebos tended to lose bone. The positive effects of active treatment were seen in women with all cycle types, including those with short luteal phase and anovulatory cycles.

Few other workers have investigated the effects of non-clinical variability in the menstrual cycle on bone. Snead *et al.* (1992) obtained 21 daily blood samples starting on day 9 of one menstrual cycle in 24 ovulatory runners who had maintained ≥ 32 km of running per week for ≥ 6 months. Bone mineral density was measured at the lumbar spine and the proximal femur. No associations were observed between integrated hormone concentrations and femoral bone density, but the integrated progesterone concentration was positively associated with density of the lumbar spine (r = 0.61, P = 0.002). Integrated estrogen concentrations were also correlated with lumbar density (r = 0.43, P = 0.04), although this correlation became nonsignificant following statistical adjustment for multiple comparisons. Although only one cycle was evaluated, these results also support the importance of a normal luteal phase in maintaining bone density in adulthood, as it is probable that women with lower integrated progesterone concentrations had experienced a short luteal phase cycle.

Another approach was taken by Sowers *et al.* (1990), who compared health and hormonal characteristics of 28 premenopausal women in the lowest 5th percentile for femoral bone mass to 25 randomly selected women with femoral bone mass within 1 SD of the mean. A single blood sample was obtained between days 18 and 22 of the menstrual cycle, and was analyzed for estradiol, progesterone, LH, FSH, thyroid stimulating hormone (TSH), prolactin and total thyroxine. Parity, relative body weight, smoking habit, oral contraceptive use and age at menarche were also assessed. The only significant difference between the two groups was the level of estradiol, 106 ± 72 pg/ml in controls and 76 ± 45 pg/ml in those with low bone density ($P < 0.05$). The authors acknowledge the significant limitations of their results consequent to use of a single measurement at one point in time. Nevertheless, their results also suggest that "normal" variation of the menstrual cycle may have a clinically significant influence on bone.

Thus, disturbances of the menstrual cycle, be they profound or more subtle, appear to have undesirable effects on bone density. While many factors have been associated with menstrual cycle disturbances, including exercise, anorexia nervosa, simple weight loss, and stresses such as moving, travel, or nursing training (Harlow and Matanoski, 1991; Schweiger *et al.*, 1988), the remainder of this article will focus on the effects of exercise and nutritional variables. It should be noted that these have not been selected as the most significant factors that could affect the cycle, but instead because they represent lifestyle variables about which women make choices, and therefore about which they require sound information on which to base their decisions.

## 6. Exercise and the Menstrual Cycle

Physical activity has many positive effects on the physical and psychological health of women (Marti, 1991). Among other benefits, weight-bearing exercise, as compared to inactivity, is reported to have favorable effects on bone (see Schoutens *et al.*, 1989, for review). The beneficial effects of activity, however, do not appear to compensate for loss of normal female reproductive hormone levels in active women who become amenorrheic; as a group, these women have lower bone density values than either menstruating athletes or sedentary women (Marti, 1991; Highet, 1989). Thus for women, especially those who may be concerned about future risk of osteoporosis, it is important to establish at what point—if any—the risk of induction of menstrual cycle abnormalities with exercise outweighs other benefits. To do this, it is important to carefully separate the effects of exercise from those of other variables (e.g., stress, weight loss, gynecologic immaturity) that could affect the susceptibility of the menstrual cycle to disruption.

In cross-sectional studies, menstrual cycle abnormalities including amenorrhea are reported to be more common among women who exercise (Carlberg *et al.*, 1983; Schweiger *et al.*, 1988; Loucks, 1990), and exercise is often reported as a "cause" of amenorrhea. The reported incidence of oligomenorrhea and amenorrhea in athletes ranges from 1 to 44% (Carlberg *et al.*, 1983; Loucks and Horvath, 1985), compared to up to 8% generally reported for control populations. Much of the variability in incidence is associated with methodological problems including variability in the definitions of amenorrhea and the method of documenting the menstrual cycle (Loucks and Horvath, 1985; Highet, 1989). Nevertheless, the incidence of amenorrhea in exercising women does appear to be higher than in control populations (Loucks and Horvath, 1985). Cross-sectional studies also reveal a higher degree of luteal phase suppression in athletes than in non-athletic women (Ellison and Lager, 1986; Loucks *et al.*, 1989).

Prospective evidence, however, does not demonstrate that exercise *alone* is a causal factor for amenorrhea or other menstrual disturbances (Bullen *et al.*, 1985; Prior *et al.*, 1990b; Bonen, 1992; Rogol *et al.*, 1992). Bullen *et al.* (1985) studied 28 untrained college women with documented normally ovulatory menstrual cycles who volunteered to participate in an eight-week exercise training program at a summer camp. During the first four weeks, running mileage was increased from 6 to 16 km per day and was maintained at that level for the following four weeks. Subjects also participated in an additional 3.5 hours per day of moderate sports activity such as playing volleyball or tennis, or bicycling. Half the subjects were fed a diet intended to maintain weight, while the remainder were placed on a diet intended to result in a weight loss of 0.45 kg per week. Results indicated that menstrual cycles were disturbed and that disturbances were more common in the weight-loss group. Only four of the 28 subjects had a normal cycle during the study, and three of these were in the weight maintenance group. In the first four weeks of the study, 63% of subjects in the weight loss group had abnormal luteal function and 31% experienced loss of the LH surge (anovulation). In the second four weeks, 75% of the women in the weight loss group had no LH surge. In contrast, women in the weight maintenance group did not experience the same progression in hormonal abnormality, although the incidence of irregular cycles was increased. Thus, rapidly intensifying severe exercise, especially when combined with weight loss, was associated with anovulation and luteal phase defects.

In contrast, a one-year prospective study, described earlier (Prior *et al.*, 1990b), revealed no effects of exercise on the menstrual cycle and bone. The women who volunteered for the study fell into one of three exercise categories: less than one hour per week of aerobic exercise; consistent running (more than one hour per week at constant pace, but not training for a specific event); and marathon training (intensifying activity over six to nine months in preparation to run a marathon). Women who were underweight (BMI $< 17$ kg/m$^2$), who had an eating disorder, whose emotional stability depended on running ("compulsive exercisers"), or who did not have two consecutive normally ovulatory cycles prior to enrollment were excluded. The participating groups of women did not differ significantly in height, weight or initial spinal bone density, although percent body fat averaged 22.8 ± 5.0% in less active women, 19.0 ± 3.4% in consistent runners and 17.2 ± 3.8% in marathon trainers. Over the year, those in the least active group ran an average of 1 km per week, consistent runners ran about 13.5 km per week, and those in the marathon training group ran an average of 35.5 km per week. Non-running exercise added an average of 17, 32 or 39 minutes of activity per week, respectively. In these women, no effect of activity was observed on the menstrual cycle or spinal bone density over the year. Short luteal phase cycles and anovulatory cycles occurred with equal frequency in all three groups: for example, more than one short luteal phase cycle occurred in 10 of 23 less active women, in 8 of 22 consistent runners, and in 10 of 21 women training for the marathon (Prior *et al.*, 1990b).

Another recent prospective study (Bonen, 1992) also failed to reveal any effect of recreational exercise training on the menstrual cycle. Volunteers first completed a control cycle, which was followed by two cycles of light calisthenics, and were then assigned to run less than 10 miles per week, 10–20 miles per week, or 20–30 miles per week for either two or four menstrual cycles. All groups were then followed during an additional two cycles of detraining. Diets were not controlled and no changes in either body weight or percent body fat occurred during the study. No consistent changes were observed in either menstrual cycle length or luteal phase length during the study.

A final prospective study is that of Rogol *et al.* (1992), who reported on a year-long training study in 17 gynecologically mature, initially untrained women. The women were assigned to run either at speeds corresponding to their lactate threshold (moderate intensity) or above their lactate threshold (high intensity). Weekly mileage progressed gradually from ~6 miles per week to ≥ 40 miles per week. Comprehensive hormonal analyses were conducted every four months. Over the course of the year, no changes in body weight occurred and a minimal decrease in body fat was detected only in the high intensity group (27.9 to 26.2%, $P < 0.05$). The only significant change in menstrual function was a slight decrease in the luteal phase length of the high intensity group by the end of the study (14.4 vs 13.0 days, $P < 0.05$).

In conclusion, it appears that exercise in and of itself does not inevitably lead to major disturbances of the menstrual cycle, particularly in gynecologically mature women. However, the finding of higher frequencies of luteal phase suppression in exercising women, and the induction of significant alterations in the cycle by rapidly escalating or very intense exercise, especially in combination with weight loss, suggest that the benefits of exercise on bone may be best realized by gradual adaptation to increased activity, without the superimposed stress of caloric restriction.

# 7. Nutritional Effects on the Menstrual Cycle

There are a number of mechanisms whereby nutritional variables could influence the menstrual cycle. In most cases, the underlying mechanism likely involves hypothalamic mechanisms that affect GnRH pulsatility and thereby the menstrual cycle, although in some cases there may also be an effect on circulating sex hormone levels. Nutrition may influence the cycle through its role in the determination of body weight and composition, as well as through the effects of specific nutrients such as fat and fiber. Recent work has also suggested that certain attitudes toward food may be associated with a greater tendency to ovulatory disturbances. Evidence associated with each of these mechanisms will be reviewed briefly.

## 7.1. Body Weight, Body Fat and Energy Intake

Frisch and McArthur (1974) hypothesized that a body fat content of 17% was necessary to initiate menses, whereas 22% body fat was needed for resumption of menses after a period of amenorrhea. This hypothesis has been widely criticized and disputed (Trussell, 1980; Scott and Johnston, 1982; Sinning and Little, 1987; Sanborn *et al.*, 1987), and there are certainly examples in the literature of groups of regularly menstruating women with less than 17% body fat. While it does seem overly simplistic to relate menstrual function to one variable—body fat—it also appears to be generally true that menstrual cycle disturbances are more common in those with low body weight or body fat percentages (Sinning and Little, 1987).

It is possible that energy (calorie) intake and balance are factors that help rationalize these findings. Caloric restriction and simple weight loss lead to alterations in the cycle, even before significant weight loss occurs (Schweiger *et al.*, 1987; Schweiger *et al.*, 1988; Lager and Ellison, 1990). If a low body fat value in a given individual is achieved and maintained by caloric restriction, menstrual disturbances may be more likely to occur than in another individual with the same low value for body fat who does not restrict energy intake. Some evidence is available in support of this suggestion, as many studies report lower energy intakes in amenorrheic women than in women with regular cycles (Drinkwater *et al.*, 1984; Marcus *et al.*, 1985; Nelson *et al.*, 1986), even when body weight is similar (Nelson *et al.*, 1986). However, discrepancies in this area of the literature have also emerged, particularly as a result of use of the doubly-labelled water method to evaluate energy balance and energy intake (Schoeller *et al.*, 1990). This non-invasive method quantitates $CO_2$ production, and thereby energy expenditure, over a period of several days. Assuming that body energy content (i.e., body weight and composition) remains stable, energy expenditure is by definition equal to energy intake. Application of this method, in association with use of food records, reveals that energy intake estimated by the food records typically underestimates expenditure as quantitated by doubly-labelled water. Moreover, the extent of underreporting of food intake (likely in association with undereating during the period in which records are kept), has been found to be greater in obese individuals and in elite athletes (Schoeller *et al.*, 1990). One could speculate that underreporting of intakes could also be more common in lean amenorrheic athletes who may be concerned about body weight, and thus the reported differences in energy intake between regularly menstruating and amenorrheic individuals can be questioned. Only the results of well-controlled studies in normally ovulatory and amenorrheic women, employing both doubly-labelled water and careful quantitation of energy intake, will allow this discrepancy to be resolved. However, irrespective of possible underreporting of intakes in some amenorrheic individuals, experimental data clearly show that restriction of energy intake results in disturbances of the menstrual cycle (Schweiger *et al.*, 1987).

## 7.2. Dietary Restraint

The concept of restrained eating, which is related to, but distinct from, caloric restriction, has recently received attention in the literature. Women with high scores on instruments designed to assess restraint are characterized by their perceptions that they consciously limit the amount of food consumed in order to prevent weight gain. In most reports, they weigh approximately the same or slightly more than women with lower restraint levels, but report eating fewer calories (Laessle *et al.*, 1989; Tuschl *et al.*, 1990a). A recent study by Schweiger *et al.* (1992) compared menstrual function in women with scores on the restraint scale above the 75th percentile (n = 9) or below the 50th percentile (n = 13) of a reference population (Tuschl *et al.*, 1990b). Individuals who had lost weight recently, had any food idiosyncrasies, participated in endurance exercise, had menstrual cycles that were abnormal in duration (i.e., < 21 or > 37 days), or who had values for BMI < 18 or > 24 kg/m$^2$, were excluded. Blood samples were taken on each weekday and urine samples and food diaries were obtained daily. The two groups of women were similar in terms of age, age at menarche and BMI, but the restrained women reported consuming 23% less food energy (approximately 1700 kcal vs 2200 kcal, P < 0.02). No differences in the percentage energy represented by carbohydrate, protein or fat were detected. Restrained eaters had significantly shorter cycles (24 vs 31 days, P < 0.01), shorter luteal phase lengths (9 vs 13 days, P < 0.02) and significantly lower mean luteal phase progesterone levels (18 vs 35 nmol/L, P < 0.05). While this study does not exclude the possibility that women with high restraint scores underreported intakes to a greater extent than women with lower scores, thus reducing or even eliminating actual differences in energy intake, the differences in menstrual cycle characteristics cannot be disputed. Perhaps the body weights maintained by the restrained women would have been considerably greater if they had not practiced restraint, and in some way this is sensed by the hypothalamus, leading to the menstrual changes. That is, the perception of needing to continuously monitor and limit food intake may be more important, or as important, as the actual absolute energy intake. Based on a study showing luteal phase defects to be associated with bone loss (Prior *et al.*, 1990b), one might speculate that women with high scores for dietary restraint would have lower values for bone density than those with lower scores. To date, however, published data comparing bone density of women with restrained and non-restrained eating patterns are not available.

## 7.3. Vegetarian Diets

Aside from energy intake or perceptions about food intake, differences in the composition of the diet also have the potential to affect the menstrual cycle. Interest in this area with respect to amenorrheic athletes arose with reports of a higher incidence of vegetarianism in athletes who were amenorrheic compared to those with regular cycles (Brooks *et al.*, 1984; Slavin *et al.*, 1984). Brooks *et al.* (1984), for example, reported that 9 of 11 amenorrheic runners were "vegetarian" (defined as eating less than 200 g of meat per week), compared to only 2 of 15 regularly menstruating runners. The two groups were comparable in terms of age, percent body fat, weekly training, age at menarche, and total energy intake, although the amenorrheic runners did eat less fat than regularly menstruating runners (68 vs 98 g/day). Similar fiber intakes were also reported, although the database for fiber at the time of the report was very incomplete. The opposite cross-sectional approach (i.e., comparing groups based on food habits rather than on menstrual status) also suggests a higher frequency of menstrual cycle disturbances in vegetarians than in omnivorous subjects (Pederson *et al.*, 1991; Lloyd *et al.*, 1991). In some cases, however, menstrual cycles were not fully characterized, or were characterized as normal or abnormal based only on the presence of flow, and more of the omnivorous subjects used oral contraceptives, which inevitably cause flow in women with an intact uterus (Pederson *et al.*, 1991). Finally, Pirke *et al.* (1986) found that anovulatory cycles developed in 7 of 9 normally ovulatory omnivorous women randomized to consume a vegetarian weight loss diet (to lose approximately 1 kg per week), compared to only 2 of 9 randomized to follow a nonvegetarian weight loss diet. Details about the composition of the diets, which were self-selected, were not provided, but weight loss was similar in the two groups. One might speculate that following an unusual diet would represent more of a stress to the women on the vegetarian diet; however, mood scores tended to be worse in the women on the nonvegetarian diet.

### 7.4. Dietary Fat and Fiber

Possible mechanisms whereby a vegetarian diet, or more accurately, the components of a vegetarian diet, could affect the menstrual cycle include: 1) high dietary fiber intake, which may decrease estrogen reabsorption by reducing the activity of beta-glucuronidase, an intestinal enzyme needed to deconjugate estrogen metabolites excreted through the bile (Adlercreutz *et al.*, 1987), and which may also bind unconjugated estrogens in the gut, impeding their reabsorption (Schultz and Howie, 1986); 2) low fat intakes, which may also affect beta-glucuronidase activity; and 3) a high intake of plants containing substances that may interfere with ovulation (Riddle and Estes, 1992). Results of controlled experimental studies in humans are available with respect to the first two mechanisms.

Rose *et al.* (1991) studied the effect of doubling dietary fiber (from about 15 to 30 g/day) in 62 premenopausal women with regular ovulatory menstrual cycles. The women were randomly assigned to supplement their diets with corn, wheat or oat bran for two months. Changes in the percentage energy from fat, which could confound the results, did not occur during the study period. Serum hormone levels were assessed between days 18 and 21 of each menstrual cycle. Compared to baseline, women receiving wheat bran had decreased levels of serum estradiol (631 ± 239 vs 536 ± 187 pmol/L, $P < 0.02$) and estrone (499 ± 152 vs 399 ± 163 pmol/L, $P < 0.002$), but no changes were seen in the corn or oat bran groups. Serum progesterone was significantly lower after one month on the wheat bran supplementation, but returned to baseline values by the second month. No changes were seen in the concentrations of sex hormone binding globulin, suggesting that the decrease in estrogen levels was physiologically significant.

Rose *et al.* (1987) have also demonstrated a significant effect of low fat intake on serum estrogen. Sixteen premenopausal, regularly menstruating women with cystic breast disease changed from their usual diet (35% of energy as fat) to a low fat diet (21% as fat). Hormone levels were again assessed in single blood samples obtained between days 17 and 20 of each cycle. Reported energy intake decreased significantly (1743 ± 397 vs 1344 ± 315 kcal/day, $P < 0.05$), but no significant change occurred in the intake of dietary fiber. After three months on the diet, during which time a small loss of weight occurred (59.4 ± 7.2 kg vs 58.0 ± 7.2 kg, $P < 0.001$), total serum estrogens decreased from 299 ± 100 pg/ml to 200 ± 63 pg/ml ($P < 0.001$). No significant changes were observed in concentrations of progesterone. In contrast, no effects on serum hormones of low fat (25% of calories) or high fat (46% of calories) diets were detected by Hagerty *et al.* (1988), who studied ovulatory lactoovovegetarian women in a crossover design. The diets were well controlled and were similar in terms of fiber, protein and the ratio of polyunsaturated to saturated fat. However, the number of subjects was small ($n = 6$) and they followed each diet for only one month.

### 7.5. Effects on Bone

Little information is available on the bone mineral density of women following vegetarian versus omnivorous diets. No difference was detected in bone mineral content of the radius, measured by single photon absorptiometry (SPA), between postmenopausal omnivores and vegetarian women (Hunt *et al.*, 1989). As noted earlier, however, cortical bone appears to be less responsive to hormonal stimuli than is trabecular bone. Dual photon absorptiometry was used by Lloyd *et al.* (1991) to compare spinal bone density in premenopausal vegetarian (n = 27) and nonvegetarian (n = 37) women. Compared to the nonvegetarians, diets of the vegetarians were higher in dietary fiber and carbohydrate, lower in protein, and similar in fat content. Menstrual disturbances (i.e., abnormal cycle length) were more common in the vegetarian women, and while their spinal bone density tended to be lower ($1.02 \pm 0.02$ g/ml vs $1.06 \pm 0.02$ g/ml), the difference was not significant. Some correlations were noted between fat intake and urinary LH levels; however, because diets were evaluated using only one 3-day record, the reliability of the findings can be questioned, and they need to be reproduced using a recording period long enough to accurately evaluate individual intakes (Basiotis *et al.*, 1987).

In summary, evidence exists for the potential influence of a number of dietary factors on the menstrual cycle, although more work is needed in this area. Because the influence of dietary factors is likely to be subtle, menstrual cycles need to be characterized by more than their length. Also, cycles should be evaluated over several months to obtain a clearer indication of an individual's menstrual status; as indicated earlier, the characteristics of one cycle do not accurately predict those of subsequent cycles. Careful evaluation of individual diets is also required in order to relate individual variables such as bone mineral density to dietary factors.

## 8. Conclusions and Directions for Research

As alluded to in the introduction to this chapter, the data presented herein demonstrate that disturbances of the menstrual cycle, whether clinically nondetectable or profound, are associated with trabecular bone loss. Given that trabecular bone is commonly involved in osteoporotic fractures, this finding may be of considerable significance for the prevention of osteoporosis in later life. Further confirmation of the importance to bone of maintaining normal ovulatory cycles during adult life is provided by the demonstration that use of cyclic medroxyprogesterone to mimic the hormonal profile of the normal ovulatory cycle prevents and/or reverses bone loss in active women with ovulatory disturbances. Although menstrual disturbances are found in higher proportions of exercising than sedentary women in cross-sectional studies, prospective data suggest that in women who are initially normally ovulatory, intensification of an exercise program over a period of several months and in the absence of weight loss is not associated with an increased prevalence of ovulatory disturbances. It appears that a number of dietary variables have the potential to influence the menstrual cycle, and of these, simple caloric restriction appears to be the most powerful.

Further work on the effects of variability of the menstrual cycle on bone are clearly needed. Priority areas include: the development of convenient methods to reliably detect anovulatory and short luteal phase cycles, and especially to identify women in whom these occur frequently; long term follow-up studies of the effects on bone of use of cyclic progesterone in women with menstrual cycle disturbances; prospective studies to assess the influence on peak bone mass of maturation of the menstrual cycle during the early postmenarcheal years; and additional study of the effect of dietary variables on the menstrual cycle, including careful documentation of the menstrual cycle over a period of time.

ACKNOWLEDGMENTS. Some of the work reported in this chapter was supported by grants from the National Health Research and Development Program, the Dairy Bureau of Canada, and the British Columbia Medical Services Foundation. We thank the women who participated in our studies.

## References

Adlercreutz, H., Hockerstedt, K., Bannwart, C., Bloiglu, S.G., Hamalainen, Fotsis, T., and Ollus, A. 1987, Effect of dietary components, including lignans and phytoestrogens, on enterohepatic circulation and liver metabolism of estrogens and on sex hormone binding globulin, *J. Steroid Biochem.* **27**:1135.

Alhava, E.M., 1991, Bone density measurements, *Calcif. Tissue Int.* **49**:S21.

American Psychiatric Association, 1987, *Diagnostic and Statistical Manual of Mental Disorders,* 3rd ed., 63 pp., American Psychiatric Association, Washington.

Bachrach, L.K, Katzman, D.K., Litt, I.F., Guido, D., and Marcus, R., 1991, Recovery from osteopenia in adolescent girls with anorexia nervosa, *J. Clin. Endocrinol. Metab.* **72**:602.

Bardawil, W.A., 1987, "Endometrium," in: *Gynecologic Endocrinology,* 4th ed., (J.J. Gold and J. B. Jasimovich, eds.), pp 185-244, Plenum Press, New York.

Basiotis, P.P., Welsh, S.O., Cronin, F.J., and Kelsey, J. L., 1987, Number of days of food intake records required to estimate individual and group nutrient intakes with defined confidence, *J. Nutr.* **117**:1638.

Biller, B.K.M., Coughlin, J.F., Saxe, V., Schoenfeld, D., Spratt, D.I., and Klibansi, A., 1991, Osteopenia in women with hypothalamic amenorrhea: a prospective study, *Obstet. Gynecol.* **78**:996.

Bonen, A., 1992, Recreational exercise does not impair menstrual cycles: a prospective study, *Int. J. Sports Med.* **13**:110.

Brooks, S.M., Sanborn, C.F., Albrecht, B.H. and Wagner, W.W., 1984, Diet in athletic amenorrhea, *Lancet* **1**:559.

Bullen, B.A., Skrinar, G.S., Beitins, I.Z., Von Mering, G., Turnbull, B.A., and McArthur, J. W., 1985, Induction of menstrual cycle disorders by strenuous exercise in untrained women, *N. Engl. J. Med.* **312**:1349.

Cann, C.E., Martin, M.C., and Jaffe, R.B., 1985, Duration of amenorrhea affects rate of bone loss in women runners: implications for therapy, *Med. Sci. Sports Exerc.* **17**:214. (abstract).

Carlberg, K.A., Buckman, M.T., Peake, G.T., and Riedesel, M.L., 1983, A survey of menstrual function in athletes, *Eur. J. Appl. Physiol.* **51**:211.

DiCarlo, C., Shoham, Z., MacDougall, J., Patel, A., Hall, M.L., and Jacobs, H.S., 1992, Polycystic ovaries as a relative protective factor for bone mineral loss in young women with amenorrhea, *Fertil. Steril.* **57**:314.

Drinkwater, B.L., Nilson, C.H., Chestnut, C.H. III, Bremner, W., Shainholtz, S., and Southworth, M., 1984, Bone mineral content of amenorrheic and eumenorrheic athletes, *N. Engl. J. Med.* **311**:277.

Drinkwater, B.L., Nilson, K., Ott, S.M., and Chestnut, C.H. III., 1986, Bone mineral density after resumption of menses in amenorrheic athletes, *JAMA* **256**:380.

Drinkwater, B.L., Bruemner, B., and Chestnut, C.H. III., 1990, Menstrual history as a determinant of current bone density in young athletes, *JAMA* **263**:545.

Ellison, P.T., 1988, Human salivary steroids: methodological considerations and applications in physical anthropology, *Yearbook of Physical Anthropology* **31**:115.

Ellison, P.T., and Lager, C., 1986, Moderate recreational running is associated with lowered salivary progesterone profiles in women, *Am. J. Obstet. Gynecol.* **154**:1000.

Ellison, P.T., Lager, C., and Calfee, J., 1987, Low profiles of salivary progesterone among college undergraduate women, *J. Adol. Health Care* **8**:204.

Frisch, R.E., and McArthur, J.W., 1974, Menstrual cycles: fatness as a determinant of minimum weight for height necessary for their maintenance and onset, *Science* **185**:949.

Fritz, M.C., and Speroff, L., 1983, Current concents of the endocrine characteristics of normal menstrual function: the key to diagnosis and management of menstrual disorders, *Clin. Obstet. Gynecol.* **26**:647.

Garn, S.M., and Hawthorne, V.M., 1985, "Calcium intake and bone loss in population context," in: *Calcium in Biological Systems* (R.P. Rubin, ed.), pp. 569-574, Plenum Press, New York.

Hagerty, M.A., Howie, B.J., Tan. S., and Schultz, T.D., 1988, Effect of low- and high-fat intakes on the hormonal milieu of premenopausal women, *Am. J. Clin. Nutr.* **47**:653.

Harlow, S.D., and Matanoski, G.M., 1991, The association between weight, physical activity, and stress and variation in the length of the menstrual cycle, *Am. J. Epidemiol.* **133**:38.

Heaney, R.P., Avioli, L.V., Chestnut, C.H. III, Lapp, J., Recker, R.R., and Brandenburger, G.H., 1989, Osteoporotic bone fragility. Detection by ultrasound transmission velocity, *JAMA* **261**:2986.

Herzog, D.B., and Copeland, P.M., 1985, Eating disorders. Medical progress, *N. Engl. J. Med.* **313**:295.

Highet, R., 1989, Athletic amenorrhoea, An update on aetiology, complications and management, *Sports Med.* 7:82.

Hunt, I.F., Murphy, N.J., Henderson, C., Clark, V.A., Jacobs, R.M., Johnston, P.K., and Coulson, A.H., 1989, Bone mineral content in postmenopausal women: comparison of omnivores and vegetarians, *Am. J. Clin. Nutr.* **50**:517.

Laessle, R.G., Tuschl, R.J., Kotthaus, B.C., and Pirke, K.M., 1989, Behavioral and biological correlates of dietary restraint in normal life, *Appetite* **12**:83.

Lager, C., and Ellison, P.T., 1990, Effect of moderate weight loss on ovarian function assessed by salivary progesterone measurements, *Am. J. Hum. Biol.* **2**:303.

Lindberg, J.S., Powell, M.R., Hund, M.M., Ducey, D.E., and Wade, C.E., 1987, Increased vertebral bone mineral in response to reduced exercise in amenorrheic women, *West. J. Med.* **146**:39.

Lloyd, T., Schaeffer, J.M., Walker, M.A., and Demers, L.M., 1991, Urinary hormonal concentrations and spinal bone densities of premenopausal vegetarian and nonvegetarian women, *Am. J. Clin. Nutr.* **54**:1005.

Loucks, A.B., 1990, Effects of exercise training on the menstrual cycle: Existence and mechanisms, *Med. Sci. Sports Exerc.* **22**:275.

Loucks, A.B., and Horvath, S.M., 1985, Athletic amenorrhea: a review, *Med. Sci. Sports Exerc.* **17**:56.

Loucks, A.B., Mortola, J.F., Girton, L., and Yen, S.S.C., 1989, Alterations in the hypothalamic-pituitary-ovarian and the hypothalamic-pituitary-adrenal axes in athletic women, *J. Clin. Endocrinol. Metab.* **68**:402.

Marcus, R., Cann, C., Madvig, P., Minkoff, J., Goddard, M., Beyer, M., Martin, M., Gaudiani, L., Haskell, W., and Genant, H., 1985, Menstrual function and bone mass in elite women distance runners: endocrine and metabolic features. *Ann. Intern. Med.* **102**:158.

Marti, B., 1991, Health effects of recreational running in women, *Sports Med.* **11**:20.

Myerson, M., Gutin, B., Warren, M.P., Wang, J., Lichtman, S., and Pierson, R.N., 1992, Total body bone density in amenorrheic runners, *Obstet. Gynecol.* **79**:973.

Nelson, M.E., Fisher, E.C., Catsos, P.D., Meredith, C.N., Turksoy, R.N., and Evans, W.J., 1986, Diet and bone status in amenorrheic runners, *Am. J. Clin. Nutr.* **43**:910.

Pederson, A.B., Bartholomew, M.J., Dolence, L.A., Aljadiar, L.P., Netteburg, K.L., and Lloyd, T., 1991, The effect of vegetarian and non-vegetarian diets on menstrual status, *Am. J. Clin. Nutr.* **53**:879.

Pettersson, F., Fries, H., and Nillius, S.J., 1973, Epidemiology of secondary amenorrhea, *Am. J. Obstet. Gynecol.* **117**:80.

Pirke, K.M., Schweiger, U., Laessle, R., Dickhaut, B., Schweiger, M., and Waechtler, M., 1986, Dieting influences the menstrual cycle: vegetarian versus nonvegetarian diet, *Fertil. Steril.* **46**:1083.

Prior, J.C., Vigna, Y.M., Schulzer, M., Hall, J.E., and Bonen, A., 1990a, Determination of luteal phase length by quantitative basal temperature methods: validation against the midcycle LH peak, *Clin. Invest. Med.* **13**:123.

Prior, J.C., Vigna, Y.M., Schechter, M.T., and Burgess, A.E., 1990b, Spinal bone loss and ovulatory disturbances, *N. Engl. J. Med.* **323**:1221.

Prior, J.C., Vigna, Y.M., Lentle, B.C., Rexworthy, C., and Connell, D., 1991, Cyclic progestin and calcium increase spinal bone density in women athletes with menstrual cycle disturbances. Abstract presented at 73rd Annual Meeting of The Endocrine Society, Washington, D.C., June 1991, page 450.

Prior, J.C., Vigna, Y.M., Barr, S.I., and Alojado, N., 1993, "Amenorrhea and anovulation—risk factors for osteoporosis that precede menopause," in: *Modern Management and Treatment of Menopause* (J. Lorrain, ed.), Springer Verlag, New York (in press).

Read, G.F., Wilson, D.W., Hughes, I.A., and Griffiths, K., 1984, The use of salivary progesterone assays in the assessment of ovarian function in postmenarcheal girls, *J. Endocrinol.* **102**:265.

Riddle, J.M., and Estes, J.W., 1992, Oral contraceptives in ancient and medieval times, *Am. Scientist* **80**:226.

Riggs, B.L., and Melton, L.J. III, 1986, Involutional osteoporosis, *N. Engl. J. Med.* **314**:1676.

Rigotti, N.A., Nussbaum, S.R., Herzog, D.B., and Neer, R.M., 1984, Osteoporosis in women with anorexia nervosa, *N. Engl. J. Med.* **311**:1601.

Rogol, A.D., Weltman, A., Weltman, J.Y., Seip, R.L., Snead, D.B., Levine, S., Haskvitz, E.M., Thompson, D.L., Schurrer, R., Dowling, D., Walberg-Rankin, J., Evans, W.S., and Veldhuis, J.D., 1992, Durability of the reproductive axis in eumenorrheic women during 1 yr of endurance training, *J. Appl. Physiol.* **72**:1571.

Rose, D.P., Boyar, A.P., Cohen, C., and Strong, L.E., 1987, Effect of a low-fat diet on hormone levels in women with cystic breast disease. I. Serum steroids and gonadotropins, *J. Natl. Cancer Inst.* **78**:623.

Rose, D.P., Goldman, M., Connolly, J.M., and Strong, L.E., 1991, High-fiber diet reduces serum estrogen concentrations in premenopausal women, *Am. J. Clin. Nutr.* **54**:520.

Rosenthal, D.I., Mayo-Smith, W., Hayes, C.W., Khurana, J.S., Biller, B.M., Neer, R.M., and Klibanski, A., 1989, Age and bone mass in premenopausal women, *J. Bone. Min. Res.* **4**:533.

Sanborn, C.F., Albrecht, B.H., and Wagner, W.W., Jr., 1987, Athletic amenorrhea: lack of association with body fat, *Med. Sci. Sports Exerc.* **19**:207.

Schoeller, D.A., Bandini, L.G., and Dietz, W.H., 1990, Inaccuracies in self-reported intake identified by comparison with the doubly labelled water method, *Can. J. Physiol. Pharmacol.* **68**:941.

Schoutens, A., Laurent, E., and Poortmans, J.R., 1989, Effects of inactivity and exercise on bone, *Sports Med.* **7**:71.

Schweiger, U., Laessle, R., Pfister, H., Hoehl, C., Schwingenschloegel, M., Schweiger, M., and Pirke, K.M., 1987, Diet-induced menstrual irregularities: effects of age and weight loss, *Fertil. Steril.* **48**:746.

Schweiger, L.L., Laessle, R., Schweiger, M., Herrmann, F., Riedel, W., and Pirke, K.M., 1988, Caloric intake, stress, and menstrual intake in athletes, *Fertil. Steril.* **49**:447.

Schweiger, U., Tuschl, R.J., Platte, P., Broocks, A,. Laessle, R.G., and Pirke, K.M., 1992, Everyday eating behavior and menstrual function in young women, *Fertil. Steril.* **57**:771.

Scott, E.C., and Johnston, F.E., 1982, Critical fat, menarche, and the maintenance of menstrual cycles: a critical review, *J. Adol. Health Care* **2**:249.

Shoupe, D., Mishell, D.R., Lacarra, M., Lobo, R.A., Horenstein, J., d'Ablaing, G., and Moyer, D., 1989, Correlation of endometrial maturation with four methods of estimating day of ovulation, *Obstet. Gynecol.* **73**:88.

Shultz, T.D., and Howie, B.J., 1986, In vitro binding of steroid hormones by natural and purified fibers, *Nutr. Cancer* **8**:141.

Singh, K., 1981, Menstrual disorders in college students, *Am. J. Obstet. Gynecol.* **3**:299.

Sinning, W.E., and Little, K.D., 1987, Body composition and menstrual function in athletes, *Sports Med.* **4**:34.

Slavin, J., Lutter, J., and Cushman, S., 1984, Amenorrhea in vegetarian athletes, *Lancet* **1**:1474.

Snead, D.B., Weltman, A., Weltman, J.Y., Evans, W.S., Veldhuis, J.D., Varma, M.M., Teates, C.D., Dowling, E.A., and Rogol, A.D., 1992, Reproductive hormones and bone mineral density in women runners, *J. Appl. Physiol.* **72**:2149.

Sowers, M.F.R., Shapiro, B., Gilbraith, M.A., and Jannausch, M., 1990, Health and hormonal characteristics of premenopausal women with lower bone mass, *Calcif. Tissue Int.* **47**:130.

Treasure, J.L., and Russell, G.F.M., 1987, Reversible bone loss in anorexia nervosa, *Br. Med. J.* **295**:474.

Trussell, J., 1980, Statistical flaws in evidence for the Frisch hypothesis that fatness triggers menarche, *Hum. Biol.* **52**:711.

Tuschl, R.J., Laessle, R.G., Platte, P., and Pirke, K.M., 1990a, Differences in food-choice frequencies between restrained and unrestrained eaters, *Appetite* **57**:772.

Tuschl, R.J., Platte, P., Laessle, R.G., Stichler, W., and Pirke, K.M., 1990b, Energy expenditure and everyday eating behaviors in healthy young women, *Am. J. Clin. Nutr.* **52**:81.

Vollman, R.F., 1977, *The Menstrual Cycle*, W.B. Saunders, Philadelphia.

## ADDENDUM

Subsequent to the submission of this manuscript, several relevant studies have been published. Associations between high scores for cognitive dietary restraint and ovulatory disturbances have now been confirmed in two groups: initially-ovulatory women with a broad range of activity levels (Barr *et al.,* 1994); and healthy vegetarian and nonvegetarian women (Barr *et al.,* 1994 in press). The latter study was designed to determine whether healthy vegetarians were actually more likely to experience abnormal menstrual cycles, as suggested by studies described elsewhere in this chapter. We recruited healthy women who reported regular cycles of normal length (compared to vegetarian women with normal cycles, those with abnormal cycles might be more likely to volunteer for a study of vegetarianism and the menstrual cycle). To avoid effects of acute dietary change, women had followed their vegetarian or nonvegetarian diets for at least two years. Using quantitative basal body temperature analysis, we prospectively assessed cycle characteristics over six months. The vegetarian women had fewer abnormal (anovulatory and short luteal phase) cycles, and also had lower scores for cognitive dietary restraint. Although we did not detect associations between dietary components (including fat, dietary fiber and energy) and ovulatory characteristics, recent studies have demonstrated acute effects of the introduction of flax seed, high in lignans, and soy protein, high in isoflavones, on the menstrual cycle. In a cross-over study, luteal phase length of 18 normally-cycling women was significantly longer during supplementation with flax seed powder (Phipps *et al.*, 1993). Conversely, addition of soy protein to the diets of six women was associated with a significant increase in follicular phase length and with higher follicular phase concentrations of estradiol (Cassidy *et al.*, 1994). Clearly, much work remains to be done in this area. Finally, a study referenced in the chapter in abstract form (as Prior *et al.*, 1991) is now published (Prior *et al.*, 1994).

Barr, S.I., Janelle, K.C., and Prior, J.C., 1994, Vegetarian versus nonvegetarian diets, dietary restraint, and subclinical ovulatory disturbances: prospective six month study, *Am. J. Clin. Nutr.* (in press).

Barr, S.I., Prior, J.C., and Vigna, Y.M., 1994, Restrained eating and ovulatory disturbances: possible implications for bone health, *Am. J. Clin. Nutr.* **59**:92.

Cassidy, A., Bingham, S., and Setchell K.D.R., 1994, Biological effects of a diet of soy protein rich in isoflavones on the menstrual cycle of premenopausal women, *Am. J. Clin. Nutr.* **60**:333.

Phipps, W.R., Martini, M.C., Lampe, J.W., Slavin, J.L., and Kurzer, M.S., 1993, Effect of flax seed ingestion on the menstrual cycle., *J. Clin. Endocrinol. Metab.* **77**:1215.

Prior, J.C., Vigna, Y.M., Barr, S.I., Rexworthy, C., and Lentle, B., 1994, Cyclic medroxyprogesterone treatment increases bone density: a controlled trial in active women with menstrual cycle disturbances. *Am. J. Med.* **96**:521.

# Index